# STP MATHEMATICS
## 3rd Edition

**9**

L BOSTOCK  S CHANDLER  A SHEPHERD

E SMITH  I BETTISON

OXFORD
UNIVERSITY PRESS

Great Clarendon Street, Oxford, OX2 6DP, United Kingdom

Oxford University Press is a department of the University of Oxford. It furthers the University's objective of excellence in research, scholarship, and education by publishing worldwide. Oxford is a registered trade mark of Oxford University Press in the UK and in certain other countries.

Text © L Bostock, S Chandler, A Shepherd and E Smith 2014
Illustrations © Oxford University Press

The moral rights of the authors have been asserted.

First published by Nelson Thornes in 1996

All rights reserved. No part of this publication may be reproduced, stored in a retrieval system, or transmitted, in any form or by any means, without the prior permission in writing of Oxford University Press, or as expressly permitted by law, by licence or under terms agreed with the appropriate reprographics rights organization.
Enquiries concerning reproduction outside the scope of the above should be sent to the Rights Department, Oxford University Press, at the address above.

You must not circulate this work in any other form and you must impose this same condition on any acquirer.

British Library Cataloguing in Publication Data
Data available

978-1-40-852380-3

14

Printed in Great Britain by Bell and Bain Ltd, Glasgow

**Acknowledgements**
Page make-up: Tech-Set Limited, UK
Illustrations: Tech-Set Limited, UK and Tony Forbes (Sylvie Poggio Agency)

The publishers would like to thank the following for the permission to reproduce photographs: **Cover:** Antishock/iStockphoto; **p30:** carlo dapino/Shutterstock; **p39:** Warren Goldswain/Shutterstock; **p51:** photobank.ch/Shutterstock; **p56:** Dream79/Shutterstock; **p63:** sweetandsour/iStockphoto; **p64:** Gjermund/Shutterstock; **p64:** cTermit/Shutterstock; **p82:** glamour/Shutterstock; **p89:** Dmitry Kalinovsky/Shutterstock; **p93:** rupbilder/Fotolia; **p98:** eans/Shutterstock; **p102:** Chris Rout/Alamy; **p102:** Roy Botterell/Corbis; **p106:** wavebreakmedia/Shutterstock; **p109:** Dirima/Shutterstock; **p112:** Brian A Jackson/Shutterstock; **p119:** Maxisport/Shutterstock; **p154:** bikeriderlondon/Shutterstock; **p155:** ollyy/Shutterstock; **p160:** Valentina Proskurina/Shutterstock; **p172:** Valadislav Gajic/Fotolia; **p180:** el lobo/Fotolia; **p183:** Monkey Business Images/Shutterstock; **p185:** Andrey_Popov/Shutterstock; **p204:** Tobik/Shutterstock; **p219:** Coprid/Shutterstock; **p232:** Bill Kennedy/Shutterstock; **p264:** Oleksii Iezhov/Shutterstock; **p269:** PLANETOBSERVER/Science Photo Library; **p305:** Sergio Stakhnyk/Shutterstock; **p340, 360:** Jorg Hackemann/Shutterstock; **p375:** FreespiritTransport/Alamy; **p384:** WitR/Shutterstock.

Although we have made every effort to trace and contact all copyright holders before publication, this has not been possible in all cases. If notified, the publisher will rectify any errors or omissions at the earliest opportunity.

Links to third party websites are provided by Oxford in good faith and for information only. Oxford disclaims any responsibility for the materials contained in any third party website referenced in this work.

# Contents

**Introduction** v

**Summary of Years 7 and 8** 1

### 1 Travel graphs 30
Finding distance from a graph. Drawing travel graphs. Calculating the time taken. Average speed. Journeys with several parts. Getting information from travel graphs.

### 2 Working with numbers 51
Range of values for a corrected number. Fractions and decimals. Addition and subtraction of fractions. Multiplication of fractions. Reciprocals. Division by a fraction. Interchanging decimals and fractions. Recurring decimals. Using numbers in standard form.

### 3 Probability 63
Mutually exclusive events. Independent events. Adding probabilities. Multiplication of probabilities. Tree diagrams. Venn diagrams.

### 4 Percentages 82
Percentage increase and percentage decrease. Income tax. Sale reductions. Finding the original quantity. Interest. Compound percentage problems.

### 5 Ratio and proportion 102
Direct proportion and inverse proportion. Ratios. Division in a given ratio. Simple direct proportion. Direct proportion. Inverse proportion.

**Summary 1** 116

### 6 Algebraic products 128
Brackets. The product of two brackets. Finding the pattern. Important products. More complex expansions. Expressions, identities and equations.

### 7 Algebraic factors 141
Finding factors. Common factors. Factorising quadratic expressions. The difference between two squares. Calculations using factorising. Mixed quadratic expressions.

### 8 Organising and summarising data 154
Analysing large sets of information. Stem-and-leaf diagrams. Finding the mean of an ungrouped frequency distribution. Finding the mean of a grouped frequency distribution. The median. Running totals. Cumulative frequency. Cumulative frequency diagrams. Finding the median of a grouped frequency distribution. Interquartile range.

### 9 Formulas 180
Constructing formulas. Substituting numbers into formulas. Changing the subject of a formula. Substituting one formula into another. The $n$th term of a sequence. Finding an expression for the $n$th term. Formula for the $n$th term of an arithmetic progression. Geometric progressions and other sequences.

### 10 Simultaneous equations 197
Solving simultaneous equations. Special cases. Solving problems. Graphical solutions.

**Summary 2** 207

# Contents

**11 Quadratic equations**     **219**
Quadratic equations. Solving quadratic equations. Solution by factorisation. Forming equations. Solving equations by trial and improvement.

**12 Graphs**     **235**
Straight lines. Parabolas. Using graphs to solve quadratic equations. Cubic graphs. The shape of a cubic curve. Reciprocal graphs. Reciprocal curves. Recognising curves. Using graphs. Inverse proportion. Gradients of curves. Interpretation of gradient.

**13 Areas and volumes**     **264**
Area of a trapezium. Units of area. Arcs and sectors of circles. Volume of a prism. Imperial units of volume. The dimensions of a formula.

**14 Transformations**     **284**
Negative scale factors. Invariant points. Reflections: finding the mirror line. Rotations. Finding the angle of rotation. Compound transformations. Vectors. Representing vectors. Vectors in the form $\binom{a}{b}$. Translations.

**15 Similar figures**     **305**
Similar figures. Similar triangles. Corresponding vertices. Finding a missing length. Using the scale factor to find the missing length. Corresponding sides. Other shapes.

**Summary 3**     **323**

**16 Trigonometry: tangent of an angle**     **340**
Tangent of an angle. Finding tangents of angles. The names of the sides of a right-angled triangle. Finding a side of a triangle. Finding a side adjacent to the given angle. Finding an angle given its tangent. Finding an angle in a right-angled triangle.

**17 Sine and cosine of an angle**     **361**
Sine of an angle. Using the sine ratio. Cosine of an angle. Using the cosine ratio. Use of all three ratios. Finding the hypotenuse. Applications.

**18 Problems in three dimensions**     **379**
Solids. Cubes, cuboids and prisms. Pyramids. Cones. Spheres.

**19 Geometry and proof**     **390**
The need for proof. Deductive proof. Showing that a hypothesis is false. Circle facts. Tangents. Tangent property.

**20 Congruent triangles**     **402**
Congruent triangles. Three pairs of sides. Two angles and a side. Two sides and an angle. Two sides and a right angle. Using congruent triangles. Properties of parallelograms. Properties of the special quadrilaterals. Parallelograms, polygons and congruent triangles. Constructions. Bisecting angles. Constructing the perpendicular bisector of a line. Dropping a perpendicular from a point to a line.

**Summary 4**     **432**

**Key words**     **449**

**Index**     **453**

# Introduction

## To the student

This is the third book of a series that helps you to learn, enjoy and progress through Mathematics in the National Curriculum. As well as a clear and concise text the book offers a wide range of practical, investigational and problem-solving work that is relevant to the mathematics you are learning.

Everyone needs success and satisfaction in getting things right. With this in mind we have provided questions of two types.

The first type are practice questions and problems to help reinforce what you have learnt.

> The second type, identified by a coloured background, are extension questions and problems for those of you who like a greater challenge.

Questions with this symbol are problems which give you a chance to apply your skills to real-life examples and practise your problem-solving skills.

Each chapter begins with a problem associated with a real-life situation entitled 'Consider'. Although some students will be able to attempt this already, it will prove to be a challenge to most of you. It should therefore act as an introductory stimulus for bringing into play the strategies for problem solving that you are already familiar with. When the problem is reintroduced at the end of the chapter in 'Consider again', you should appreciate how your skills have expanded.

Class discussions may be used by your teacher to promote discussion of key concepts.

Most chapters have a 'Mixed exercise' after the main work of the chapter has been completed. This will help you to revise what you have done, either when you have finished the chapter or at a later date.

Practical or investigational work is shown in the book by the icon. For this work you are encouraged to share your ideas with others, to use any mathematics you are familiar with, and to try to solve each problem in different ways, appreciating the advantages and disadvantages of each method.

The book starts with a summary of the main results from Years 7 and 8. After every five chapters you will find a further summary. This lists the most important points that have been studied in the previous chapters and concludes with revision exercises that test the work you have studied up to that point. This is followed by a mental arithmetic test.

# Introduction

 Hints are included throughout to help you avoid common errors and misconceptions.

The first time key words appear, they are highlighted in **bold blue** text. A definition can be found in the 'Key words' section at the back of the book so that you can check the meanings of words.

At this stage you will find that you use a calculator frequently but it is still unwise to rely on a calculator for work that you should do in your head. Remember, whether you use a calculator or do the working yourself, always estimate your answer and always ask yourself the question, 'Is my answer a sensible one?'

Mathematics is an exciting and enjoyable subject when you understand what is going on. Remember, if you don't understand something, ask someone who can explain it to you. If you still don't understand, ask again. Good luck with your studies.

## To the teacher

This is the third book of the STP Mathematics Third Edition. It is based on the ST(P) Mathematics series but has been extensively rewritten and is now firmly based on the 2014 Programme of Study for Key Stage 3.

This series of books aims to prepare students for a high level of achievement at Key Stage 3 leading on to the higher tier at GCSE.

The majority of scientific calculators now on sale use direct keying sequences for entering functions such as $\tan 33$, $\sqrt{2}$. This is the order used in this book.

## Kerboodle

This textbook is supported by online resources on Kerboodle, including a wide range of assessment activities and teacher support to help develop problem-solving skills. Kerboodle also includes an online version of this student book for independent use at home.

# Summary of Years 7 and 8

## Types of number

A *factor* of a number will divide into the number exactly.

A *multiple* of a number has that number as a factor,

e.g.    3 is a factor of 12   and   12 is a multiple of 3.

A *prime number* has only 1 and itself as factors, e.g. 7. Remember that 1 is not a prime number.

*Square numbers* can be drawn as a square grid of dots, e.g. 9:

*Rectangular numbers* can be drawn as a rectangular grid of dots, e.g. 6:

*Triangular numbers* can be drawn as a triangular grid of dots, e.g. 6:

## Operations of ×, ÷, +, −

The sign in front of a number refers to that number only.

When a calculation involves a *mixture of operations*, start by calculating anything inside brackets, then follow the rule 'do the multiplication and division first'.

## Fractions

*Equivalent fractions* are formed by multiplying or by dividing the top and the bottom of a fraction by the same number,

e.g.    $\frac{1}{2} = \frac{3}{6}$   (multiplying top and bottom of $\frac{1}{2}$ by 3)

Fractions can be *added* or *subtracted* when they have the *same* denominator, e.g. to add $\frac{1}{2}$ to $\frac{1}{3}$, we first change them into equivalent fractions with the same denominators,

i.e.    $\frac{1}{2} + \frac{1}{3} = \frac{1 \times 3}{2 \times 3} + \frac{1 \times 2}{3 \times 2} = \frac{3}{6} + \frac{2}{6} = \frac{5}{6}$

To *multiply one fraction by another fraction,* we multiply their numerators and multiply their denominators,

e.g.    $\frac{1}{2} \times \frac{5}{3} = \frac{1 \times 5}{2 \times 3} = \frac{5}{6}$

To *divide by a fraction*, we multiply by the fraction turned upside down,

e.g. $\quad \frac{1}{2} \div \frac{5}{3} = \frac{1}{2} \times \frac{3}{5} = \frac{3}{10}$

To multiply or divide with *mixed numbers*, e.g. $1\frac{3}{4}$, first change the mixed numbers to improper fractions.

A *fraction can be changed to a decimal* by dividing the bottom number into the top number,

e.g. $\quad \frac{3}{8} = 3 \div 8 = 0.375$

A *fraction can be expressed as a percentage* by multiplying the fraction by 100,

e.g. $\quad \frac{2}{5} = \frac{2}{5} \times 100\% = \frac{2}{5} \times \frac{100}{1}\% = 40\%$

To find *a fraction of a quantity*, we multiply the fraction by the quantity,

e.g. $\quad \frac{1}{2}$ of $\frac{3}{4}$ means $\frac{1}{2} \times \frac{3}{4}$, and $\frac{3}{8}$ of £24 $= £\left(\frac{3}{8} \times 24\right)$

To express *one quantity as a fraction of another*, first make sure that both quantities are in the same unit, then place the first quantity over the second,

e.g. $\quad$ 24 p as a fraction of £2 is $\frac{24}{200}\left(=\frac{3}{25}\right)$

## Decimals

*Decimals can be added or subtracted* using the same methods as for whole numbers, provided that the decimal points are placed in line.

*To multiply a decimal by 10, 100, 1000, ...,* we move the point 1, 2, 3, ... places to the right,

e.g. $\quad 2.56 \times 10 = 25.6, \quad$ and $\quad 2.56 \times 1000 = 2560(.0)$

*To divide a decimal by 10, 100, 1000, ...,* we move the point 1, 2, 3, ... places to the left (equivalent to moving the figures to the right),

e.g. $\quad 2.56 \div 10 = 0.256, \quad$ and $\quad 2.56 \div 1000 = 0.00256$

*To multiply decimals* without using a calculator, first ignore the decimal point and multiply the numbers. Then add together the number of decimal places in each of the decimals being multiplied together; this gives the number of decimal places in the answer,

e.g. $\quad 7.5 \times 0.5 = 3.75 \quad (75 \times 5 = 375)$
$\quad\quad\quad [(1) + (1) = (2)]$

*To divide by a decimal*, move the point in *both* numbers to the right until the number we are dividing by is a whole number,

e.g. $\quad 2.56 \div 0.4 = 25.6 \div 4 = 6.4$

*A decimal can be expressed as a percentage* by multiplying the decimal by 100,

e.g.    $0.325 = 32.5\%$

## Significant figures

The first significant figure in a number is the first non-zero figure when reading from left to right.

The second significant figure is the next figure to the right, whether or not it is zero, and so on.

For example, in 0.0205,    the first significant figure is 2, and the second significant figure is 0.

## Rounding numbers

To round (i.e. to correct) a number to a specified place value or number of significant figures, look at the digit in the next place: if it is 5 or more, add 1 to the specified digit, otherwise leave the specified digit as it is,

e.g.    $1\,3|7 \;\;\;\; = 140$ to the nearest 10, or 2 s.f.

$2.|5\,6\,4 \;= 3$ to the nearest whole number, or 1 s.f.

$2.5\,6|4 \;= 2.56$ correct to 2 decimal places, or 3 s.f.

## Standard form

A number written in standard form is a number between 1 and 10 multiplied by a power of ten,

e.g.    $1.2 \times 10^5$

## Percentages

'Per cent' means 'out of one hundred'.

Hence *a percentage can be expressed as a fraction* by placing the percentage over 100,

e.g.    $33\% = \dfrac{33}{100}$,

and *a percentage can be expressed as a decimal* by dividing the percentage by 100, that is, by moving the decimal point two places to the left,

e.g.    $33\% = 0.33$

To find *one quantity as a percentage of another quantity*, we place the first quantity over the second quantity and multiply this fraction by 100,

e.g.    24 p as a percentage of £2 is $\dfrac{24}{200} \times \dfrac{100}{1}\% = 12\%$

To find *a percentage of a quantity*, change the percentage to a decimal and multiply it by the quantity,

e.g.    $32\%$ of £18 $= 32\% \times £18 = 0.32 \times £18 = £5.76$

To *increase* a quantity by 15%,
we find *the increase* by finding 15% of the quantity;
we find *the new quantity* directly by finding 100% + 15%, i.e. 115%, of the original quantity; so we multiply it by 1.15.

To *decrease* a quantity by 15%,
we find *the decrease* by finding 15% of the quantity,
we find *the new quantity* directly by finding 100% − 15%, i.e. 85%, of the original quantity; so we multiply it by 0.85.

## Directed numbers

Positive and negative numbers are collectively known as directed numbers. They can be represented on a number line.

$$-3 \quad -2 \quad -1 \quad 0 \quad 1 \quad 2 \quad 3 \quad 4 \quad 5 \quad 6$$

The *rules for multiplying and dividing pairs of directed numbers* are

when the signs are the same, the result is positive.
when the signs are different, the result is negative.

For example, $\quad (+2) \times (+3) = +6 \quad$ and $\quad (-2) \times (-3) = +6$
$\quad\quad\quad\quad\quad\quad (+2) \times (-3) = -6 \quad$ and $\quad (-8) \div (+2) = -4$

Similar rules apply to adding and subtracting directed numbers,
e.g. $\quad 3 + (+2) = 3 + 2 = 5 \quad$ and $\quad 3 - (-2) = 3 + 2 = 5$
$\quad\quad 3 + (-2) = 3 - 2 = 1 \quad$ and $\quad 3 - (+2) = 3 - 2 = 1$

## Square roots

A square root of 20 is a number which, when multiplied by itself, gives 20. A number has two square roots, one positive and one negative, for example if $x^2 = 4$, then $x = \pm 2$

2 is the positive square root of 4 and −2 is the negative square root of 4.

$\sqrt{20}$ means the positive square root of 20.

## Ratio

Ratios are used to compare the relative sizes of quantities.

For example, if a model car is 20 cm long and the real car is 200 cm long, we say that their lengths are in the ratio 20 : 200.

Ratios can be simplified by dividing the parts of the ratio by the same number,
e.g. $\quad$ 20 : 200 = 1 : 10 (dividing 20 and 200 by 20).

A *map ratio* is the ratio of a length on a map to the length it represents on the ground. When expressed as a fraction (or sometimes as a ratio), it is called the representative fraction.

## Direct proportion

Two quantities are directly proportional when they are always in the same ratio.

For example, if 1 kilogram of apples costs 96 p, then $n$ kg of apples cost $96n$ pence, so the ratio of weight to cost is $n:96n = 1:96$, that is, the weight of these apples is directly proportional to their cost.

## Indices

When a number is written in the form $3^4$, 3 is called the *base* and 4 is called the *index* or *power*. $3^4$ means $3 \times 3 \times 3 \times 3$.

$a^2$ is called $a$ squared, $a^3$ is called $a$ cubed.

### Negative index

$3^{-4}$ means $\frac{1}{3^4}$

### Zero index

$3^0 = 1$. In fact, $a^0 = 1$, whatever number $a$ stands for, except when $a = 0$.

### Rules of indices

We can multiply different powers of the same base by adding the indices,

e.g. $\quad 3^4 \times 3^2 = 3^{4+2} = 3^6$

We can divide different powers of the same base by subtracting the indices,

e.g. $\quad 3^4 \div 3^2 = 3^{4-2} = 3^2$

## Units

*Metric units of length* in common use are the kilometre, metre, centimetre and millimetre, where

$\quad$ 1 km = 1000 m $\quad$ 1 m = 100 cm $\quad$ 1 cm = 10 mm

*Metric units of mass* are the tonne (t), kilogram (kg), gram (g) and milligram (mg), where

$\quad$ 1 tonne = 1000 kg $\quad$ 1 kg = 1000 g $\quad$ 1 g = 1000 mg

*Imperial units of length* in common use are the mile, yard (yd), foot (ft) and inch (in), where

$\quad$ 1 mile = 1760 yards $\quad$ 1 yard = 3 feet $\quad$ 1 foot = 12 inches

*Imperial units of mass* still in common use are the ton, hundredweight (cwt), stone, pound (lb) and ounce (oz), where

$\quad$ 1 ton = 2240 lb $\quad$ 1 stone = 14 lb $\quad$ 1 lb = 16 ounces

For a *rough conversion* between metric and imperial units, use

$\quad$ 1 km $\approx \frac{1}{2}$ mile $\quad$ 1 yard $\approx$ 1 m $\quad$ 1 kg $\approx$ 2 lb $\quad$ 1 tonne $\approx$ 1 ton

For a more accurate conversion, use

5 miles ≈ 8 km    1 inch ≈ 2.5 cm    1 kg ≈ 2.2 lb

*Area* is measured by standard-sized squares.

$1\,cm^2 = 10 \times 10\,mm^2 = 100\,mm^2$
$1\,m^2 = 100 \times 100\,cm^2 = 10\,000\,cm^2$
$1\,km^2 = 1000 \times 1000\,m^2 = 1\,000\,000\,m^2$

*Volume* is measured by standard-sized cubes.

$1\,cm^3 = 10 \times 10 \times 10\,mm^3 = 1000\,mm^3$
$1\,m^3 = 100 \times 100 \times 100\,cm^3 = 1\,000\,000\,cm^3$

The *capacity* of a container is the volume of liquid it can hold. The main *metric units of capacity* are the litre and the millilitre (ml), where

1 litre = 1000 ml    and    1 litre = 1000 cm³    so    1 ml = 1 cm³

The main *Imperial units of capacity* are the gallon and the pint, where

1 gallon = 8 pints

Rough conversions between metric and Imperial units of capacity are given by

1 litre ≈ 1.75 pints    and    1 gallon ≈ 4.5 litres

## Circles

The *diameter* of a circle is twice the *radius*.

The *circumference* is given by    $C = 2\pi r$,
where $r$ units is the radius of the circle and    $\pi = 3.1415...$

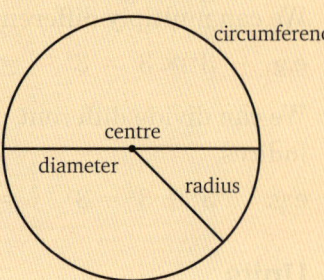

## Area

The *area of a square* = (length of a side)²

The *area of a rectangle* = length × breadth

The *area of a circle* is given by    $A = \pi r^2$

The *area of a parallelogram* is given by    $A$ = length × height

The *area of a triangle* is given by

$A = \frac{1}{2}$ base × height

When we talk about the height of a triangle or of a parallelogram, we mean the perpendicular height.

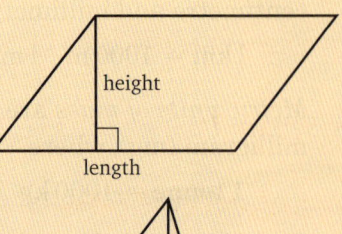

## Volume and capacity

The *volume of a cuboid* = length × breadth × height

A solid with a constant cross-section is called a *prism*.

The *volume of a prism* is given by    area of cross-section × length

The *volume of a cylinder* is given by $\quad V = \pi r^2 h$

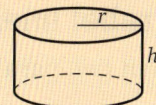

## Density

The *density* of a material is the mass of one unit of volume of the material, for example, the density of silver is 10.5 g/cm³, that is, 1 cm³ of silver weighs 10.5 g.

## Sets

A *set* is a collection of objects that have something in common.

The objects are called *members* or *elements*. The members are listed inside curly brackets and separated by commas. For example, the set of even numbers from 1 to 9 is written as {2, 4, 6, 8}.

## Venn diagrams

*Venn diagrams* represent sets using circles inside a rectangle. This Venn diagram represents the sets A and B:

The rectangle represents the *universal set*, U. This is a set that includes the members of A and B and possibly other members not in A or in B.

The *intersection* of the sets A and B is the set containing the members that are in both A and B and is written $A \cap B$. This is represented by the shaded region in this Venn diagram.

The *union* of the sets A and B is the set containing the members of A and the members of B. The union of A and B is written $A \cup B$ and is represented by the shaded region in this Venn diagram.

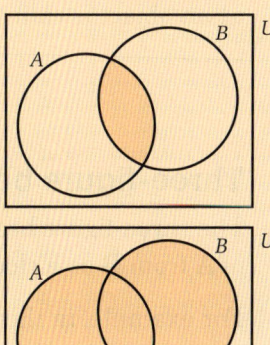

## Angles

One complete revolution = 4 right angles = 360°.

An *acute angle* is less than 90°.

An *obtuse angle* is larger than 90° but less than 180°.

A *reflex angle* is larger than 180°.

*Vertically opposite* angles are equal.

*Angles on a straight line* add up to 180°. Two angles that add up to 180° are called *supplementary angles*.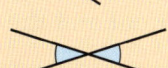

Angles at a point add up to 360°.

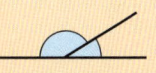
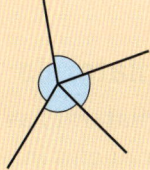

## Parallel lines

When two parallel lines are cut by a transversal,

the *corresponding angles* are equal

the *alternate angles* are equal

the *interior angles* add up to 180°.

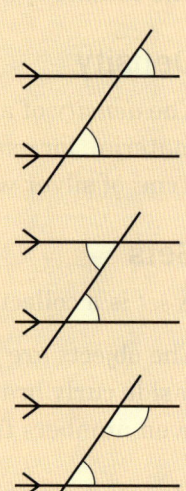

## Angles of elevation and depression

If you start by looking straight ahead, the angle that you turn your eyes through to look *up* at an object is called the angle of elevation, the angle you turn your eyes through to look *down* at an object is called the angle of depression.

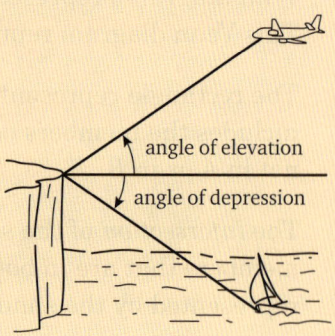

## Three-figure bearings

The three-figure bearing of a point A from a point B gives the direction of A from B as a clockwise angle measured from the north.

For example, in this diagram, the bearing of A from B is 140°.

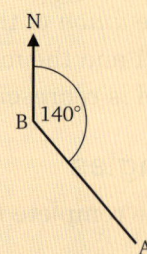

## Triangles

The three angles in a triangle add up to 180°.

An *equilateral triangle* has all three sides equal and each angle is 60°.

An *isosceles triangle* has two sides equal and the angles at the base of these sides are equal.

## Quadrilaterals

A quadrilateral has four sides.
The four angles in a quadrilateral add up to 360°.

## Special quadrilaterals

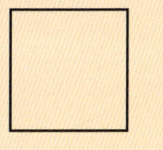

In a square
- all four sides are the same length
- both pairs of opposite sides are parallel
- all four angles are right angles.

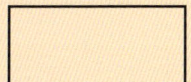

In a rectangle
- both pairs of opposite sides are the same length
- both pairs of opposite sides are parallel
- all four angles are right angles.

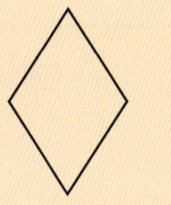

In a rhombus
- all four sides are the same length
- both pairs of opposite sides are parallel
- the opposite angles are equal.

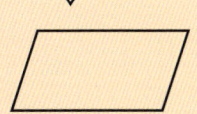

In a parallelogram
- the opposite sides are the same length
- the opposite sides are parallel
- the opposite angles are equal.

In a trapezium
- just one pair of opposite sides are parallel.

## Polygons

A polygon is a plane figure bounded by straight lines, e.g.

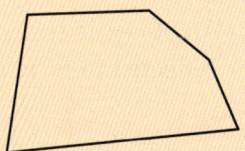

A *regular polygon* has all angles equal and all sides the same length. This is a regular hexagon.

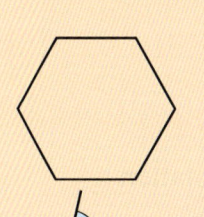

The *sum of the exterior angles* of any polygon is 360°.

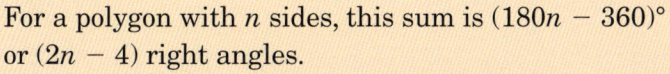

The *sum of the interior angles* of any polygon depends on the number of sides.

For a polygon with $n$ sides, this sum is $(180n - 360)°$ or $(2n - 4)$ right angles.

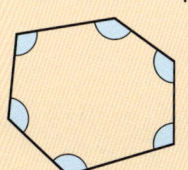

## Pythagoras' theorem

Pythagoras' theorem states that in any right-angled triangle ABC with $\hat{C} = 90°$,

$$AB^2 = AC^2 + BC^2$$

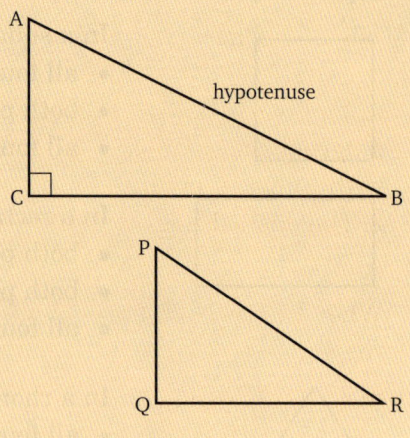

Conversely if, in a triangle PQR, $PR^2 = PQ^2 + QR^2$

then $\hat{Q} = 90°$

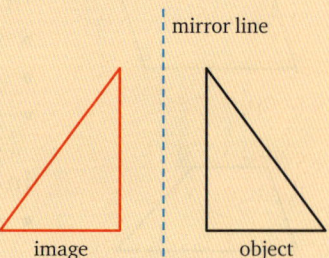

## Congruence

Two figures are congruent when they are exactly the same shape and size.

## Transformations

### Reflection in a mirror line

When an object is reflected in a mirror line, the object and its image form a symmetrical shape, with the mirror line as the axis of symmetry.

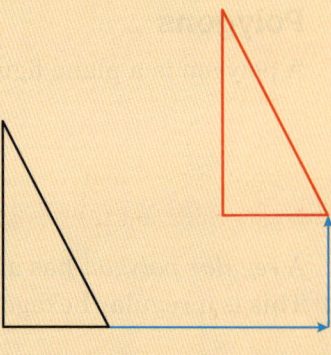

### Translation

An object is translated when it moves without being turned or reflected to form an image.

### Rotation

When an object is rotated about a point to form an image, the point about which it is rotated is called the *centre of rotation* and the angle it is turned through is called the *angle of rotation*.

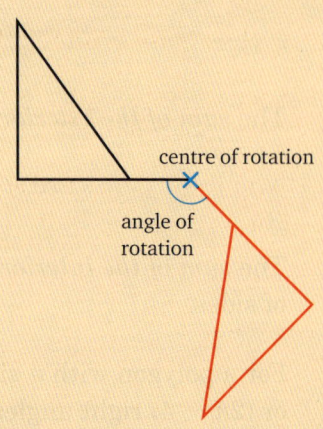

## Enlargement

When an object is enlarged by scale factor 2, each line on the image is twice the length of the corresponding line on the object. The diagram shows an enlargement of a triangle, with centre of enlargement X and scale factor 2. The dashed lines are guide-lines.

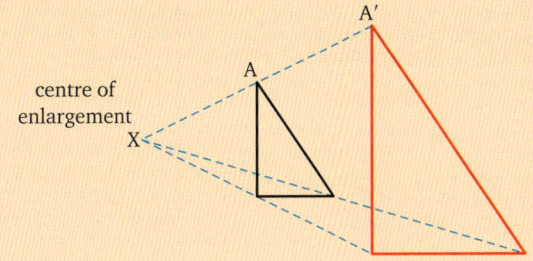

centre of enlargement

$XA' = 2XA$

When the scale factor is less than one, the image is smaller than the object.

This diagram shows an enlargement with scale factor $\frac{1}{4}$ and centre of enlargement O.

$OA' = \frac{1}{4}OA$

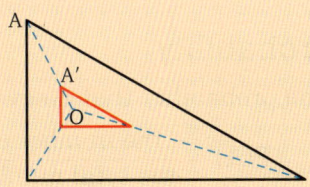

## Statistics

A *hypothesis* is a statement that is not known to be true or untrue.

*Discrete values* are exact and distinct, for example, the number of people in a queue.

*Continuous values* can only be given in a range of a continuous scale, for example, the length of a piece of wood.

For a list of values
- the *range* is the difference between the largest value and the smallest value
- the *mean* is the sum of all the values divided by the number of values
- the *median* is the middle value when the values have been arranged in order of size, (when the middle of the list is halfway between two values, the median is the average of these two values)
- the *mode* is the value that occurs most frequently.

For a grouped frequency distribution
- the *range* is estimated as the higher end of the last group minus the lower end of the first group
- the *modal group* is the group with the largest number of items in it.

We get a *scatter graph* when we plot values of one quantity against corresponding values of another quantity.

When the points are scattered about a straight line, we can draw that line by eye. It is called the *line of best fit*.

We use the word *correlation* to describe the amount of scatter about this line.

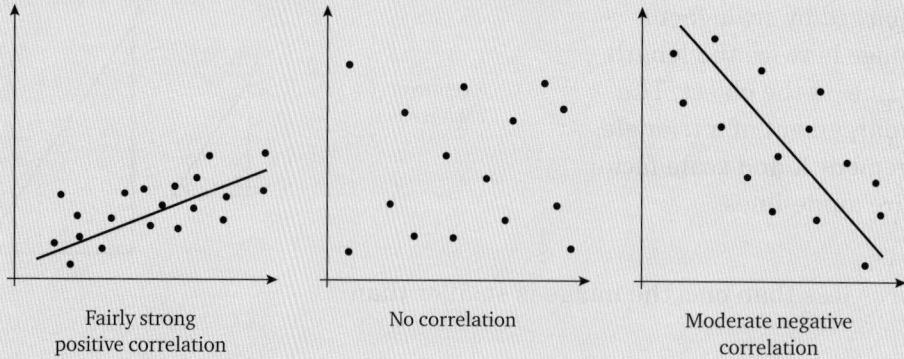

Fairly strong positive correlation

No correlation

Moderate negative correlation

## Probability

The probability that an event $A$ happens is $P(A)$, where

$$P(A) = \frac{\text{the number of ways in which } A \text{ can occur}}{\text{the total number of equally likely outcomes}}$$

If $A$ cannot happen, $P(A) = 0$. If $A$ is certain to happen, $P(A) = 1$. In general, $P(A)$ lies between 0 and 1.

The probability that an event $A$ does not happen, $P(\overline{A})$, is equal to one minus the probability that it does happen, i.e. $P(\overline{A}) = 1 - P(A)$.

If $p$ is the probability that an event happens on one occasion, then we expect it to happen $np$ times on $n$ occasions. For example, if we toss an unbiased coin 50 times, we expect

$\frac{1}{2} \times 50 = 25$ heads

When we perform experiments to find out how often an event occurs, the *relative frequency* of the event is given by

$$\frac{\text{the number of times the event occurs}}{\text{the number of times the experiment is performed}}$$

Relative frequency is used to give an approximate value for probability.

## Formulas

A formula is a general rule for finding one quantity in terms of other quantities. For example, the formula for finding the area of a rectangle is given by    Area = length × breadth

When letters are used for unknown numbers, the formula can be written more concisely, so the area, $A$ cm$^2$, of a rectangle measuring $l$ cm by $b$ cm, is given by the formula    $A = l \times b$

## Algebraic expressions

Terms such as $5n$ mean    $5 \times n = n + n + n + n + n$
Similarly $ab$ means $a \times b$.
$2x + 5x$ can be simplified to $7x$.

## Multiplication and division of algebraic fractions

The same rules apply to fractions with letter terms as to fractions with numbers only,

i.e. to multiply fractions, multiply the numerators and multiply the denominators,

e.g. $\dfrac{2x}{3} \times \dfrac{5x}{7} = \dfrac{2x \times 5x}{3 \times 7} = \dfrac{10x^2}{21}$

and to divide by a fraction, turn it upside down and multiply,

e.g. $\dfrac{x}{6} \div \dfrac{2x}{3} = \dfrac{\cancel{x}^1}{\cancel{6}_2} \times \dfrac{\cancel{3}^1}{2\cancel{x}_1} = \dfrac{1}{4}$

## Simplification of brackets

$x(2x - 3)$ means $x \times 2x + x \times (-3)$.

Therefore $x(2x - 3) = 2x^2 - 3x$

## Solving equations

An equation is a relationship between an unknown number, represented by a letter, and other numbers, for example, $2x - 3 = 5$

Solving the equation means finding the unknown number.

Provided that we do the same thing to both sides of an equation, we keep the equality. This can be used to solve the equation.

When an equation contains brackets, first multiply out the brackets,

e.g. $3x - 2(3 - x) = 6$

gives $3x - 6 + 2x = 6$

When an equation contains fractions, multiply each term in the equation by the lowest number that each denominator divides into exactly. This will eliminate all fractions from the equation,

e.g. if $\dfrac{x}{2} + 1 = \dfrac{2}{3}$

multiplying each term by 6 gives

$\dfrac{6}{1} \times \dfrac{x}{2} + 6 \times 1 = \dfrac{6}{1} \times \dfrac{2}{3}$ which simplifies to $\quad 3x + 6 = 4$

which can be solved easily.

Two equations with two unknown quantities are called *simultaneous equations*. A pair of simultaneous equations can be solved algebraically by eliminating one of the letters. If two letter terms are the same size:

when the signs are different, we add the equations;

when the signs are the same we subtract the equations.

For example, to eliminate $y$ from $\quad 2x + y = 5 \quad$ [1]

and $\quad 3x - y = 7 \quad$ [2]

we add [1] and [2] to give $\quad 5x = 12 \quad$ [3]

The value of $x$ can be found from [3]. This value is then substituted for $x$ in [1] or [2] to find $y$.

*Polynomial equations* in one unknown contain terms involving powers of $x$,

e.g. $x^3 - 2x - 4 = 2$ and $2x^2 = 5$ are polynomial equations.

Equations containing an $x^2$ term and a number only may be solved by finding square roots.

More complex equations can be solved by *trial and improvement*. That means trying possible values for $x$ until we find a value that fits the equation.

Equations can also be solved by *drawing a graph*, for example to solve $x^3 - x = 10$, we draw the graph of $y = x^3 - x - 10$. The solutions are the values of $x$ where this graph crosses the $x$-axis (i.e. where $y = 0$).

## Inequalities

An inequality remains true when the same number is added to, or subtracted from, both sides,

e.g. if $x > 5$ then $x + 2 > 5 + 2$
and $x - 2 > 5 - 2$

An inequality also remains true when both sides are multiplied, or divided, by the same *positive* number,

e.g. if $x > 5$ then $2x > 10$
and $\frac{x}{2} > \frac{5}{2}$

However, multiplication or division by a negative number must be avoided because this destroys the inequality.

## Graphs

The equation of a line or curve gives the $y$-coordinate of a point on the line or curve in terms of its $x$-coordinate. This relationship between the coordinates is true only for points on the line or curve.

### Straight lines

The *gradient* of a straight line can be found from any two points, P and Q, on the line, by calculating

$$\frac{\text{increase in } y \text{ in moving from P to Q}}{\text{increase in } x \text{ in moving from P to Q}}$$

$$= \frac{(y\text{-coordinate of Q}) - (y\text{-coordinate of P})}{(x\text{-coordinate of Q}) - (x\text{-coordinate of P})}$$

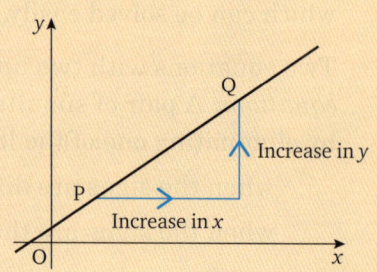

When the gradient is positive, the line slopes uphill when moving from left to right.

When the gradient is negative, the line slopes downhill when moving from left to right.

*The equation of a straight line* is of the form $y = mx + c$

where     $m$ is the gradient of the line

and       $c$ is the $y$-intercept,

e.g.      the line whose equation is $y = 2x - 3$ has gradient 2 and $y$-intercept $-3$.

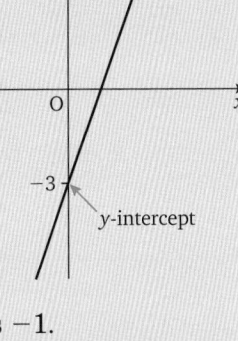

Two lines are parallel when they have the same gradient,

e.g.      $y = 2x + 1$  and  $y = 2x - 5$  are parallel.

Two lines are perpendicular when the product of their gradients is $-1$.

e.g      $y = 2x + 4$  and  $y = -\frac{1}{2}x + 6$  are perpendicular.

An equation of the form    $y = c$    gives a line parallel to the $x$-axis.

An equation of the form    $x = b$    gives a line parallel to the $y$-axis.

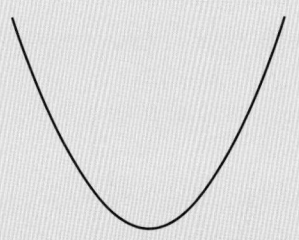

## Curves

A *parabola* is a curve whose equation is in the form $y = ax^2 + bx + c$.

The shape of this curve looks like this:

When the $x^2$ term is positive, the curve is as shown.

When the $x^2$ term is negative, the curve is upside down.

The exercises that follow are *not* intended to be worked through before starting the main part of this book. They are here for you to use when you need practice of the basic techniques.

## REVISION EXERCISE 1 (Fractions and decimals – the four rules)

**Do not use a calculator for any questions in this exercise.**

1  Express as a single fraction in its simplest form

   **a**  $\frac{2}{3} + \frac{7}{8}$        **d**  $\frac{3}{16} + \frac{3}{4} + \frac{5}{12}$       **g**  $\frac{3}{4} + \frac{13}{20} + \frac{4}{5}$

   **b**  $\frac{3}{5} + \frac{3}{30}$       **e**  $\frac{1}{4} + \frac{2}{3}$                      **h**  $\frac{2}{7} + \frac{1}{9} + \frac{1}{6}$

   **c**  $\frac{2}{7} + \frac{1}{2} + \frac{3}{14}$   **f**  $\frac{5}{6} + \frac{2}{3} + \frac{1}{4}$   **i**  $\frac{8}{21} + \frac{1}{2} + \frac{2}{3}$

**2** Express as a single fraction in its simplest form

  **a** $\frac{7}{9} - \frac{5}{12}$      **d** $\frac{1}{10} - \frac{1}{20}$      **g** $2\frac{3}{4} + 1\frac{1}{2} - 1\frac{1}{3}$

  **b** $\frac{1}{4} - \frac{2}{9}$      **e** $2\frac{2}{5} - 1\frac{3}{8}$      **h** $1\frac{3}{8} + 1\frac{1}{4} - 2\frac{1}{2}$

  **c** $\frac{3}{10} - \frac{1}{15}$      **f** $5\frac{5}{6} - 2\frac{5}{9}$      **i** $3\frac{1}{5} - 4\frac{1}{8} + 1\frac{7}{10}$

**3** Express as a single fraction in its simplest form

  **a** $\frac{2}{3} \times \frac{5}{6}$      **d** $\frac{2}{5}$ of $1\frac{3}{7}$      **g** $\frac{2}{9} \div 1\frac{2}{7}$

  **b** $\frac{2}{3} \times \frac{1}{4} \times \frac{3}{5}$      **e** $\frac{2}{5} \times \frac{7}{8} \times \frac{3}{7} \times \frac{10}{11}$      **h** $\frac{3}{7} \div 1\frac{3}{4}$

  **c** $\frac{1}{2}$ of $1\frac{1}{3} \times \frac{5}{7}$      **f** $1\frac{2}{3} \div \frac{5}{6}$      **i** $3 \div \frac{2}{3}$

**4** Express as a single fraction in its simplest form

  **a** $1\frac{2}{3} \times \frac{1}{2} - \frac{3}{5}$

  **b** $\frac{2}{7} + \frac{1}{4} \times 1\frac{1}{3}$

  **c** $4\frac{1}{2} \div 3 + \frac{3}{4}$

  **d** $\left(2\frac{1}{5} + 1\frac{2}{3}\right) \div 5\frac{4}{5}$

  **e** $\left(2\frac{1}{2} - 1\frac{1}{3}\right) \div 4\frac{2}{3}$

  **f** $\frac{4}{5} \div \frac{1}{4} + \frac{1}{3} \times 4\frac{1}{2}$

  **g** $2\frac{2}{5} \times 1\frac{7}{8} - 1\frac{2}{3}$

  **h** $2\frac{1}{7} - \frac{1}{3}$ of $1\frac{2}{7}$

  **i** $\left(\frac{1}{2} - \frac{1}{3}\right) \div \left(\frac{3}{4} - \frac{1}{3}\right)$

  **j** $\left(4\frac{1}{2} - 3\frac{3}{8}\right) \times 1\frac{1}{3}$

  **k** $\left(\frac{11}{12} - \frac{1}{2}\right) \times \frac{2}{5} + \frac{1}{2}$

  **l** $2\frac{1}{2} \div 1\frac{3}{7} + 1\frac{1}{3}$

**5** Find

  **a** $1.26 + 3.75$

  **b** $12.4 + 6.7$

  **c** $5.82 + 0.35$

  **d** $0.04 + 8.86$

  **e** $4.002 + 0.83$

  **f** $0.000\,32 + 0.0017$

  **g** $5.3 - 2.1$

  **h** $0.16 - 0.08$

  **i** $1.07 - 0.58$

  **j** $0.37 - 0.009$

  **k** $2 - 0.17$

  **l** $0.0127 - 0.0059$

**6** Find
   **a** $1.2 \times 0.8$
   **b** $0.7 \times 0.06$
   **c** $0.4 \times 0.02$
   **d** $0.5 \times 0.5$
   **e** $3.0501 \times 1.1$
   **f** $1.002 \times 0.36$
   **g** $1.08 \div 0.4$
   **h** $0.2 \div 2.5$
   **i** $0.18 \div 1.2$
   **j** $42.8 \div 200$
   **k** $0.01 \div 0.5$
   **l** $0.0013 \div 1.3$

**7** Fill in the blanks in the following calculations.
   **a** $3.7 \times \square = 15.54$
   **b** $22.96 \div \square = 8.2$
   **c** $0.374 \times 0.06 = \square$
   **d** $\square \div 0.45 = 1.44$
   **e** $0.37 \times 1.92 - \square = 0$
   **f** $\square \times 0.85 = 1.105$
   **g** $5.9 \times \square = 0.2537$
   **h** $\square - 4.08 \div 1.7 = 0$
   **i** $0.00162 \div 0.045 = \square$
   **j** $\square \div 0.026 = 1.4$

## REVISION EXERCISE 2 (Using fractions, decimals, ratios and percentages)

**1 a** Express as a decimal
   **i** $\frac{9}{25}$   **ii** $\frac{19}{20}$   **iii** $54\%$   **iv** $82\frac{1}{2}\%$
   **b** Express as a fraction in its lowest terms
   **i** $85\%$   **ii** $0.42$   **iii** $65\%$   **iv** $0.125$
   **c** Express as a percentage
   **i** $0.44$   **ii** $\frac{7}{25}$   **iii** $1.38$   **iv** $\frac{37}{40}$

**2** Copy and complete the table.

|   | Fraction | Percentage | Decimal |
|---|---|---|---|
| a | $\frac{17}{20}$ | | |
| b | | $37\frac{1}{2}\%$ | |
| c | | | $0.625$ |
| d | | $5\frac{3}{4}\%$ | |
| e | | | $1.15$ |
| f | $4\frac{3}{4}$ | | |

**3 a** Put either < or > between each of the following pairs of fractions.

    **i** $\frac{5}{8}$   $\frac{7}{10}$     **ii** $\frac{2}{5}$   $\frac{1}{3}$     **iii** $\frac{1}{5}$   $\frac{4}{15}$     **iv** $\frac{4}{7}$   $\frac{5}{9}$

**b** Change into an improper fraction

    **i** $2\frac{3}{7}$      **ii** $5\frac{4}{9}$      **iii** $3\frac{3}{5}$      **iv** $9\frac{3}{4}$

**c** Give as a mixed number

    **i** $\frac{42}{5}$      **ii** $\frac{17}{4}$      **iii** $\frac{46}{7}$      **iv** $\frac{26}{17}$

**4 a** Find

    **i** 36% of 50 kg      **ii** 4.5% of 440 g      **iii** 84% of 15 m

**b**  **i** Increase £480 by 45%

    **ii** Decrease £320 by 55%

    **iii** Increase 150 cm² by 56%

    **iv** Decrease £44 by 35%

**c** Find, giving your answer correct to 3 significant figures

    **i** 37% of 46 km      **ii** $4\frac{3}{4}$% of 12.6 m      **iii** $13\frac{1}{2}$% of 245 mm

**d** Express

    **i** 12 mm as a percentage of 6 cm

    **ii** 650 m as a fraction of 2 km

    **iii** 56 cm² as a percentage of 1 m²

    **iv** 6 pints as a fraction of 4 gallons.

**5 a** Give the following ratios in their simplest form.

    **i** 12 : 18      **iii** 320 : 480      **v** $\frac{1}{2} : \frac{5}{6} : \frac{2}{3}$

    **ii** 3 : 6 : 9      **iv** 3.5 : 2.5      **vi** 288 : 128 : 144

**b** Simplify the following ratios.

    **i** 45 cm : 0.1 m      **iii** 340 m : 1.2 km      **v** 450 mg : 1 g

    **ii** 42 p : £1.05      **iv** 32 g : 2 kg      **vi** 2.2 t : 132 kg

**c** Find $x$ if

    **i** $x : 5 = 2 : 9$      **iii** $x : 6 = 5 : 4$      **v** $3 : 8 = 9 : x$

    **ii** $x : 3 = 1 : 7$      **iv** $5 : x = 7 : 2$      **vi** $15 : 2 = x : 3$

**d**  **i** Divide £45 into two parts in the ratio 4 : 5.

    **ii** Divide 96 m into two parts in the ratio 9 : 7.

    **iii** Divide 5 kg into three parts in the ratio 1 : 2 : 5.

    **iv** Divide seven hours into three parts in the ratio 1 : 5 : 8.

**6 a** Find the map ratio of a map on which 10 cm represents 1 km.

**b** The map ratio of a map is 1 : 200 000. The distance between two factories is 8 km. What distance is this on the map?

**7 a** In a sale a pair of trainers priced £35 is reduced by 30%. What is the sale price?

**b** Sally and Tim bought a portable CD player between them for £44.94. Sally paid $\frac{4}{7}$ of the cost and Tim paid the remainder.

    **i** What fraction did Tim pay?

    **ii** How much did Sally pay?

**8** At a concert 64% of the audience were females.
  **a** What fraction of the audience were females?
  **b** Express the part of the audience that was male as
   **i** a percentage
   **ii** a decimal
   **iii** a fraction in its lowest terms.

**9** Estimate the value of 236.4 ÷ 48.7, and then use a calculator to find its value correct to 2 decimal places.

**10** A popular leisure club has 2750 members. Of these, 42% are girls, 0.3 are boys, $\frac{4}{25}$ are men and the remainder are women.
  **a** What fraction of the members are girls?
  **b** What percentage of the members are women?
  **c** What decimal fraction of the members are male?
  **d** How many of the members are females?

**11 a** Find
   **i** $\frac{2}{3}$ of £36
   **ii** $\frac{3}{4}$ of 34 cm
   **iii** $\frac{4}{9}$ of 54 kg.
  **b** Which is the smaller, $\frac{5}{12}$ of 10 or $\frac{3}{4}$ of 5?
  **c** Which is the larger, $\frac{3}{5}$ of $\frac{9}{10}$ or $\frac{5}{7}$ of $\frac{3}{4}$?

## REVISION EXERCISE 3 (Number work)

**1** Without using a calculator, find
  **a** 349 + 276
  **b** 723 − 584
  **c** 66 × 80
  **d** 48 × 500
  **e** 336 ÷ 6
  **f** 560 ÷ 80
  **g** 7 × 63 − 249
  **h** 6421 − 236 × 7
  **i** (19 + 6) × 5 − 96
  **j** 429 ÷ 21, giving the remainder
  **k** (36 − 14) × 3 − 49
  **l** 339 ÷ 23, giving the remainder

**2 a** Find the value of  **i** $2^6$  **ii** $3^5$  **iii** $2^3 \times 3^2 \times 7$
  **b** Express in index form
   **i** 128  **ii** 343  **iii** 625  **iv** 729
  **c** Express as the product of prime factors in index form
   **i** 1080  **ii** 3276  **iii** 1800

**3 a** Find the lowest number that is a multiple of all the numbers in each set (called the lowest common multiple, LCM).
   **i** 3, 7   **iii** 2, 8, 10   **v** 26, 3
   **ii** 2, 9  **iv** 3, 4, 6    **vi** 5, 4
  **b** Find the highest whole number that divides exactly into all the given numbers (called the highest common factor, HCF).
   **i** 4, 6     **iii** 22, 44, 55   **v** 24, 8, 16
   **ii** 3, 6, 12  **iv** 18, 6, 9    **vi** 14, 28, 56

**4** Find the value of
- **a** $5^2$
- **b** $3^4$
- **c** $2^5$
- **d** $5^3$
- **e** $2^4 \times 3^2$
- **f** $8^2 \times 5^2$
- **g** $6^3 \times 2^2$
- **h** $7^3 \times 2^3$
- **i** $3.25 \times 10^2$
- **j** $8.01 \times 10^3$
- **k** $0.072 \times 10^4$
- **l** $1.1 \times 10^6$

**5** Write, where possible, as a single expression in index form
- **a** $2^3 \times 2^4$
- **b** $5^2 \times 3^5$
- **c** $5^1 \times 5^3$
- **d** $2 \times 2^4$
- **e** $7^3 \div 7^2$
- **f** $3^6 \div 3^2$
- **g** $3^4 \div 3$
- **h** $(2^3)^2$
- **i** $(5^3)^2$

**6** Find, as a fraction, the value of
- **a** $2^{-1}$
- **b** $10^{-1}$
- **c** $\left(\frac{1}{3}\right)^{-1}$
- **d** $\left(\frac{2}{3}\right)^{-1}$
- **e** $2^{-3}$
- **f** $6^{-2}$
- **g** $\left(\frac{1}{5}\right)^{-3}$
- **h** $\left(\frac{1}{6}\right)^{-2}$
- **i** $4^{-2}$
- **j** $\left(\frac{2}{7}\right)^{-2}$
- **k** $\left(\frac{3}{10}\right)^{-4}$
- **l** $5^0$

**7 a** Write the following numbers in standard form.
- **i** 265
- **ii** 0.18
- **iii** 76 700
- **iv** 0.000 007
- **v** 450 000
- **vi** 0.092

**b** Write the following numbers as ordinary numbers.
- **i** $3.45 \times 10^{-2}$
- **ii** $5.01 \times 10^{-2}$
- **iii** $7.3 \times 10^{-1}$
- **iv** $6.37 \times 10^{-4}$
- **v** $1.4 \times 10^5$
- **vi** $2.83 \times 10^5$

**8** Give each of the following numbers correct to
  **i** 3 decimal places  **ii** 3 significant figures.
- **a** 2.7846
- **b** 0.1572
- **c** 0.073 25
- **d** 0.150 76
- **e** 254.1627
- **f** 7.8196
- **g** 3.2994
- **h** 0.000 925 8
- **i** 0.009 638

**9** For each calculation, first make a rough estimate of the answer, then use your calculator to give the answer correct to 3 significant figures.
- **a** $78.4 \times 0.527$
- **b** $842 \times 284$
- **c** $9.827 \div 4.731$
- **d** $(5.09)^3$
- **e** $(0.185)^{-2}$
- **f** $3000 \div 48.66$
- **g** $\dfrac{7.21 \times 5.93}{13.74}$
- **h** $\dfrac{849 \times 0.773}{16.34}$
- **i** $\dfrac{0.515}{6.37 \times 0.717}$

**10** Show on a sketch the range in which each of the following measurements lies.
- **a** 8 m to the nearest metre

b  15 minutes to the next complete minute

c  123 cm rounded down to the nearest cm

## REVISION EXERCISE 4 (Sets)

1  $A$ = {prime factors of 15} and $B$ = {prime factors of 12}
   a  List the members of $A \cup B$.
   b  List the members of $A \cap B$.

2  $A$ = {all factors of 36} and $B$ = {all factors of 42} with
   $U$ = {whole numbers from 1 to 42 inclusive}
   a  Illustrate $A$, $B$ and $U$ in a Venn diagram.
   b  List the members of     i $A \cup B$    ii $A \cap B$.
   c  Give the number of members in $U$ that are not in $A$ or in $B$.

3  $U$ = {all triangles} and $A$ = {isosceles triangles},
   $B$ = {right-angled triangles}
   a  State whether $A$ and $B$ are finite or infinite sets.
   b  Illustrate the information on a Venn diagram.
   c  Describe the set $A \cap B$.

4  $A$ and $B$ are two sets such that $A$ contains 8 members, $B$ contains
   5 members and $A \cap B$ contains 3 members.
   Illustrate this information in a Venn diagram and hence find the
   number of members in $A \cup B$.

## REVISION EXERCISE 5 (Shape and space)

1  Find the size of each marked angle.

a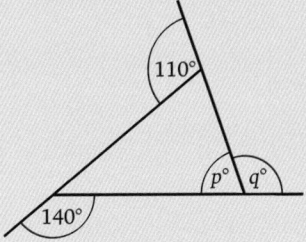

c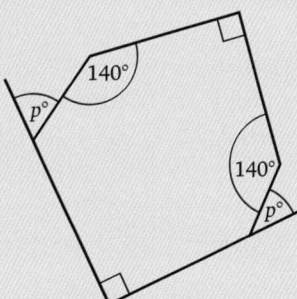

b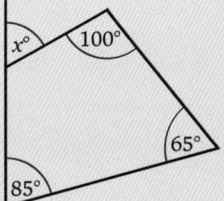

d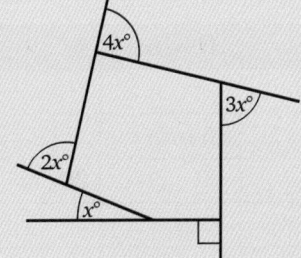

**2 a** Find the size of each exterior angle of a regular polygon with
  **i** 15 sides    **ii** 20 sides.
**b** Find the size of each interior angle of a regular polygon with
  **i** 8 sides    **ii** 18 sides.
**c** How many sides has a regular polygon
  **i** if each exterior angle is 15°
  **ii** if each interior angle is 162°?
**d** Is it possible for each exterior angle of a regular polygon to be
  **i** 40°    **ii** 70°?
  If it is, give the number of sides.
**e** Is it possible for each interior angle of a regular polygon to be
  **i** 120°    **ii** 160°?
  If it is, give the number of sides.

**3** Find the area of each shape.

**a**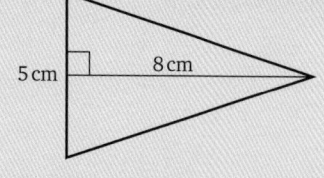

**c**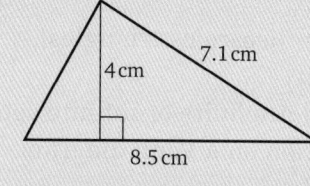

**b**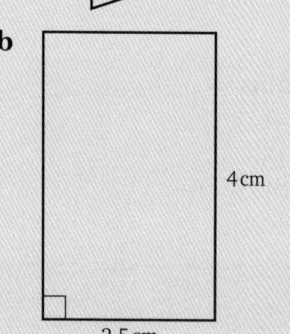

**d**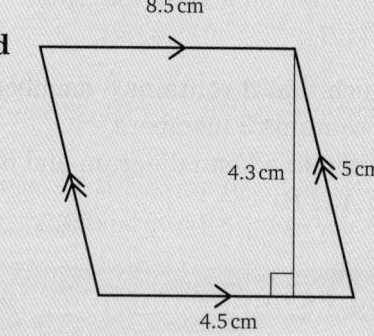

**4** Use squared paper and draw axes for $x$ and $y$ in the ranges
$-6 \leq x \leq 6$, $-6 \leq y \leq 6$ using 1 square to 1 unit. Draw the figure and find its area in square units.
  **a** Triangle ABC with A (0, 6), B (6, 6) and C (5, 2)
  **b** Parallelogram ABCD with A (0, 1), B (0, 6), C (6, 4) and D (6, −1)
  **c** Rectangle ABCD with A (−4, 2), B (0, 2) and C (0, −1)
  **d** Square ABCD with A (0, 0), B (0, 4) and C (4, 4)
  **e** Triangle ABC with A (−5, −4), B (2, −4), C (−2, 3)

**5** For each of the following figures, find the missing measurement. Draw a diagram in each case.

| | Figure | Base | Height | Area |
|---|---|---|---|---|
| a | Triangle | 8 cm | | 16 cm² |
| b | Rectangle | 3 cm | 15 mm | |
| c | Parallelogram | 4 cm | | 20 cm² |
| d | Square | 5 m | | |
| e | Triangle | 70 mm | | 14 cm² |

**6** Find the area of the following shapes. Draw a diagram for each question and mark in all the measurements.

**a**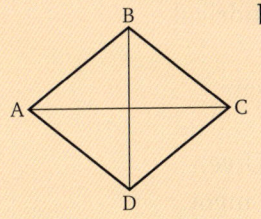
ABCD is a rhombus
AC = 15 cm and
BD = 8 cm

**b**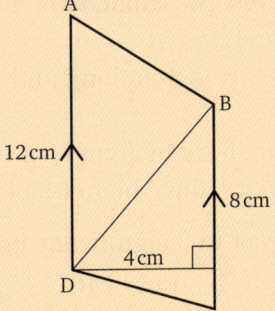

**c**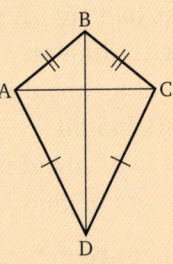
AC = 6 cm and
BD = 10 cm

**7 a**   **i** Find, giving your answer in metres, 137 cm + 234 mm + 3.2 m
  **ii** Find, giving your answer in grams, 645 g + 0.37 kg + 960 mg
  **iii** Find, giving your answer in inches, 3 feet + 2 yards + 8 inches
**b** Express
  **i** 45 mm in cm
  **ii** 0.56 km in m
  **iii** 48 inches in feet
  **iv** 13 yards in feet
  **v** 5 cm$^2$ in mm$^2$
  **vi** 4000 cm$^2$ in m$^2$
  **vii** 0.6 m$^2$ in cm$^2$
  **viii** 3 square feet in square inches
  **ix** 5000 mm$^3$ in cm$^3$
  **x** 0.002 m$^3$ in cm$^3$
  **xi** 4 000 000 cm$^3$ in m$^3$
  **xii** $\frac{5}{12}$ cubic feet in cubic inches

**8** For the blue part of each diagram, find **i** the perimeter  **ii** the area. Give answers that are not exact correct to 3 significant figures.

**a**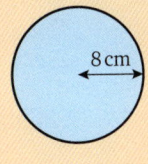

**b**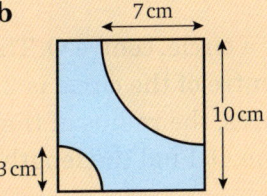

**c**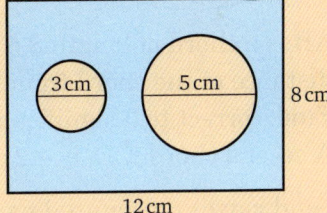

**9** In this question, the cross-sections of the prisms and their lengths are given. Find their volumes.

**a**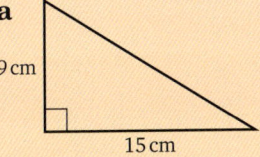
Length 30 cm

**b**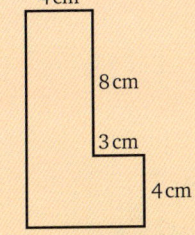
Length 15 cm

**c**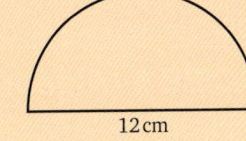
Length 40 cm

**10 a** Find the volume of a rectangular metal block measuring 4.2 cm by 3.8 cm by 1.5 cm.

**b** Find, correct to 3 significant figures, the capacity of a cylindrical metal can with diameter 12 cm and height 10 cm.

**c** The volume of a cuboid is 136 cm³. It is 8 cm long and 3.4 cm wide. Find its height.

**d** A gold ingot is a cuboid measuring 5 cm by 5 cm by 3 cm. Given that the density of gold is 19.3 g/cm³, find the mass of the ingot.

**11** Give answers that are not exact correct to 3 significant figures.

**a** Find AC. 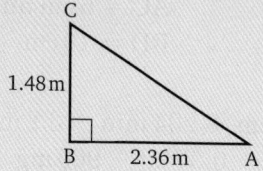   **b** Find PR.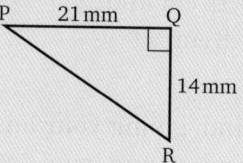

**c**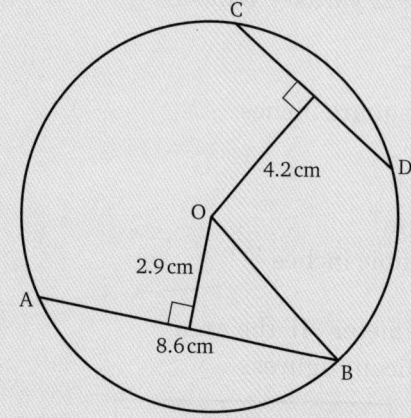

  **i** AB is a chord of length 8.6 cm in a circle, centre O. The distance of the chord from the centre of the circle is 2.9 cm. Find, correct to 3 significant figures, the radius of the circle.

  **ii** A second chord, CD, is 4.2 cm from O. Find the length of CD.

**12** Draw $x$- and $y$-axes on squared paper for values of $x$ and $y$ from 0 to 10. Mark the points A (2, 1) and B (8, 9) on your graph.

Find **a** the coordinates of the midpoint of AB

**b** the length of the line AB.

## REVISION EXERCISE 6 (Algebra)

**1** Simplify the following expressions.

**a** $3(x + 7) + 2x$

**b** $5 + (3a - 7)$

**c** $3(2x + 3) + 4(x - 2)$

**d** $5x - 3(x + 2)$

**e** $4x - 2(3 + x)$

**f** $8a - (2a + 3)$

**g** $-3(x - 2)$

**h** $-2(4 - 3x)$

**i** $-5(2x + 7)$

**j** $4(a + 2) - 3(a - 4)$

**k** $15x - 3x - 2(4x - 3)$

**l** $6a - 3(2a - 5) + 3$

2. Simplify
   a. $2a \times 3b \times 4c$
   b. $5x \times 2x \times 4y$
   c. $3a \times 4b \times 5b$
   d. $\dfrac{4a}{3} \times \dfrac{7}{20}$
   e. $\dfrac{3x}{5} \div \dfrac{9x}{10}$
   f. $\dfrac{5x}{3} \div \dfrac{3}{10x}$
   g. $(-3a) \times (-2a)$
   h. $(-x) \div (-y)$
   i. $12x \div (-3x)$

3. a. If $P = 2(a - 6b)$, find $P$ when
      i. $a = 3$ and $b = 1$
      ii. $a = -5$ and $b = 2$
      iii. $a = -4$ and $b = -2$
   b. If $A = xy$, find $A$ when
      i. $x = 3$ and $y = \tfrac{1}{2}$
      ii. $x = 4$ and $y = -2$
      iii. $x = -3$ and $y = -6$

4. a. Apples cost $x$ pence each. Write down a formula for $C$ if $C$ pence is the cost of 6 apples.
   b. Find a formula for $u_n$ in terms of $n$.

   | $n$ | 1 | 2 | 3 | 4 | 5 |
   |---|---|---|---|---|---|
   | $u_n$ | 4 | 7 | 10 | 13 | 16 |

5. Solve the following equations.
   a. $3x = 15$
   b. $4x + 1 = 17$
   c. $7x - 3 = 2x + 7$
   d. $\dfrac{x}{5} = 2$
   e. $3x - 4 = 11 - 2x$
   f. $4(x - 2) = 5(2x + 5)$
   g. $\dfrac{x}{4} - 3 = 5$
   h. $\dfrac{x}{2} - \dfrac{1}{4} = \dfrac{7}{12}$
   i. $0.8x = 5.6$
   j. $0.03x = 0.42$

6. a. Solve the following inequalities and illustrate your solutions on a number line.
      i. $x - 6 < 4$
      ii. $9 - x \geqslant 4$
      iii. $8 > 3 - x$
      iv. $4x - 1 \leqslant 15$
      v. $2 > 8 + x$
      vi. $7 > 3 - x$
      vii. $4x + 1 > 6$
      viii. $3x + 4 \geqslant 5 - 2x$
      ix. $2x + 3 \leqslant 9 - 4x$

**b** Solve each pair of inequalities and hence find the range of values of $x$ which satisfy both of them.

  **i** $x - 3 < 5$ and $x > 2 - x$    **ii** $2x + 1 > 5$ and $3x - 10 < 2$

**c** Find the range of values of $x$ for which the following inequalities are true.

  **i** $x + 3 > 2x - 1 > 3$
  **ii** $x - 1 < 2x + 3 \leqslant 7$
  **iii** $4x + 1 < x - 2 < 4$

**7** Solve the simultaneous equations.

  **a** $x + y = 4$
  $3x + y = 10$
  
  **b** $7x - 2y = 22$
  $3x + 2y = 18$
  
  **c** $5x + 3y = 25$
  $8x - 3y = 1$
  
  **d** $2x - 3y = 15$
  $2x - y = 9$
  
  **e** $x + 5y = 9$
  $x - y = 21$
  
  **f** $2a + 3b = 9$
  $2a + 7b = 13$

**8** Solve the following equations, giving your answers correct to 3 significant figures.

  **a** $x^2 = 23$    **b** $x^2 = 0.47$

## REVISION EXERCISE 7 (Graphical work)

**1**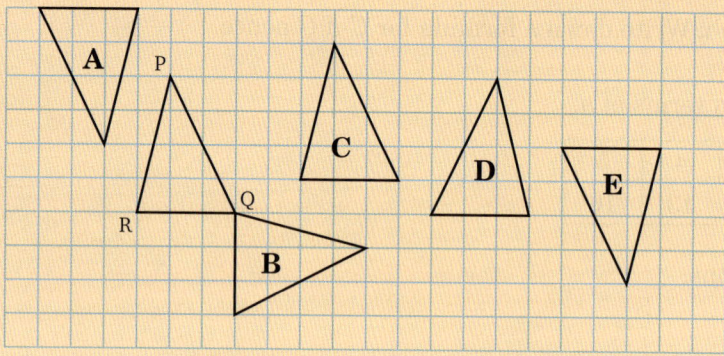

In the diagram, which images of $\triangle PQR$ are given by
**a** a translation
**b** a reflection
**c** a rotation
**d** none of these?

**2** Draw axes for $x$ and $y$ for values from 0 to 10 using 1 cm as 1 unit. Draw $\triangle XYZ$ with X (4, 2), Y (4, 4) and Z (5, 1) and $\triangle X'Y'Z'$ with X' (6, 4), Y' (6, 10) and Z' (9, 1). Find the centre of enlargement and the scale factor that maps $\triangle XYZ$ to $\triangle X'Y'Z'$.

**3 a** Write down the gradient and $y$-intercept of the line whose equation is

  **i** $y = 3x$    **ii** $y = 2x + 6$    **iii** $y = 3 - \frac{1}{2}x$

**b** Write down the equation of the line that is parallel to the line $y = 4x$ that goes through the point    **i** (0, 2)    **ii** (0, −3)

**c** A (2, $a$) and B ($b$, 10) are points on the line $y = -3x + 7$. Find $a$ and $b$.

4   The table gives the coordinates of three points on a straight line. What is the equation of the line?

| x | −3 | 0 | 2 |
|---|---|---|---|
| y | −4 | 2 | 6 |

5   Determine whether each of the following straight lines makes an acute angle or an obtuse angle with the positive $x$-axis.
   a   $y = -x + 2$
   b   $y = 3 - 7x$
   c   $y = 0.6x$
   d   $y = 2 + 3x$

6   Determine whether each pair of lines is parallel, perpendicular or neither.
   a   $y = 3x + 1$, $y = 4 - 3x$
   b   $y = 3x - 1$, $y = 3x + 5$
   c   $y = -\frac{1}{2}x + 1$, $y = 2x + \frac{1}{2}$
   d   $y = -x + 3$, $y = x - 3$

7   The equation of a curve is $y = 3x^2$. Which of these sketches could be the curve?

   A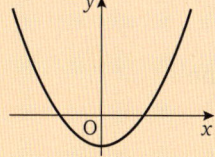
   C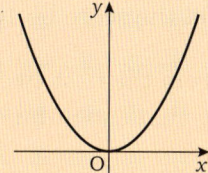
   B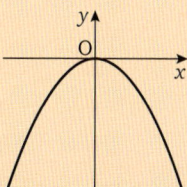
   D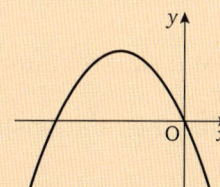

8   Sketch the curve whose equation is   a   $y = -2x^2$   b   $y = 5x^2$

## REVISION EXERCISE 8 (Probability and statistics)

1   The pie chart shows the breakdown of John Peters' bill for £120 when his car was serviced.
   a   What fraction of the bill was labour?
   b   How much was the charge for VAT?
   c   What percentage of the bill was for parts?

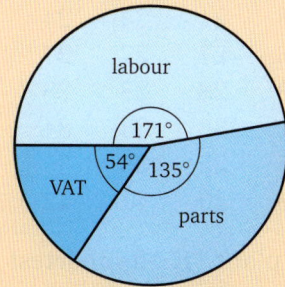

**2 a** Find the mode, median and range of the numbers

9, 8, 11, 8, 9, 13, 12, 11, 8, 7, 9, 8, 12

**b** Find the mean of the numbers given in part **a**. Give your answer correct to 3 significant figures.

**3** In a sale 80 dresses are reduced. Of these 16 are size 8, 28 are size 10, 24 are size 12 and the remainder are size 14. Aleisha takes a dress from the rail at random. What is the probability that the dress she takes is
**a** size 10
**b** not size 12
**c** size 12 or larger?

**4** Two fair six-sided dice are rolled together 360 times.
**a** Draw a table to show the equally likely outcomes when the dice are rolled once.
**b** About how many double threes are there likely to be?
**c** About how many times should the score be 10?

**5** A number is chosen at random from the first twelve non-zero whole numbers. What is the probability that the number is
**a** a prime number
**b** exactly divisible by 4
**c** not exactly divisible by 3
**d** a rectangular number but not a square number?

**6** For each scatter diagram, describe the relationship between the quantities.

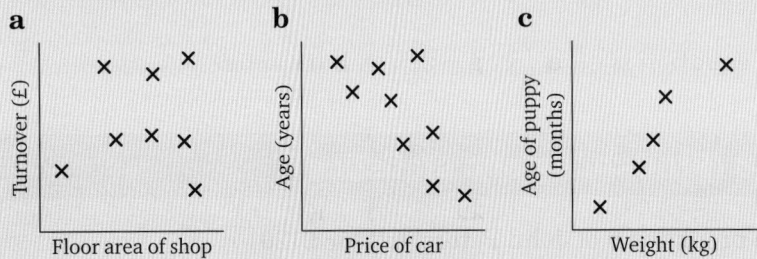

**7** Given below is a list of the heights, in centimetres rounded up to the nearest centimetre, of 60 tomato plants. The list is in numerical order.

| 20 | 20 | 21 | 22 | 22 | 22 | 23 | 24 | 24 | 24 |
|----|----|----|----|----|----|----|----|----|----|
| 24 | 25 | 25 | 25 | 26 | 26 | 26 | 27 | 27 | 27 |
| 27 | 27 | 27 | 28 | 28 | 28 | 28 | 28 | 28 | 28 |
| 28 | 29 | 30 | 30 | 31 | 32 | 32 | 33 | 33 | 34 |
| 34 | 34 | 34 | 34 | 35 | 35 | 35 | 35 | 35 | 35 |
| 36 | 36 | 36 | 37 | 37 | 38 | 40 | 40 | 41 | 42 |

**a** What is the height of **i** the tallest plant **ii** the shortest plant?

**b** Copy and complete this frequency table.

| Height, $h$ (cm) | Frequency |
|---|---|
| $18 < h \leq 23$ | |
| $23 < h \leq 28$ | |
| $28 < h \leq 33$ | |
| $33 < h \leq 38$ | |
| $38 < h \leq 43$ | |

**c** How many plants have a height that is
   **i** greater than 28 cm
   **ii** 33 cm or less?
**d** What is the modal group?
**e** Illustrate this information with
   **i** a bar chart
   **ii** a frequency polygon.

**8**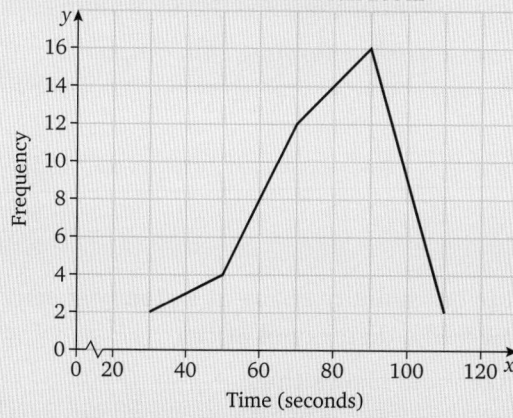

Use the frequency polygon to answer the following questions. If you cannot give an answer, say why.
**a** How many times were measured?
**b** What is the range of times?
**c** Is it true to say that half the times were less than 80 seconds?
**d** How many students ran 200 metres?

# 1 Travel graphs

## Consider

From the bus stop Isobel travels 1 mile along the bus route to school, and she must be in school by 8.50 a.m. She can walk at 4 mph and, if she catches the bus, it will travel at an average speed of 20 mph. The buses arrive at unpredictable times so Isobel is often undecided whether to wait for a bus or to walk.

- What is the latest time that Isobel can leave the bus stop on foot to be certain of arriving at school on time?
- If she could be sure of catching a bus as soon as she gets to the bus stop, what is the latest time she could arrive there and still get to school in time?
- One morning she decides to walk but sees a bus pass her after she has been walking for 5 minutes. How much longer would she have had to wait before the bus arrived? If she had waited, how much earlier would she have got to school than she did by walking?

*You should be able to solve this problem after you have worked through this chapter.*

## Finding distance from a graph

When we went on holiday in the car we travelled to our holiday resort at a steady speed of 30 kilometres per hour (km/h), so each hour we covered a distance of 30 km.

This graph plots distance covered against time taken and shows that

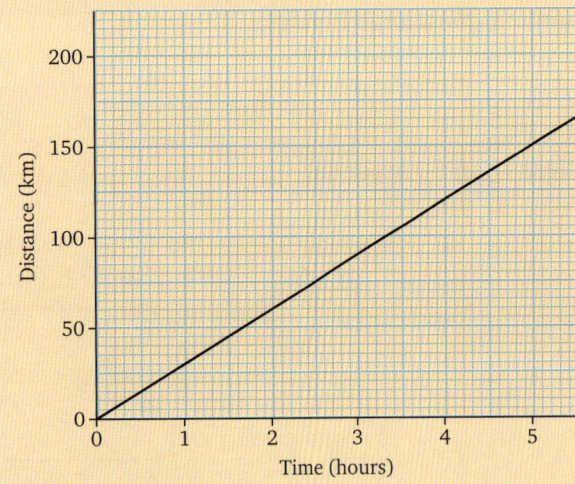

in 1 hour we travelled 30 km

in 2 hours we travelled 60 km

in 3 hours we travelled 90 km

in 4 hours we travelled 120 km

in 5 hours we travelled 150 km.

# 1 Travel graphs

## Exercise 1a

The graphs in questions **1** to **5** show five different journeys.
For each journey find

a the distance travelled
b the time taken
c the distance travelled: in 1 hour (questions **1**, **2** and **3**)
in 1 second (questions **4** and **5**).

**1**
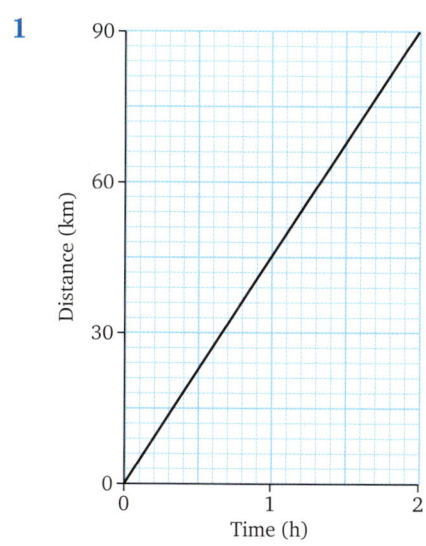

**3**
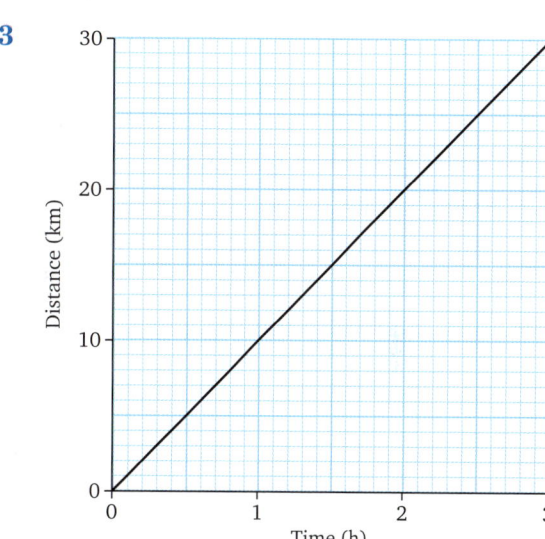

**2**
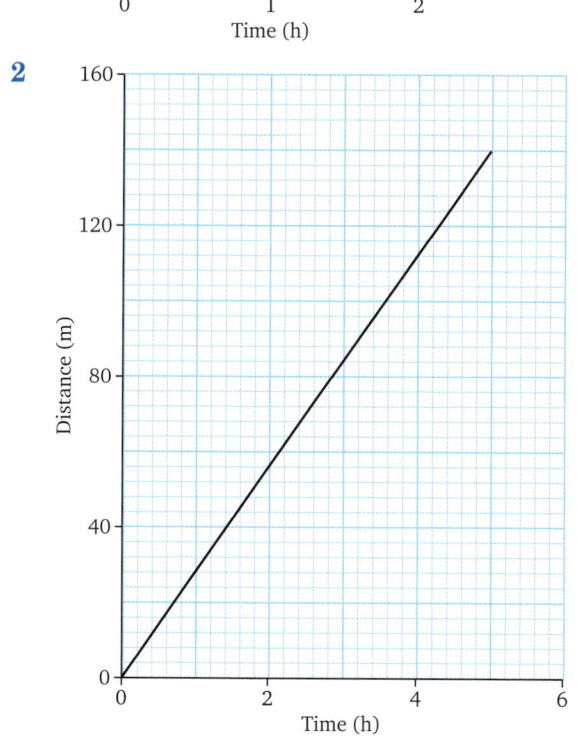

**4**
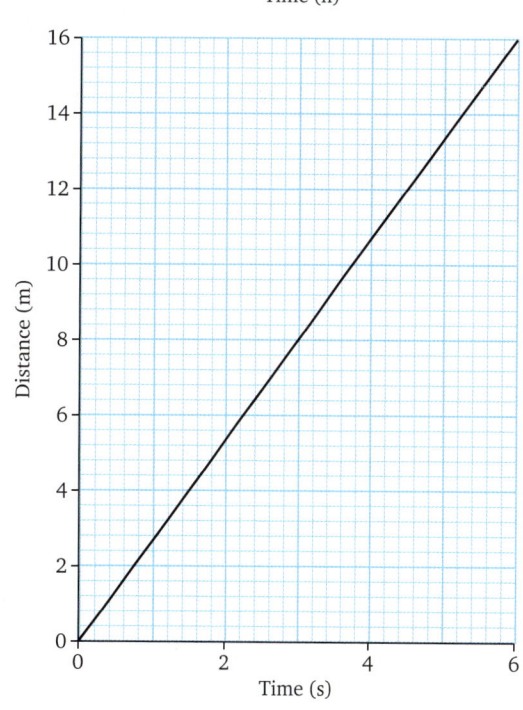

5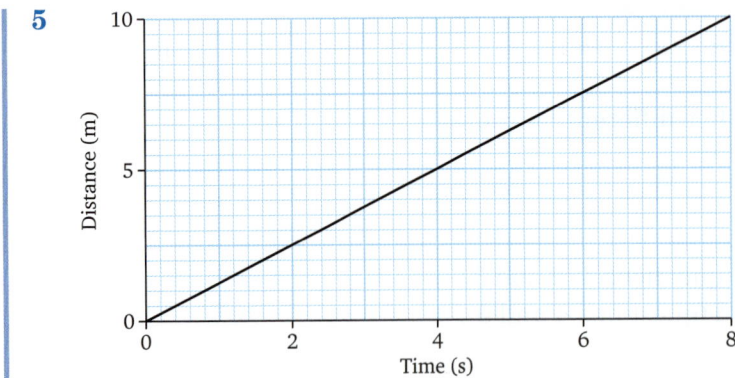

## Drawing travel graphs

If Ethan walks at 6 km/h, we can draw a graph to show this, using 2 cm to represent 12 km on the distance axis and 2 cm to represent 1 hour on the time axis.

Plot the point which shows that in 1 hour he has travelled 6 km. Join the origin to this point and produce the straight line to give the graph shown.

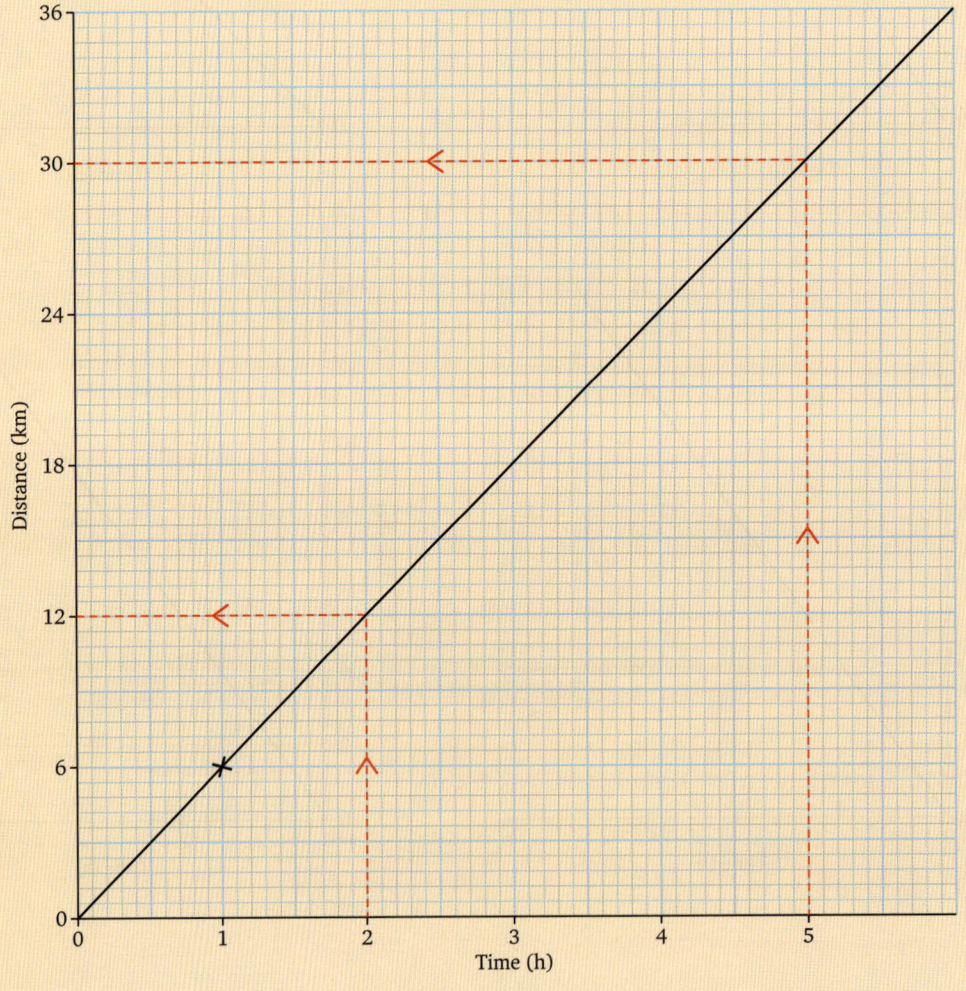

From this graph we can see that in 2 hours Ethan travels 12 km and in 5 hours he travels 30 km.

Alternatively we could say that

if he walks 6 km in 1 hour,

he will walk 6 × 2 km, i.e. 12 km, in 2 hours

and he will walk 6 × 5 km, i.e. 30 km, in 5 hours.

The distance walked is found by multiplying the speed by the time,

i.e.     **distance = speed × time**

## Exercise 1b

### Worked example

→ Draw a travel graph to show Sally's journey of 150 km in 3 hours. Plot distance along the vertical axis and time along the horizontal axis. Let 4 cm represent 1 hour and 2 cm represent 50 km.

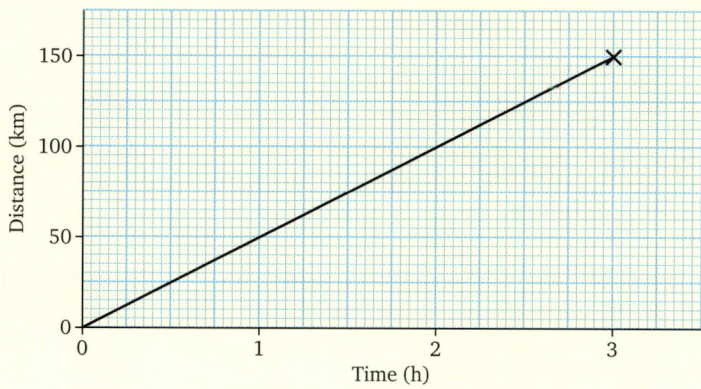

For questions **1** to **9**, draw travel graphs to show the following journeys. Plot distance along the vertical axis and time along the horizontal axis. Use the scales given in brackets.

1  60 km in 2 hours (4 cm ≡ 1 hour, 1 cm ≡ 10 km)

2  180 km in 3 hours (4 cm ≡ 1 hour, 2 cm ≡ 50 km)

3  100 km in $2\frac{1}{2}$ hours (2 cm ≡ 1 hour, 2 cm ≡ 25 km)

4  75 miles in $1\frac{1}{4}$ hours (8 cm ≡ 1 hour, 2 cm ≡ 25 miles)

5  240 m in 12 seconds (1 cm ≡ 1 second, 2 cm ≡ 50 m)

**6** Callum walks at 5 km/h. Draw a graph to show him walking for 3 hours. Use 4 cm to represent 5 km and 4 cm to represent 1 hour. Use your graph to find how far he walks in
  **a** $1\frac{1}{2}$ hours
  **b** $2\frac{1}{4}$ hours.

**7** Georgia can jog at 10 km/h. Draw a graph to show her jogging for 2 hours. Use 1 cm to represent 2 km and 8 cm to represent 1 hour. Use your graph to find how far she jogs in
  **a** $\frac{3}{4}$ hour
  **b** $1\frac{1}{4}$ hours.

**8** Jo drives at 35 mph. Draw a graph to show her driving for 4 hours. Use 1 cm to represent 10 miles and 4 cm to represent 1 hour. Use your graph to find how far she drives in
  **a** 3 hours
  **b** $1\frac{1}{4}$ hours.

**9** Luke walks at 4 mph. Draw a graph to show him walking for 3 hours. Use 1 cm to represent 1 mph and 4 cm to represent 1 hour. Use your graph to find how far he walks in
  **a** $\frac{1}{2}$ hour
  **b** $3\frac{1}{2}$ hours.

Solve questions **10** to **18** by calculation.

**10** An express train travels at 200 km/h. How far will it travel in
  **a** 4 hours
  **b** $5\frac{1}{2}$ hours?

**11** Kieran cycles at 24 km/h. How far will he travel in
  **a** 2 hours
  **b** $3\frac{1}{2}$ hours
  **c** $2\frac{1}{4}$ hours?

**12** An aeroplane flies at 300 mph. How far will it travel in
  **a** 4 hours
  **b** $5\frac{1}{2}$ hours?

**13** A bus travels at 60 km/h. How far will it travel in
  **a** $1\frac{1}{2}$ hours
  **b** $2\frac{1}{4}$ hours?

**14** Sophie can cycle at 12 mph. How far will she ride in
  **a** $\frac{3}{4}$ hour
  **b** $1\frac{1}{4}$ hours?

**15** An athlete can run at 10.5 metres per second. How far will he travel in
  **a** 5 seconds
  **b** 8.5 seconds?

**16** A boy cycles at 12 mph. How far will he travel in
  **a** 2 hours 40 minutes
  **b** 3 hours 10 minutes?

**17** A Boeing 777 travels at 540 mph. How far does it travel in
  **a** 3 hours 15 minutes
  **b** 7 hours 45 minutes?

**18** A racing car travels around a 2 km circuit at 120 km/h. How many laps will it complete in
  **a** 30 minutes
  **b** 1 hour 12 minutes?

## Calculating the time taken

Greta walks at 6 km/h so we can find out how long it will take her to walk 24 km. As she walks 6 km in 1 hour, it will take her $\frac{24}{6}$ hours to walk 24 km, i.e. 4 hours.

In the same way, if Greta walks 15 km, it will take her $\frac{15}{6}$ hours, i.e. $2\frac{1}{2}$ hours.

This demonstrates that

$$\text{time} = \frac{\text{distance}}{\text{speed}}$$

### Exercise 1c

1. How long will Zena, walking at 5 km/h, take to walk
   - **a** 10 km
   - **b** 15 km?

2. How long will a car, travelling at 80 km/h, take to travel
   - **a** 400 km
   - **b** 260 km?

3. How long will it take David, running at 10 mph, to run
   - **a** 5 miles
   - **b** $12\frac{1}{2}$ miles?

4. How long will it take an aeroplane flying at 450 mph to fly
   - **a** 1125 miles
   - **b** 2400 miles?

5. A cowboy rides at 14 km/h. How long will it take him to ride
   - **a** 21 km
   - **b** 70 km?

6. A rally driver drives at 50 mph. How long does it take him to travel
   - **a** 75 miles
   - **b** 225 miles?

7. An athlete runs at 8 m/s. How long does it take him to cover
   - **a** 200 m
   - **b** 1600 m?

8. A dog runs at 20 km/h. How long does the dog take to travel
   - **a** 8 km
   - **b** 18 km?

9. A cruise ship travels at 20 nautical miles per hour. How long will it take to travel
   - **a** 6048 nautical miles
   - **b** 3528 nautical miles?
   
   Give your answers to the nearest hour.

10. A car travels at 56 mph. How long does it take to travel
    - **a** 70 miles
    - **b** 154 miles?

11. Anna cycles at 12 mph. How long will it take her to cycle
    - **a** 30 miles
    - **b** 64 miles?

12. How long will it take a car travelling at 64 km/h to travel
    - **a** 48 km
    - **b** 208 km?

## Average speed

Mr Compton left home at 8 a.m. to travel the 50 km to his place of work. He arrived at 9 a.m. Although he had travelled at many different speeds during his journey, he covered the 50 km in exactly 1 hour. We say that his *average speed* for the journey was 50 kilometres per hour, or 50 km/h. If he had travelled at the same speed all the time, he would have travelled at 50 km/h.

Amy travelled the 135 miles from her home to London in 3 hours. If she had travelled at the same speed all the time, she would have travelled at $\frac{135}{3}$ mph = 45 mph. We say that her average speed for the journey was 45 mph.

In each case:  average speed = $\dfrac{\text{distance travelled}}{\text{time taken}}$

This formula can also be written:

distance travelled = average speed × time taken

and  time taken = $\dfrac{\text{distance travelled}}{\text{average speed}}$

A useful way to remember these relationships is from the triangle: (Cover up the one you want to find.)

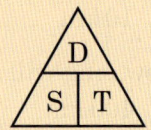

Suppose that a car travels 35 km in 30 minutes and we want to find its speed in kilometres per hour. To do this, we must express the time taken in hours instead of minutes,

i.e.  time taken = 30 min = $\frac{1}{2}$ hour

Then  average speed = $\dfrac{35}{\frac{1}{2}}$ km/h = $35 \times \frac{2}{1}$ km/h

= 70 km/h

We must take care with units. If we want a speed in kilometres per hour, we need the distance in kilometres and the time in hours. If we want a speed in metres per second, we need the distance in metres and the time in seconds.

### Exercise 1d

In questions **1** to **12**, find the average speed for each journey.

1  80 km in 1 hour
2  120 km in 2 hours
3  60 miles in 1 hour
4  480 miles in 4 hours
5  80 m in 4 seconds
6  135 m in 3 seconds
7  150 km in 3 hours
8  520 km in 8 hours
9  245 miles in 7 hours
10  104 miles in 13 hours
11  252 m in 7 seconds
12  255 m in 15 seconds

# 1 Travel graphs

## Worked example

→ Tony drives 39 km in 45 minutes. Find his average speed.

$45 \text{ min} = \frac{45}{60} \text{ hour} = \frac{3}{4} \text{ hour}$

average speed $= \dfrac{\text{distance travelled}}{\text{time taken}}$

$= \dfrac{39 \text{ km}}{\frac{3}{4} \text{ hour}}$

$= 39 \times \frac{4}{3} \text{ km/h} = 52 \text{ km/h}$

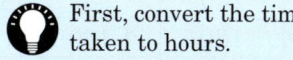
First, convert the time taken to hours.

Find the average speed in km/h for a journey of

13  40 km in 30 minutes

14  60 km in 40 minutes

15  48 km in 45 minutes

16  66 km in 33 minutes.

## Worked example

→ Find the average speed in km/h for a journey of 5000 m in $\frac{1}{2}$ hour.

$5000 \text{ m} = \frac{5000}{1000} \text{ km} = 5 \text{ km}$

average speed $= \dfrac{\text{distance travelled}}{\text{time taken}}$

$= \dfrac{5 \text{ km}}{\frac{1}{2} \text{ hour}}$

$= 5 \times \frac{2}{1} \text{ km/h} = 10 \text{ km/h}$

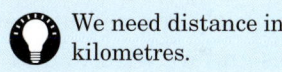
We need distance in kilometres.

Find the average speed in km/h for a journey of

17  4000 m in 20 minutes

18  6000 m in 45 minutes

19  40 m in 8 seconds

20  175 m in 35 seconds.

Find the average speed in mph for a journey of

21  27 miles in 30 minutes

22  18 miles in 20 minutes

23  25 miles in 25 minutes

24  28 miles in 16 minutes.

# STP Maths 9

The following table shows the distances in kilometres between various places in the United Kingdom.

|            | London | Bradford | Cardiff | Leicester | Manchester | Oxford | Reading |
|------------|--------|----------|---------|-----------|------------|--------|---------|
| Bradford   | 320    |          |         |           |            |        |         |
| Cardiff    | 250    | 332      |         |           |            |        |         |
| Leicester  | 160    | 160      | 224     |           |            |        |         |
| Manchester | 310    | 55       | 277     | 138       |            |        |         |
| Oxford     | 90     | 280      | 172     | 120       | 230        |        |         |
| Reading    | 64     | 320      | 192     | 164       | 264        | 45     |         |
| York       | 315    | 53       | 390     | 174       | 103        | 290    | 210     |

Use this table to find the average speeds for journeys between

**25** London, leaving at 1025, and Manchester, arriving at 1625

**26** Oxford, leaving at 0330, and Cardiff, arriving at 0730

**27** Leicester, leaving at 1914, and Oxford, arriving at 2044

**28** Reading, leaving at 0620, and London, arriving at 0750

**29** Bradford, leaving at 1537, and Oxford, arriving at 1907

**30** Cardiff, leaving at 1204, and York, arriving at 1624

**31** Bradford, leaving at 1014, and Reading, arriving at 1638.

## Journeys with several parts

Problems frequently occur if different parts of a journey are travelled at different speeds in different times and we wish to find the average speed for the whole journey.

For example, Roger travels the first 50 miles of a journey at an average speed of 25 mph and the next 90 miles at an average speed of 30 mph.

Using $\text{time in hours} = \dfrac{\text{distance in miles}}{\text{speed in mph}}$

gives $\text{time to travel 50 miles at 25 mph} = \dfrac{\text{distance}}{\text{speed}}$

$= \dfrac{50 \text{ miles}}{25 \text{ mph}} = 2 \text{ hours}$

$\text{time to travel 90 miles at 30 mph} = \dfrac{\text{distance}}{\text{speed}}$

$= \dfrac{90 \text{ miles}}{30 \text{ mph}} = 3 \text{ hours}$

∴ the total distance of 140 miles is travelled in 5 hours

i.e. $\text{average speed for whole journey} = \dfrac{\text{total distance}}{\text{total time}}$

$= \dfrac{140 \text{ miles}}{5 \text{ hours}} = 28 \text{ mph}$

*Note:* never add or subtract average speeds.

Often it is more convenient to enter this information in a table like the one given below. You may need to do some calculations like those shown above, before you can complete the table.

|  | Speed in mph | Distance in miles | Time in hours |
|---|---|---|---|
| First part of journey | 25 | 50 | 2 |
| Second part of journey | 30 | 90 | 3 |
| Whole journey |  | 140 | 5 |

We can add the distances to give the total length of the journey, and add the times to give the total time taken for the journey.

Average speed for whole journey $= \dfrac{\text{total distance}}{\text{total time}}$

$= \dfrac{140 \text{ miles}}{5 \text{ hours}} = 28 \text{ mph}$

## Exercise 1e

1. I walk for 24 km at 8 km/h and then jog for 12 km at 12 km/h.
   Find my average speed for the whole journey.

2. A cyclist rides for 23 miles at an average speed of $11\frac{1}{2}$ mph before his cycle breaks down, forcing him to push his cycle the remaining distance of 2 miles at an average speed of 4 mph.
   Find his average speed for the whole journey.

3. An athlete runs 6 miles at 8 mph, then walks 1 mile at 4 mph.
   Find his average speed for the whole journey.

4. A woman walks 3 miles at an average speed of $4\frac{1}{2}$ mph and then runs 4 miles at 12 mph.
   Find her average speed for the whole journey.

5. A motorist travels the first 30 km of a journey at an average speed of 120 km/h, the next 60 km at 60 km/h, and the final 60 km at 80 km/h.
   Find the average speed for the whole journey.

6. Phil Sharp walks the 2 km from his home to the bus stop in 15 minutes, and catches a bus immediately which takes him the 9 km to the railway station at an average speed of 36 km/h.
   He arrives at the station just in time to catch the London train which takes him the 240 km to London at an average speed of 160 km/h.
   Calculate his average speed for the whole journey from home to London.

**7** A cruise ship travelling at 24 knots takes 18 days to travel between two ports.

By how much must it increase its speed to reduce the length of the voyage by 2 days?

(A knot is a speed of 1 nautical mile per hour.)

## Getting information from travel graphs

If we are given a graph representing a journey, which shows the distance travelled plotted against the time taken, we can get a lot of information from it. The worked example in the next exercise shows how we can extract such information.

### Exercise 1f

### Worked example

→ The graph below shows the journey of a coach, which passes three service stations A, B and C on a motorway. B is 60 km north of A and C is 20 km north of B. Use the graph to answer the following questions.

**a** At what time does the coach leave A?

**b** At what time does the coach arrive at C?

**c** At what time does the coach pass B?

**d** What is the average speed of the coach for the journey from A to C?

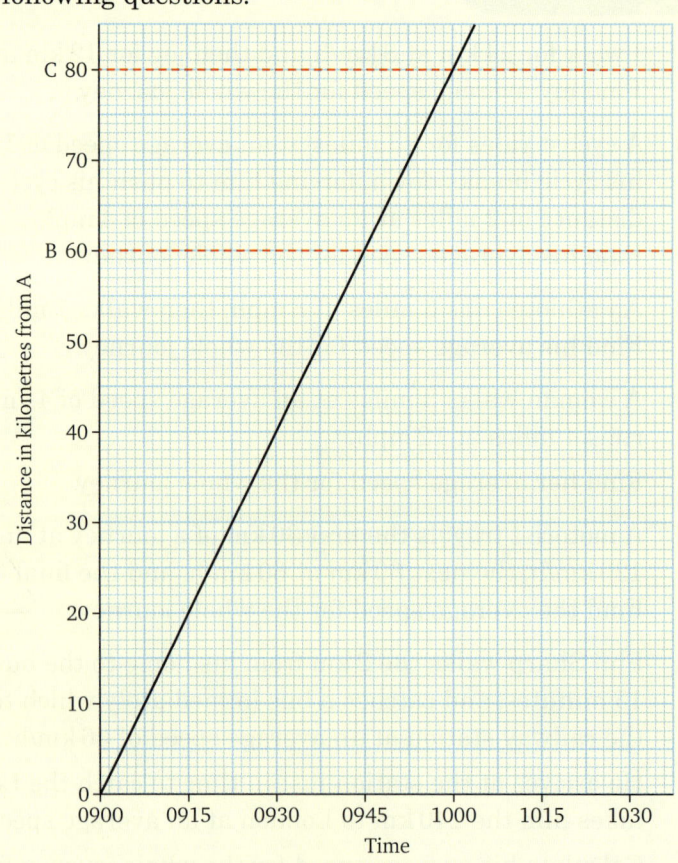

**a** The coach leaves A at 0900.

**b** It arrives at C at 1000.

**c** It passes through B at 0945.

**d** Distance from A to C = 80 km.
Time taken to travel from A to C is 1000 − 0900, i.e. 1 hour.

$$\text{Average speed} = \frac{\text{distance travelled}}{\text{time taken}}$$

$$= \frac{80\,\text{km}}{1\,\text{hour}} = 80\,\text{km/h}$$

> Find the point on the graph level with C, then use a ruler as a guide to find the point on the time axis that is immediately below this point.

**1** The graph shows the journey of a car through three towns, Axeter, Bexley and Canton. Axeter is 100 km south of Bexley and Canton is 60 km north of Bexley. Use the graph to answer the following questions.

**a** At what time does the car

   **i** leave Axeter

   **ii** pass through Bexley

   **iii** arrive at Canton?

**b** How long does the car take to travel from Axeter to Canton?

**c** How long does the car take to travel

   **i** the first 80 km of the journey

   **ii** the last 80 km of the journey?

**d** What is the average speed of the car for the journey from Axeter to Canton?

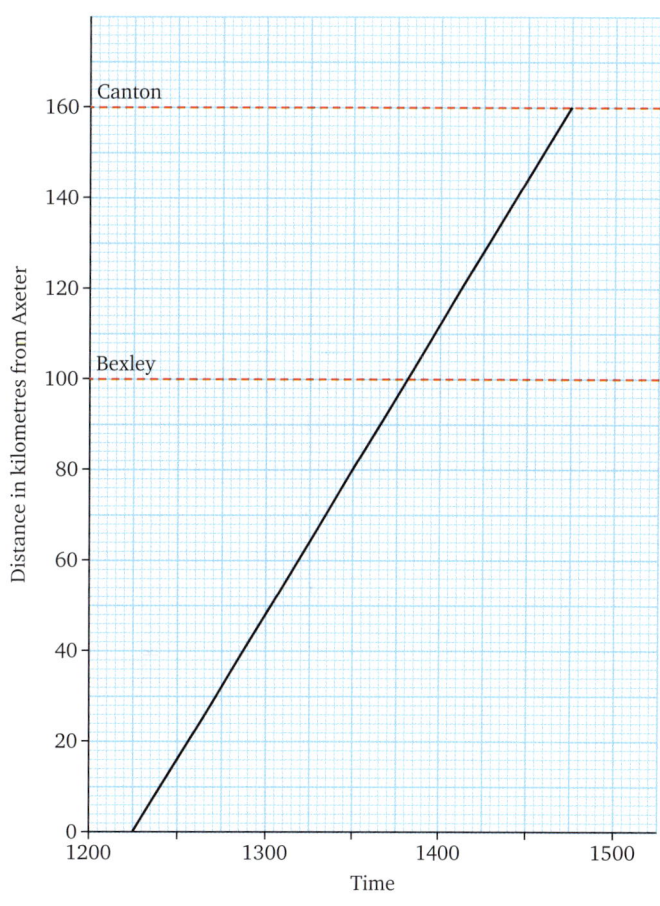

2  A coach leaves Newcombe at noon on its journey to Lee via Manley. The graph shows its journey.

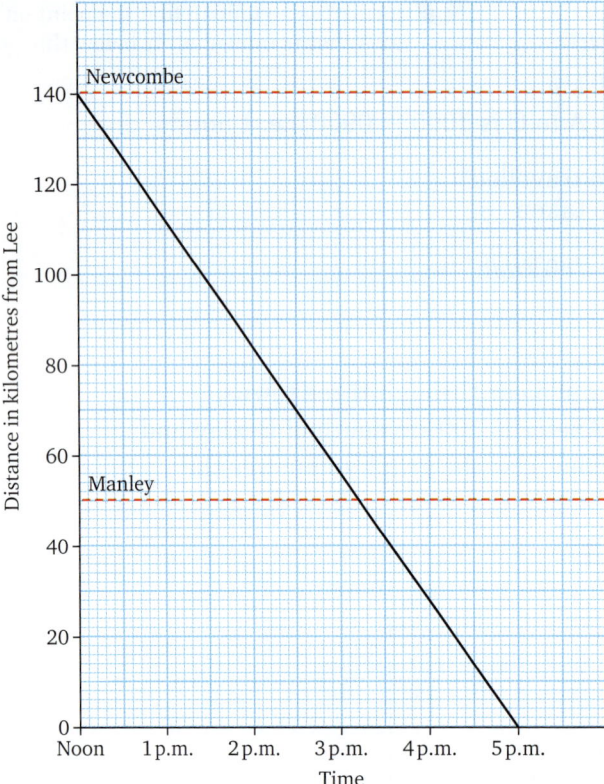

a  How far is it
   i  from Newcombe to Manley
   ii from Manley to Lee?
b  How long does the coach take to travel from Newcombe to Lee?
c  What is the coach's average speed for the whole journey?
d  How far does the coach travel between 1.30 p.m. and 2.30 p.m.?
e  After travelling for $1\frac{1}{2}$ hours, how far is the coach from
   i  Newcombe
   ii Manley?

3   Mr Brown used the family car to drive his children from their home to the nearest railway station and then returned home. The graph shows his journey.

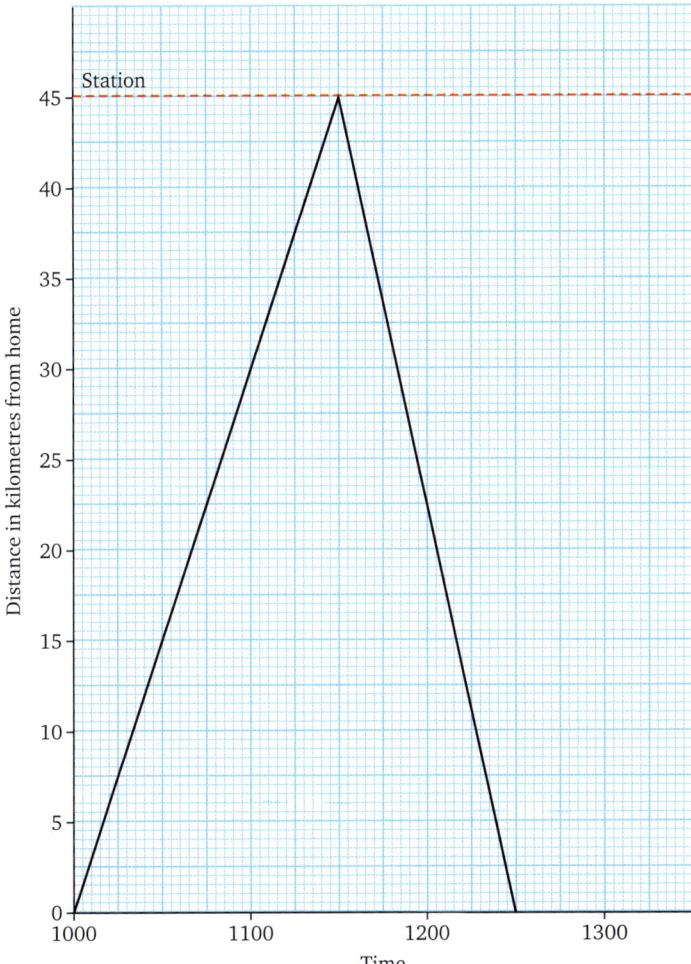

a   How far is it from home to the station?
b   How long did it take the family to get to the station?
c   What was the average speed of the car on the journey to the station?
d   How long did the car take for the return journey?
e   What was the average speed for the return journey?
f   What was the car's average speed for the round trip?

4   The graph shows the journey of a car through three service stations, A, B and C, on a motorway.

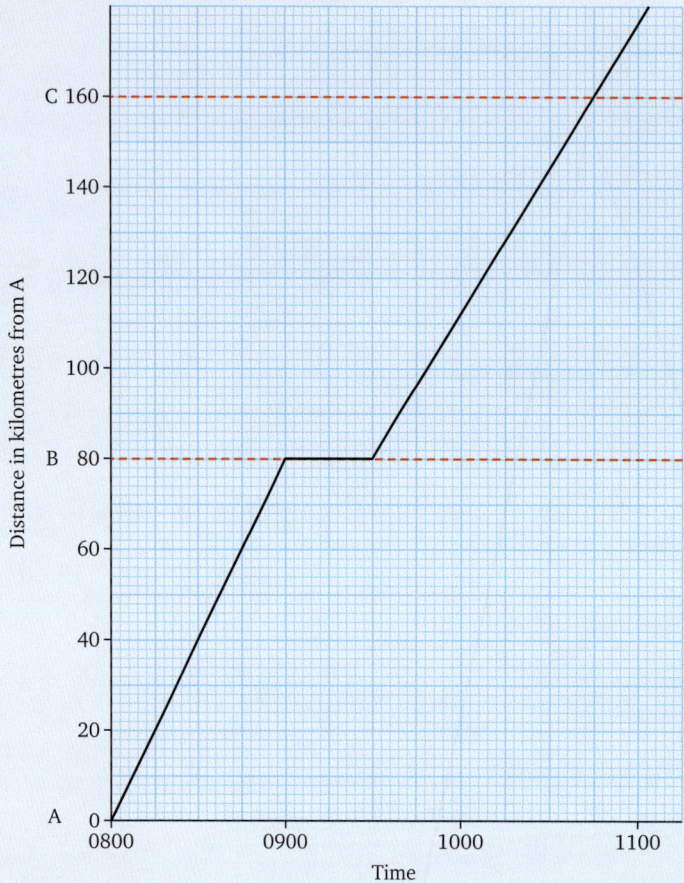

a   Where was the car at
    i   0900
    ii  0930?
b   What was the average speed of the car between
    i   A and B
    ii  B and C?
c   For how long does the car stop at B?
d   How long did the journey from A to C take?
e   What was the average speed of the car from A to C?
    Give your answer correct to 1 significant figure.

5   The graph shows Anand's journey on a sponsored walk.

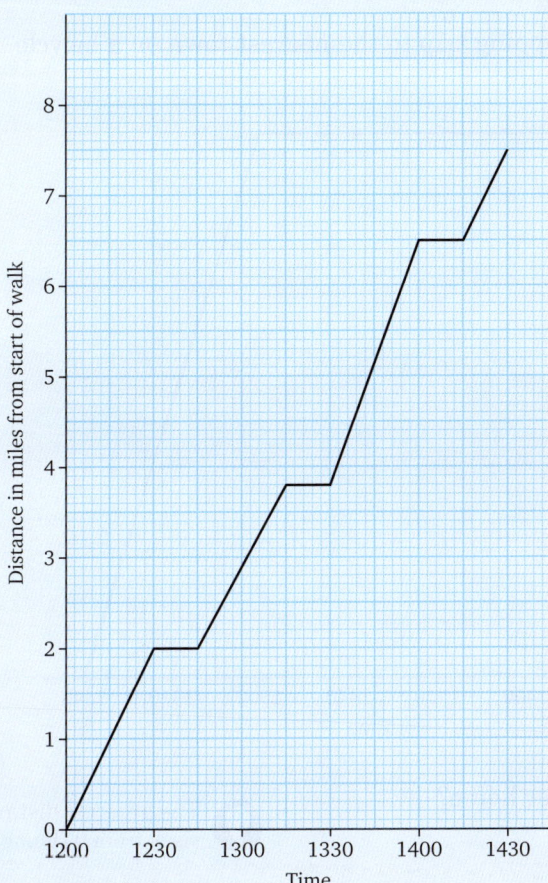

a   How far did he walk?
b   How many times did he stop?
c   What was the total time he spent resting?
d   How long did he actually spend walking?
e   How long did the walk take him?
f   What was his average speed for the whole journey?
g   Over which of the four stages did he walk fastest?
h   Over which two stages did he walk at the same speed?

## Worked example

The graph shows Mrs Webb's shopping trip to the nearest town on a bicycle.

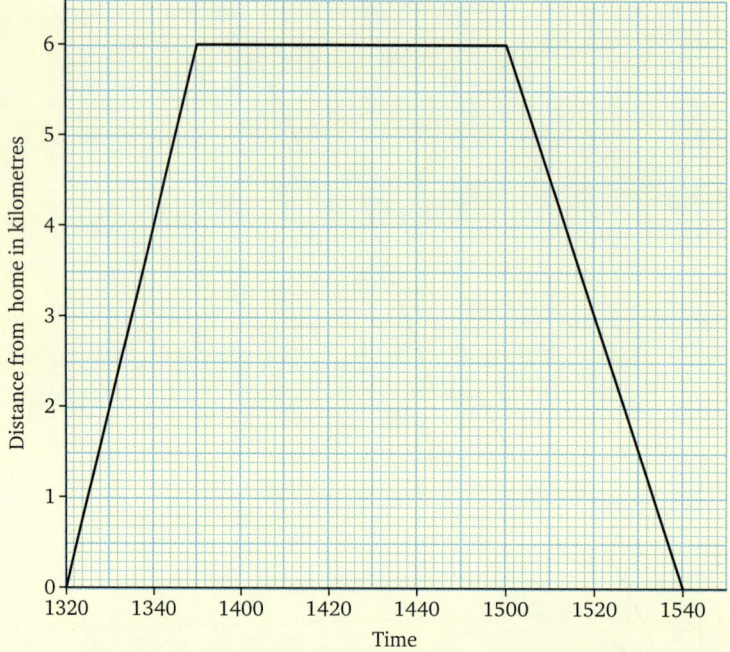

→ How far is it to town from her home?

It is 6 km to town from her home.

*The greatest distance the graph rises from 0 (home) is 6 km.*

→ How long did she take to get to town?

Mrs Webb left home at 1320 and arrived in town at 1350.
She therefore took 30 minutes to get to town.

*She arrived in town when she stopped going further away from home. This is where the graph stops going up.*

→ How long did she spend in town?

She arrived in town at 1350 and left at 1500. She therefore spent 1 hour and 10 minutes there.

*She left for home when the graph starts to go down. i.e. she stayed in town for the time that the graph is parallel to the time axis.*

→ What was her average speed on the outward journey?

On the outward journey:

$$\text{average speed} = \frac{\text{distance travelled}}{\text{time taken}} = \frac{6 \text{ km}}{\frac{1}{2} \text{ hour}}$$

$$= \frac{6}{1} \times \frac{2}{1} \text{ km/h} = 12 \text{ km/h}$$

6  The graph shows the journey of a train from Newpool to London and back again. Use the graph to answer the questions that follow.

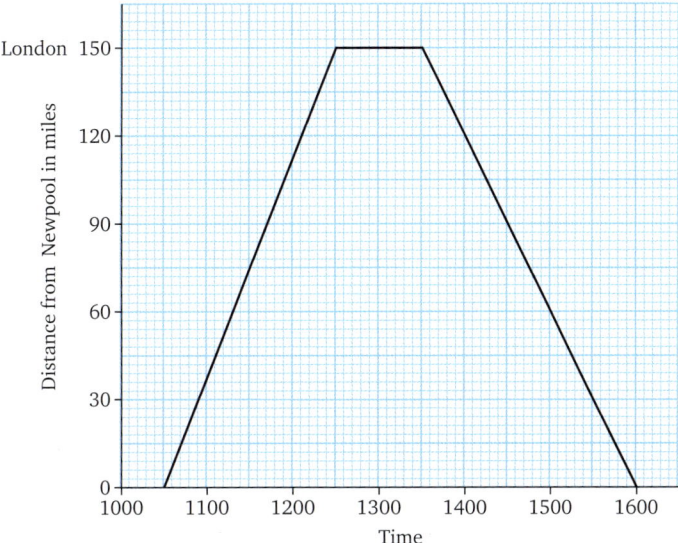

a  How far is Newpool from London?
b  How long did the outward journey take?
c  What was the average speed for the outward journey?
d  How long did the train remain in London?
e  At what time did the train leave London, and how long did the return journey take?
f  What was the average speed on the return journey?

7  The graph represents the journey of a motorist from Leeds to Manchester and back again.

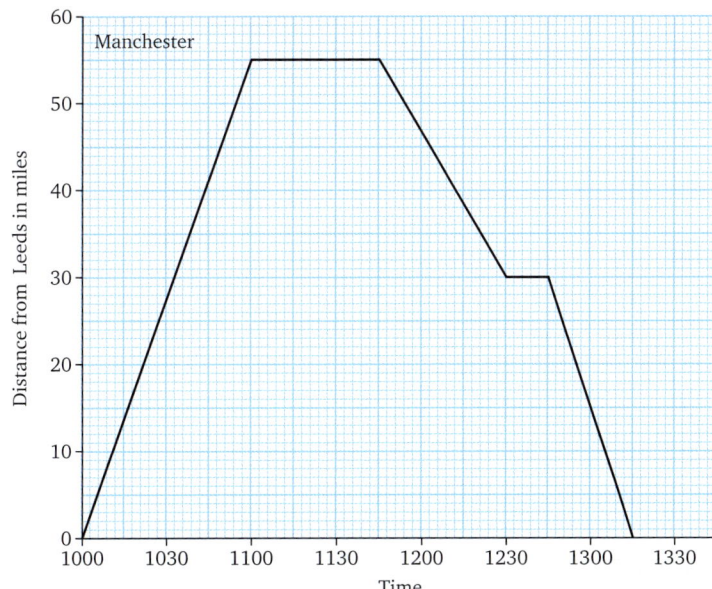

Use this graph to find
a  the distance between the two cities
b  the time the motorist spent in Manchester

**c** his average speed on the outward journey

**d** his average speed on the homeward journey (including the stop).

**8** The graph below shows Caitlin's journeys between home and school in one day.

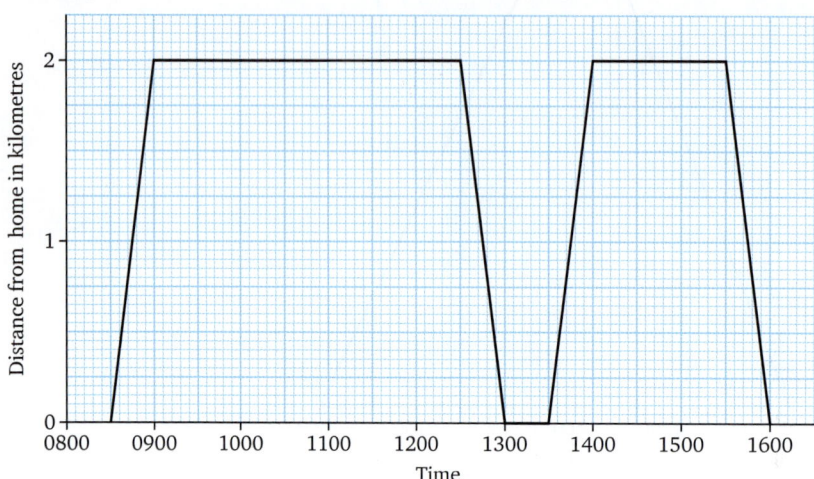

**a** At what time did she leave home in
  **i** the morning  **ii** the afternoon?
**b** How long was she at school during the day?
**c** How long was she away from school for her midday break?
**d** What was her average speed for each of these journeys?
**e** Find the total time for which she was away from home.

**9** The graph below shows the journeys of two cars between two service stations, A and B, which are 180 km apart. Use the graph to find

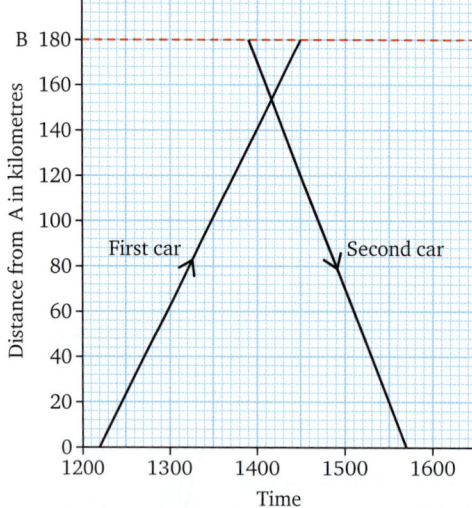

**a** the average speed of the first motorist and his time of arrival at B
**b** the average speed of the second motorist and the time at which she leaves B
**c** when and where the two motorists pass
**d** their distance apart at 1427.

10 The school fête is always held on the first Saturday in July. The graphs show the journeys from home to school and back again of three students on the day of the fête last year.

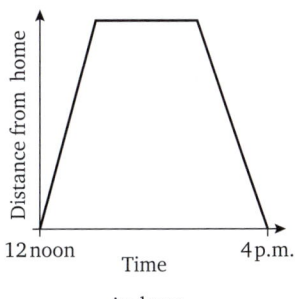
Andrew

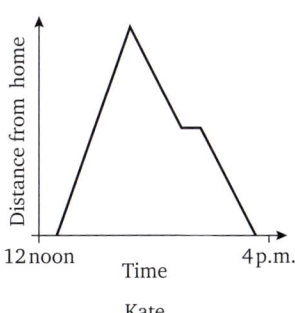
Kate

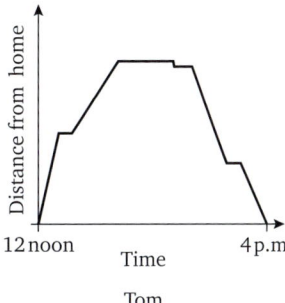
Tom

Use these graphs to answer the following questions.
a Who got to the fête first?
b Who stayed there the longest?
c Who left for home first?
d Who took the longest time to get home?
e Who had the slowest journey to the fête?
f What did Kate do when she got to the fête?
g Who lived nearest to the school?

11 Molly leaves home at 1 p.m. to walk at a steady 4 mph towards Cornforth, which is 6 miles away, to meet her boyfriend Rob. Rob leaves Cornforth at 2.00 p.m. and jogs at a steady 6 mph to meet her. Draw a graph for each of these journeys, using 4 cm ≡ 1 hour on the time axis and 1 cm ≡ 1 mile on the distance axis. From your graph find
a when and where they meet
b their distance apart at 2.10 p.m.

## Mixed exercise

### Exercise 1g

1 Una runs at $12\frac{1}{2}$ km/h. Draw a graph to show her running for $2\frac{1}{2}$ hours. Use your graph to find
  a how far she has travelled in $1\frac{3}{4}$ hours
  b how long she takes to run the first 20 km.

2 A ship travels at 18 nautical miles per hour. How long will it take to travel
  a 252 nautical miles      b 1026 nautical miles?

3 Find the average speed in km/h of a journey of 48 km in 36 minutes.

4 I left London at 1147 to travel the 315 miles to York. I arrived at 1717. What was my average speed?

5 I walk $\frac{2}{3}$ mile in 10 minutes and then run $\frac{1}{3}$ mile in 2 minutes. What is my average speed for the whole journey?

**6** The graph shows Paul's journey in a sponsored jog from A to B. On the way his sister, who is travelling by car in the opposite direction from B to A, passes him.
  **a** How far does Paul jog?
  **b** How long does he take?
  **c** How much of this time does he spend resting?
  **d** What is his average speed for the whole journey?
  **e** What is his sister's average speed?
  **f** How far from A did Paul and his sister pass each other?

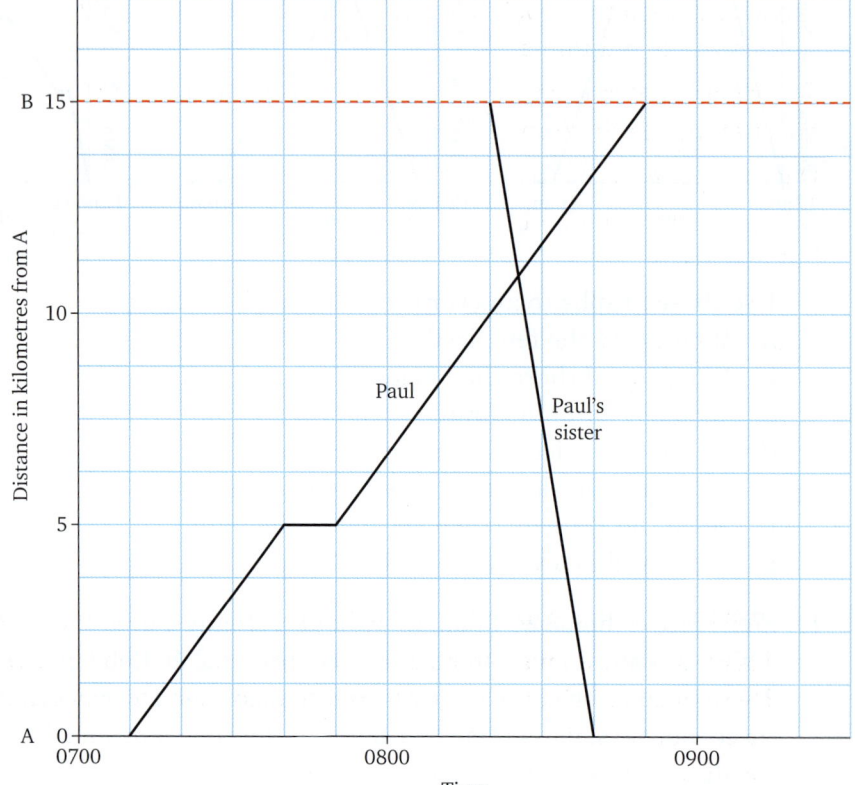

## Consider again

From the bus stop Isobel travels 1 mile along the bus route to school, and she must be in school by 8.50. She can walk at 4 mph and, if she catches the bus, it will travel at an average speed of 20 mph. The buses arrive at unpredictable times so Isobel is often undecided whether to wait for a bus or to walk.
- What is the latest time that Isobel can leave the bus stop on foot to be certain of arriving at school on time?
- If she could be sure of catching a bus as soon as she gets to the bus stop, what is the latest time she could arrive there and still get to school in time?
- One morning she decides to walk but sees a bus pass her after she has been walking for 5 minutes. How much longer would she have had to wait before the bus arrived? If she had waited, how much earlier would she have got to school than she did by walking?

Now can you answer these questions?

# 2 Working with numbers

### Consider

David and his father sometimes have communication problems, especially when numbers are involved.

For example, on one occasion, David's father asked him to measure the mat well inside the front door.

David wrote down '54 cm by 32 cm'.

David's father then had a piece of matting cut to these measurements but, when he got it home, he found that it was too large for the hole. This resulted in an argument in which each unfairly blamed the other.

What reasons could there be for this error?

*You should be able to solve this problem after you have worked through this chapter.*

### Class discussion

Discuss the possible consequences of using rounded numbers in the following situations.

1. Hasib wanted enough topsoil to cover an area of 12 m² to a depth of 50 cm. He worked out the area correct to the nearest square metre. He ordered 6 cubic metres of topsoil.

2. Hannah measured the length of her garden path as 38.5 metres correct to 1 decimal place. She ordered some square paving tiles of side 50 cm, which she then placed edge-to-edge in a single line to pave the path.

3. The operating equipment on a long-range shell launcher needs to be set with the distance and bearing of the target. A soldier worked out these measurements from maps as 2558 m on a bearing of 026.79°. He assumed that they were accurate to the number of figures given but, because of various reading and calculation errors, they were accurate only to 2 significant figures.

The examples above show that working with numbers that have been rounded can give results that are less accurate than we need for the situation. Unless we are aware of this, the consequences can be anything from mildly annoying to disastrous.

The following points may have arisen from your discussions.

- Some numbers are exact and others are rounded.
- It is impossible to give exact numerical values for some quantities, for example lengths, so we need to use and work with corrected numbers.
- We need to be aware that some numbers have been rounded and, where possible, to know how they have been rounded.
- When we use rounded numbers in calculations, we need to appreciate that the results contain errors.

# STP Maths 9

## Range of values for a corrected number

Suppose we are told that, correct to the nearest ten, 250 people boarded a particular train. People are counted in whole numbers only. Hence in this case, 245 is the lowest number that gives 250 when corrected to the nearest 10 and 254 is the highest number that can be corrected to 250. We can therefore say that the actual number of people who boarded the train is any whole number from 245 to 254.

Now suppose that we are given a nail and are told that its length is 25 mm correct to the nearest millimetre.

Look at the reading on this measuring gauge.

The lowest number that can be rounded up to 25 is 24.5. The highest number that can be rounded down to 25 is not so easy to determine. All we can say is that any number up to, but not including, 25.5 can be rounded down to 25.

The length of the nail is therefore in the range from 24.5 mm up to, but not including, 25.5 mm.

If $l$ mm is the length of the nail, we can write

$$24.5 \leqslant l < 25.5$$

To illustrate this on a number line we use a line segment with a solid circle at the lower end to show that 24.5 is included in the range and an open circle at the upper end to show that 25.5 is not included in the range.

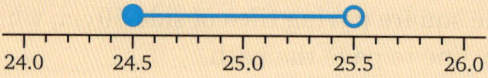

24.5 is called the **lower bound** of $l$ and 25.5 is called the **upper bound** of $l$.

## Exercise 2a

### Worked example

→ Illustrate on a number line the range of values of $x$ given by $0.1 < x \leqslant 0.8$

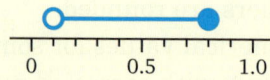

Use a number line like this for questions **1** to **6**.

In each question, illustrate the range on your number line.

1  $5 \leq x \leq 10$
2  $0 < x \leq 15$
3  $-2 \leq x \leq 6$
4  $5 \leq x < 15$
5  $0 \leq x < 10$
6  $-5 < x \leq 5$

Use a number line like this for questions **7** to **12**. In each question, illustrate the range on your number line.

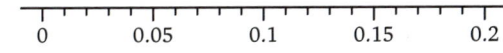

7  $0 < x < 0.1$
8  $0.1 < x < 0.2$
9  $0.05 < x < 0.15$
10  $0.08 < x < 0.16$
11  $0.02 < x < 0.08$
12  $0.03 < x < 0.13$

## Worked example

→ A number is given as 3.15 correct to 2 decimal places. Illustrate on a number line the range in which this number lies.

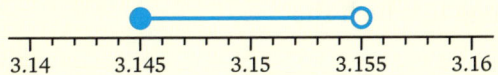 

3.15 is between 3.14 and 3.16 so we will use just that part of the number line.

In questions **13** to **24**, illustrate on a number line the range of possible values for each corrected number.

13  1.5 correct to 1 d.p.
14  0.6 correct to 1 d.p.
15  0.2 correct to 1 d.p.
16  1.3 correct to 1 d.p.
17  0.1 correct to 1 d.p.
18  6.2 correct to 1 d.p.
19  0.25 correct to 2 d.p.
20  0.52 correct to 2 d.p.
21  1.15 correct to 2 d.p.
22  6.89 correct to 2 d.p.
23  12.26 correct to 2 d.p.
24  0.05 correct to 2 d.p.

## Worked example

→ It is stated that a packet of pins contains 500 pins to the nearest 10. Find the range in which the actual number of pins lies.

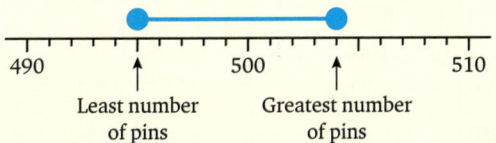 

There must be a whole number of pins in the packet. 504 is the largest whole number that rounds to 500 to the nearest 10.

If $n$ is the number of pins in the packet, then $n$ is a whole number such that

$495 \leq n \leq 504$

### Worked example

→ A copper tube is sold as having an internal bore (diameter) of 10 mm to the nearest millimetre. Find the range in which the actual bore lies.

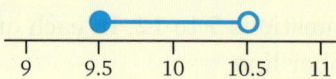

If $d$ mm is the diameter of the tube, then
$$9.5 \leq d < 10.5$$

In some of the following questions you are asked to find a range of values for a quantity that can only have whole number values. When this is the case, you must clearly state this in your answer.

25  The weight, $w$ kg, of a bag of sand is given as 5.6 kg correct to 1 decimal place. Find the range of values in which $w$ lies.

26  A shop is said to make a profit of £5400 a month. If this figure is given correct to the nearest £100, find the range in which the actual monthly figure, £$x$, lies.

27  On a certain model of bicycle, the brake pads have to be 12.5 mm thick to work efficiently. If the brake pads are $x$ mm thick, find the range in which $x$ lies when the measurement given is correct to 1 decimal place.

28  Referring to a football match, a newspaper headline proclaimed '75 000 watch England win'. Assuming that this figure for the number of spectators is correct to the nearest 1000, find the range in which $x$, the number of people who actually attended the match, lies. If, in fact, the figure given was a guess, what can you say about $x$?

29  One of the component parts of a metal hinge is a pin. In order to work properly, this pin must have a diameter of 1.25 mm correct to 2 decimal places. If $d$ mm is the diameter, find the range in which $d$ must lie.

30  Alan measured the width of a space between two kitchen units in metres correct to 1 decimal place, and wrote down 1.6 m.
   a  Find the range within which the width of the space lies.
   b  The cupboard bought to go into the space is 1.62 m wide correct to 2 decimal places. Will it fit the space?
   c  Comment on Alan's measuring.

31  The length of a car is given as 455 cm. If this is correct to the nearest 5 cm, find the range in which the actual length lies.

32 Knitting yarn is sold by weight. It is found that 10 g of double knitting pure wool has a length of 20 m correct to the nearest metre. What is the minimum length you would expect in a 50 g ball of this wool?

33 The weight of a wall tile is given as 40 g to the nearest gram. A DIY store sells these tiles in polythene-wrapped packs of ten. Ignoring the weight of the wrapping, find the range in which the weight of one of these packs lies.

34 The length of a side of a square carpet tile is 29.9 cm correct to 1 decimal place. One hundred of these carpet tiles are laid end-to-end on the floor of a shop.
   a  Find the range in which the length of this line of tiles lies.
   b  Hence give the difference between the length of the longest possible and the shortest possible line of 100 tiles.

35 Toy bricks are cubes of side 34 mm correct to the nearest millimetre. Eight of these cubes are placed side by side in a row. What is the upper bound of the length of this row?

36 A cube of side 34 mm, correct to the nearest millimetre, is to be packed in a cubical box with internal sides of 34.2 mm correct to 3 significant figures. Explain why the cube may not fit in the box.

37 Connor and Jessica both said that they were 15 years old. What is the greatest possible difference in their ages?

38 A wooden pole is 450 mm long and a 200 mm length is cut from it. Both of these measurements are correct to the nearest millimetre. What can you say about the length of the remaining section of the pole?

39 A rectangular sheet of stainless steel, measuring 35 mm by 23 mm correct to the nearest millimetre, needs edging with wire.
   a  What length of wire is needed to make sure that there is enough to edge one of these sheets?
   b  The wire costs £18.20 per metre. What is the cost of enough wire to ensure that 500 000 of these sheets can be edged? (Assume that there is no waste.)

40 A pair of digital scales gives weights correct to the nearest gram. When one nail is weighed on these scales, the reading is 8 grams. When 100 identical nails are weighed, the reading is 775 grams.
   a  Explain the apparent contradiction in the readings.
   b  Give a more accurate weight for one nail than the scales are capable of showing when only one nail is weighed.

## Fractions and decimals

The questions in the last exercise show that an answer worked out using corrected numbers is not exact but lies within a range. If we are not aware of this, we may get errors when we use corrected numbers, as the following situation shows.

A caterer was asked to provide 800 filled baguettes for a local function.

One long French loaf is enough for three filled baguettes.

The caterer reasoned that he needed 800 thirds of French loaves and worked this number out as follows.

$$800 \text{ thirds} = 800 \times \tfrac{1}{3}$$
$$= 800 \times 0.33 = 264$$

He ordered 264 loaves, made them up, then found that he was 8 baguettes short.

The caterer could have avoided this situation by checking his answer:
264 loaves makes $264 \times 3$ baguettes. This is 792 baguettes, which are not enough.

He would not have made the error in the first place if

- he had not replaced $\tfrac{1}{3}$ with 0.33. Although 0.33 is correct to 2 decimal places, it is not exactly equal to $\tfrac{1}{3}$.

  In fact $\tfrac{1}{3} = 0.333\,33...$ so 0.33 is less than $\tfrac{1}{3}$; hence the shortage.

- he had worked with fractions and had calculated $800 \times \tfrac{1}{3}$ as $\tfrac{800}{1} \times \tfrac{1}{3} = \tfrac{800}{3} = 266\tfrac{2}{3}$.

  He would then have seen that he needed 267 loaves for 800 baguettes with 1 baguette to spare.

This situation shows, firstly, that we need to be aware that some fractions do not have exact decimal equivalents. Secondly, if we choose to use the decimal form of a fraction, it is important to use enough decimal places to ensure the accuracy we need.

It also illustrates that we need to be comfortable working with fractions so that we can choose to do so when it is appropriate rather than always resorting to decimals.

You can use the revision exercises at the front of this book to practise the basic operations with fractions. In the next section we introduce a different way of interpreting division by a fraction and use fractions in less simple calculations.

## Addition and subtraction of fractions

We cannot add apples to oranges unless we reclassify them both as, say, fruit. In much the same way, we cannot add tenths to quarters unless we change them both into the same kind of fraction. That means changing them so that they have a common denominator. To do this we use the following fact.

The value of a fraction is unaltered if both numerator and denominator are multiplied by the same number.

## Multiplication of fractions

To multiply fractions, multiply the numerators together and multiply the denominators together.

Any mixed numbers must be changed into improper fractions, and factors that are common to the numerator and denominator should be cancelled before multiplication.

## Reciprocals

**If the product of two numbers is 1, then each number is called the reciprocal of the other.**

We know that $\frac{1}{3} \times 3 = 1$

so $\frac{1}{3}$ is the reciprocal of 3 and 3 is the reciprocal of $\frac{1}{3}$.

To find the reciprocal of $\frac{3}{4}$ we require the number which when multiplied by $\frac{3}{4}$ gives 1.

Now $\quad \frac{4}{3} \times \frac{3}{4} = 1 \quad$ so $\quad \frac{4}{3}$ is the reciprocal of $\frac{3}{4}$.

In all cases the reciprocal of a fraction is obtained by turning the fraction upside down.

A number can be written as a fraction, for example, $3 = \frac{3}{1}$ and $2.5 = \frac{2.5}{1}$.

so $\quad$ the reciprocal of $\quad \frac{3}{1}$ is $\frac{1}{3} \quad$ or $\quad 1 \div 3$

and $\quad$ the reciprocal of $\quad \frac{2.5}{1}$ is $\frac{1}{2.5} \quad$ or $\quad 1 \div 2.5 \, (= 0.4)$

**The reciprocal of a number is 1 divided by that number.**

## Division by a fraction

Consider $\quad \frac{2}{5} \div \frac{3}{7} \quad$ This can be interpreted as $\quad \frac{2}{5} \times 1 \div \frac{3}{7}$

Now $\quad 1 \div \frac{3}{7}$ is the reciprocal of $\frac{3}{7}$, i.e. $\frac{7}{3}$

Therefore $\quad \frac{2}{5} \div \frac{3}{7} = \frac{2}{5} \times \frac{7}{3} = \frac{14}{15}$

**To divide by a fraction we multiply by its reciprocal.**

## Exercise 2b

Write down the reciprocals of the following numbers.

**1** 4

**2** $\frac{1}{2}$

**3** $\frac{2}{5}$

**4** 10

**5** $\frac{1}{8}$

**6** $\frac{3}{11}$

**7** 100

**8** $\frac{2}{9}$

**9** $\frac{15}{4}$

**10** 0.25

**11** 3.2

**12** 1.6

Find, without using a calculator

**13** $\frac{2}{3} \div \frac{1}{2}$  **16** $5 \div \frac{4}{5}$  **19** $\frac{3}{7} \div 1\frac{3}{4}$

**14** $1\frac{2}{3} \div \frac{5}{6}$  **17** $\frac{2}{9} \div 1\frac{2}{7}$  **20** $\frac{5}{9} \div 10$

**15** $2\frac{1}{2} \div 4$  **18** $\frac{1}{2} \div \frac{3}{4}$  **21** $3 \div \frac{2}{3}$

## Worked example

→ Find $2\frac{1}{2} + \frac{3}{5} \div 1\frac{1}{2} - \frac{1}{2}\left(\frac{3}{5} + \frac{1}{3}\right)$

 Remember to work out the brackets first, then the multiplication and division and lastly the addition and subtraction.

$$2\frac{1}{2} + \frac{3}{5} \div 1\frac{1}{2} - \frac{1}{2}\left(\frac{3}{5} + \frac{1}{3}\right) = 2\frac{1}{2} + \frac{3}{5} \div 1\frac{1}{2} - \frac{1}{2}\left(\frac{9+5}{15}\right)$$

$$= 2\frac{1}{2} + \frac{3}{5} \div \frac{3}{2} - \frac{1}{2} \times \frac{14}{15}$$

$$= 2\frac{1}{2} + \frac{\cancel{3}^1}{5} \times \frac{2}{\cancel{3}_1} - \frac{1}{\cancel{2}_1} \times \frac{\cancel{14}^7}{15}$$

$$= 2\frac{1}{2} + \frac{2}{5} - \frac{7}{15}$$

$$= 2 + \frac{15 + 12 - 14}{30}$$

$$= 2 + \frac{13}{30} = 2\frac{13}{30}$$

Find, without using a calculator

**22** $1\frac{2}{3} \times \frac{1}{2} - \frac{2}{5}$  **29** $\frac{3}{5}\left(1\frac{1}{4} - \frac{2}{3}\right)$  **36** $\dfrac{\frac{9}{10}}{\frac{5}{6}}$

**23** $\frac{3}{7} + \frac{1}{4} \div 1\frac{1}{3}$  **30** $3\frac{1}{2} - \frac{2}{3} \times 6$  **37** $\left(\frac{2}{3} - \frac{1}{2}\right) \div \left(\frac{3}{4} - \frac{1}{3}\right)$

**24** $\frac{2}{5} \div \left(\frac{1}{2} + \frac{3}{4}\right)$  **31** $2\frac{1}{3} + \frac{1}{2}\left(2 \div \frac{4}{5}\right)$  **38** $\frac{7}{9} - \frac{1}{3}$ of $1\frac{2}{7}$

**25** $5\frac{1}{2} \div 3 + \frac{2}{9}$  **32** $\frac{3}{4}\left(5\frac{1}{3} - 2\frac{1}{5}\right) \div \frac{7}{9}$  **39** $\frac{2}{3} \times \frac{6}{7} - \frac{5}{8} \div 1\frac{1}{4}$

**26** $\frac{4}{5} \div \frac{1}{6} + \frac{1}{3} \times 1\frac{1}{2}$  **33** $\frac{1}{2} + \left(\frac{3}{4} \div \frac{1}{6}\right)$ of $3$  **40** $\dfrac{\frac{7}{8}}{3\frac{1}{2} - \frac{2}{3}}$

**27** $\frac{9}{11} - \frac{2}{5} \times \frac{3}{4}$  **34** $\dfrac{\frac{1}{3} + \frac{1}{4}}{\frac{5}{6} - \frac{3}{4}}$  **41** $\dfrac{3\frac{1}{4}}{2\frac{3}{5}}$

**28** $2\frac{1}{2} \div \frac{7}{9} + 1\frac{1}{3}$  **35** $\dfrac{1\frac{1}{5} - \frac{3}{4}}{2\frac{1}{2}}$  **42** $\dfrac{\frac{2}{3} \times \frac{3}{4}}{\frac{5}{6} \times \frac{3}{10}}$

## Interchanging decimals and fractions

When a fraction is written in the form $\frac{2}{3}$, it is called a **common fraction**, or **vulgar fraction**.

Another way of representing fractions is by placing a point after the number of units and continuing with digits to the right. Fractions written in the form 0.75 are called **decimal fractions**.

Usually we refer to common fractions simply as fractions and to decimal fractions simply as decimals.

### Exercise 2c

**Worked example**

→ Express 0.705 as a fraction.

$0.705 = \frac{7}{10} + \frac{5}{1000}$    This step is usually omitted.

$= \frac{705}{1000} = \frac{141}{200}$

Express the following decimals as fractions.

| 1 | 0.35 | 4 | 1.36 | 7 | 0.005 | 10 | 2.05 |
| 2 | 0.216 | 5 | 0.03 | 8 | 1.01 | 11 | 1.104 |
| 3 | 0.204 | 6 | 0.012 | 9 | 0.11 | 12 | 0.0001 |

**Worked example**

→ Express $\frac{7}{8}$ as a decimal.

$\frac{7}{8} = 0.875$    $\frac{7}{8}$ means $7 \div 8$

$\phantom{8)}0.875$
$8\overline{)7.000}$

Express the following fractions as decimals. Do not use a calculator.

| 13 | $\frac{3}{20}$ | 16 | $\frac{6}{25}$ | 19 | $1\frac{3}{4}$ | 22 | $\frac{5}{16}$ |
| 14 | $\frac{1}{8}$ | 17 | $\frac{1}{16}$ | 20 | $\frac{5}{32}$ | 23 | $2\frac{3}{8}$ |
| 15 | $\frac{3}{5}$ | 18 | $\frac{27}{50}$ | 21 | $\frac{4}{25}$ | 24 | $\frac{1}{500}$ |

## Recurring decimals

If we try to change $\frac{1}{6}$ to a decimal, i.e. $6\overline{)1.0000...}$ giving $0.1666...$, we discover that

- we cannot write $\frac{1}{6}$ as an exact decimal
- from the second decimal place, the 6 recurs for as long as we have the patience to continue the division.

Similarly, if we convert $\frac{2}{11}$ to a decimal by dividing 2 by 11, we get 0.181 818 18... and we see that

- $\frac{2}{11}$ cannot be expressed as an exact decimal
- the pair of digits '18' recurs indefinitely.

Decimals like these are called **recurring decimals**. To save time and space, we place a dot over the digit that recurs. In the case of a group of digits recurring, we place a dot over the first and last digit in the group.

Therefore we write  0.166 666...  as  $0.1\dot{6}$
and  0.181 818...  as  $0.\dot{1}\dot{8}$
and  0.316 316 316...  as  $0.\dot{3}1\dot{6}$

### Exercise 2d

Use the dot notation to write the following fractions as decimals.

1  $\frac{1}{3}$     3  $\frac{5}{6}$     5  $\frac{1}{7}$     7  $\frac{1}{11}$     9  $\frac{5}{12}$     11  $\frac{7}{30}$
2  $\frac{2}{9}$     4  $\frac{1}{15}$    6  $\frac{1}{12}$    8  $\frac{1}{18}$    10  $\frac{1}{14}$    12  $\frac{1}{13}$

Some recurring decimals can be expressed as fractions by recognition.
For example

$\frac{1}{3} = 0.\dot{3}$     $\frac{1}{9} = 0.\dot{1}$     $\frac{1}{11} = 0.\dot{0}\dot{9}$

#### Worked example

→ Express as a fraction    **a** $0.0\dot{3}$    **b** $0.\dot{2}$

**a**  $0.0\dot{3} = 0.\dot{3} \div 10$
      $= \frac{1}{3} \div 10 = \frac{1}{30}$

**b**  $0.\dot{2} = 0.\dot{1} \times 2$
      $= \frac{1}{9} \times 2 = \frac{2}{9}$

13  Express as a fraction    **a** $0.\dot{5}$    **b** $0.0\dot{2}$    **c** $0.0\dot{5}$    **d** $0.00\dot{4}$

14  **a**  Express $0.\dot{9}$ as a fraction.
    **b**  What do you deduce about the value of $0.\dot{9}$?
    **c**  Express $0.0\dot{9}$ as a fraction.

## Using numbers in standard form

Calculations involving numbers in **standard form** can be done on a calculator.

For example, to enter $1.738 \times 10^{-6}$, the number 1.738 is entered normally followed by the $10^x$ button and then the power of 10, i.e. $-6$.

However, as with all calculations, it is important to know whether the answer is about right. This means that we need to be able to estimate results using non-calculator methods. These are illustrated in the worked example below.

### Exercise 2e

**Worked example**

→ If $a = 1.2 \times 10^{-2}$ and $b = 6 \times 10^{-4}$,

find **a** $ab$ **b** $\dfrac{a}{b}$ **c** $a + b$

**a** $ab = (1.2 \times 10^{-2}) \times (6 \times 10^{-4}) = 7.2 \times 10^{-6}$

**b** $\dfrac{a}{b} = \dfrac{1.2 \times 10^{-2}}{6 \times 10^{-4}} = \dfrac{1.2}{6} \times 10^{-2-(-4)} = 0.2 \times 10^2 = 2 \times 10^1$

**c** $a + b = 1.2 \times 10^{-2} + 6 \times 10^{-4}$

$= 0.012 + 0.0006$

$= 0.0126 = 1.26 \times 10^{-2}$

 Multiplication must be done before addition, so each number must be written in full.

**1** Without using a calculator, write down the value of $ab$ in standard form if

    **a** $a = 2.1 \times 10^2$, $b = 4 \times 10^3$      **d** $a = 5 \times 10^{-4}$, $b = 2.3 \times 10^{-2}$

    **b** $a = 5.4 \times 10^4$, $b = 2 \times 10^5$      **e** $a = 1.6 \times 10^{-2}$, $b = 2 \times 10^4$

    **c** $a = 7 \times 10^{-2}$, $b = 2.2 \times 10^{-3}$      **f** $a = 6 \times 10^5$, $b = 1.3 \times 10^{-7}$

**2** Without using a calculator, write down the value of $\dfrac{p}{q}$ in standard form if

    **a** $p = 6 \times 10^5$, $q = 3 \times 10^2$      **c** $p = 7 \times 10^{-3}$, $q = 5 \times 10^2$

    **b** $p = 9 \times 10^3$, $q = 3 \times 10^5$      **d** $p = 1.8 \times 10^{-3}$, $q = 6 \times 10^{-4}$

**3** Without using a calculator, write down the value of $x + y$ in standard form if

    **a** $x = 2 \times 10^2$, $y = 3 \times 10^3$      **c** $x = 2.1 \times 10^4$, $y = 3.1 \times 10^5$

    **b** $x = 3 \times 10^{-2}$, $y = 2 \times 10^{-3}$      **d** $x = 1.3 \times 10^{-4}$, $y = 4 \times 10^{-3}$

For questions **4** and **5**, first estimate the answer and then use a calculator to give the answer correct to 3 significant figures.

**4** The special theory of relativity states that a mass $m$ is equivalent to a quantity of energy $E$, where $E = mc^2$.
$c$ m/s is the speed of light and $c = 2.998 \times 10^8$.
Find $E$ when $m = 1.66 \times 10^{-27}$.

**5** The quantity of nitrate in one bottle of mineral water is $1.5 \times 10^{-3}$ g. The quantity of nitrate in another bottle is $7.3 \times 10^{-4}$ g.

The two bottles are emptied into the same jug. How much nitrate is there in the water in the jug?

### Consider again

David's father asked him to measure the mat well inside the front door.
David wrote down '54 cm by 32 cm'.
David's father then had a piece of matting cut to these measurements but, when he got it home, he found that it was too large for the hole.
Now can you suggest what reasons there could be for this error?

 **Practical work**

You will need access to a set of scales that give weights in kilograms and grams. They do not need to be able to measure small masses accurately; a set of kitchen scales is ideal.
You also need a bag of uncooked rice.
  **a** Try weighing one grain of rice and report on the result.
  **b** Describe a method by which it is possible to estimate the mass of one grain of rice in grams to 2 decimal places.
  **c** Use your method to estimate the mass of one grain.
  **d** Suggest a way to judge the accuracy of your estimate.

 **Investigation**

  **a** Express $\frac{1}{99}$ as a decimal.   **b** Use the result from part **a** to express $0.1\dot{5}$ as a fraction.
  **c** Find a rule for expressing any recurring decimal with a two-digit repeating pattern as a fraction.
  **d** Express $\frac{1}{999}$ as a decimal.
  **e** Explain how the result from part **d** can be used to express any recurring decimal with a three-digit repeating pattern as a fraction.
  **f** Find a rule for expressing any recurring decimal as a fraction and test your rule with examples of four-digit repeating patterns and five-digit repeating patterns.

# 3 Probability

## Consider

Fruit machines, also known as 'one-armed bandits', are a popular gambling game. You can win money by playing on fruit machines, but it is more likely that you will lose money. You put a coin in the machine and pull the lever.

This makes three drums rotate quickly, then slow down and finally stop. Each drum has pictures of several fruits. When the drums stop, three fruits are shown in the centre of the display. Depending on what they are there may be a prize.

Max has been interested in fruit machines, but he has never played them. He would win a prize on the machine he is looking at if cherries appear on all three drums, or lemons show on all three drums. Before he takes the plunge, he would like to assess the risk of losing money.

To do this, he needs to know how to find the probability of combined events such as
a getting either a cherry or a lemon on the first drum
b getting a cherry on both the first and the second drum
c getting a cherry on all three drums.

There are 10 different types of fruit on each drum.

Can you give Max the answers he needs?

## Consider also

An ordinary unbiased dice is rolled. What is the probability of throwing a prime number or an even number?

*You should be able to solve these problems after you have worked through this chapter.*

## Class discussion

Each sentence describes a situation where two events are involved. Discuss what the events are and whether they fall into the 'either ... or' category or into the 'both ... and' category.

1 An ordinary six-sided dice is rolled and scores five or six.
2 Two dice are rolled and a double six is scored.
3 Tim picks a box from a lucky dip. Some boxes contain a prize and the others are empty.
4 The England cricket captain tosses a coin to find out who has the choice to bat or to field.

## Mutually exclusive events

When an ordinary dice is rolled, it is possible to score either a five or a six. It is not possible to score both a five and a six. Such events are called **mutually exclusive**.

## Independent events

When two ordinary dice are rolled, it is possible to score a six on the first dice and a six on the second dice Also, the score obtained on the second dice is not affected in any way by the score on the first dice. Such events, where both can happen but each has no influence on whether the other event happens or not, are called **independent events**.

Not all events are independent. There are two green sweets and two red sweets in a bag. Bethany takes one of these sweets then Tom takes one. If Bethany's sweet is red, there is only one out of three ways in which Tom can choose a red sweet. But if Bethany's sweet is green, there are two out of three ways in which Tom can choose a red sweet. So the probability that the second sweet is red depends on the colour of the first sweet taken.

### Exercise 3a

Decide whether the events described are mutually exclusive, independent or dependent.

1. Jasmine and Oliver each buy a ticket for a raffle and one of them wins first prize.

2. Two coins are tossed.
    a. The first coin lands heads up or tails up.
    b. Both coins land head up.

3. A 10 pence coin is tossed and a dice is rolled.
    a. The coin lands heads up and an even number is scored on the dice.
    b. A three or a six is scored on the dice.

4. A blue bag and a red bag each contain a large number of coins, some of which are counterfeit. One coin is selected at random from each bag.
    a. The coin taken from the blue bag is counterfeit or not counterfeit.
    b. Both coins are counterfeit.

5 Hartfield Airport has 100 scheduled flights due to depart on Saturday.
   a  Two or three flights are cancelled.
   b  One flight is cancelled because the plane is faulty and another flight is cancelled because of a hurricane at its destination.

6 A box contains six blue pens and three red pens. One pen is removed at random.
   a  The pen is put back then a pen is removed again.
   b  The pen is not put back and another pen is removed.

## Adding probabilities

If we select a card at random from an ordinary pack of 52 playing cards, the probability of selecting an ace is $\frac{4}{52}$ and the probability of selecting a black king is $\frac{2}{52}$.

Now selecting either an ace or a black king involves two events that are mutually exclusive since it is impossible to draw one card which is both an ace and a black king.

There are 4 aces and 2 black kings, so if we want to find the probability of selecting either an ace or a black king, there are 6 cards that we would count as 'successful', therefore

$$P(\text{ace or a black king}) = \frac{6}{52}$$

 Remember that the probability that an event $A$ happens is $P(A)$, where
$$P(A) = \frac{\text{the number of ways in which } A \text{ can occur}}{\text{the total number of equally likely outcomes}}$$

$$P(\text{ace}) = \frac{4}{52} \quad \text{and} \quad P(\text{black king}) = \frac{2}{52}$$

Since $\frac{6}{52} = \frac{4}{52} + \frac{2}{52}$ it follows that

$$P(\text{ace or black king}) = P(\text{ace}) + P(\text{black king})$$

Now consider the probability of scoring 5 or 6 when one dice is rolled.

$$P(\text{score 5 or 6}) = \frac{2}{6}$$

From one roll of a dice, a score of 5 and a score of 6 are mutually exclusive, where

$$P(\text{score 5}) = \frac{1}{6} \quad \text{and} \quad P(\text{score 6}) = \frac{1}{6}$$

i.e. $P(\text{score 5 or 6}) = \frac{2}{6} = \frac{1}{6} + \frac{1}{6}$

$$= P(\text{score 5}) + P(\text{score 6})$$

From these examples we see that

**if $A$ and $B$ are mutually exclusive events, then $P(A \text{ or } B) = P(A) + P(B)$**

# STP Maths 9

Now consider the probability of scoring 1 or 2 or 3 or 4 or 5 or 6 when one dice is rolled. These events are mutually exclusive and they cover all the possible outcomes. The set of all possible outcomes is called **exhaustive**.

Now $\quad$ P(score 1 or 2 or 3 or 4 or 5 or 6) $= \frac{1}{6} + \frac{1}{6} + \frac{1}{6} + \frac{1}{6} + \frac{1}{6} + \frac{1}{6} = \frac{6}{6} = 1$

This illustrates the general rule that

**the sum of the probabilities of an exhaustive set of mutually exclusive outcomes is 1**

For example, a bag contains some black discs, some red discs and some white discs. When one disc is removed at random, the outcome is either black, red or white. These outcomes are mutually exclusive and exhaustive.

Given that the probability that the disc is black is $\frac{1}{3}$ and the probability that it is white is $\frac{1}{5}$, we can use the fact above to find the probability that the disc is red.

$\quad$ P(black) + P(red) + P(white) = 1

giving $\quad \frac{1}{3}$ + P(red) + $\frac{1}{5}$ = 1

Therefore $\quad$ P(red) = $1 - \left(\frac{1}{3} + \frac{1}{5}\right) = 1 - \frac{8}{15} = \frac{7}{15}$

## Exercise 3b

1. A card is selected at random from an ordinary pack of 52. What is the probability that the card is
   a. a red ace
   b. a black king
   c. a red ace or a black king?

2. Emily rolls an ordinary dice once. What is the probability that the number shown is
   a. 2
   b. 3 or 4
   c. 2, 3 or 4?

3. A card is drawn at random from the 12 court cards (jacks, queens and kings). What is the probability that the card is
   a. a black jack
   b. a red queen
   c. either a black jack or a red queen?

4. Graham is looking for his house key. The probability that it is in his pocket is $\frac{5}{9}$, while the probability that it is in his car is $\frac{1}{13}$. What is the probability that
   a. the key is either in his pocket or in his car
   b. the key is somewhere else?

5  When Mrs George goes shopping, the probability that she returns by bus is $\frac{3}{7}$, in a taxi $\frac{1}{7}$ and on foot $\frac{5}{14}$. What is the probability that she returns
   a  by bus or taxi
   b  by bus or on foot
   c  by none of these three ways?

6  Abigail has a bag containing discs of four different colours. One disc is removed at random. The table shows the probabilities of choosing three of the four colours.

   | Colour | red | white | blue | pink |
   |---|---|---|---|---|
   | Probability | $\frac{2}{7}$ | $\frac{2}{9}$ | $\frac{1}{4}$ | |

   Abigail removes one disc at random. What is the probability that this disc is
   a  red or white
   b  white or blue
   c  red, white or blue
   d  pink?

7  Rajev has a pack of playing cards with some cards missing. There are 45 cards in the pack. He knows that all the clubs and hearts are in his pack. One card is drawn at random from the pack. What is the probability that this card is not a club or a heart?

8  Maya rolls an ordinary dice. What is the probability that the number on the dice is
   a  an even number
   b  a prime number
   c  either even or prime?

   Your answer to part **c** should not be the sum of the answers to parts **a** and **b**. Why not?

## Multiplication of probabilities

When a coin is tossed and a dice is rolled, we can use a table to list all the possible outcomes.
A set of all the possible outcomes of an experiment is called a **possibility space**.

|  |  | \multicolumn{6}{c}{Dice} |
|---|---|---|---|---|---|---|---|
|  |  | 1 | 2 | 3 | 4 | 5 | 6 |
| Coin | H | H, 1 | H, 2 | H, 3 | H, 4 | H, 5 | H, 6 |
|  | T | T, 1 | T, 2 | T, 3 | T, 4 | T, 5 | T, 6 |

From the table we can see that

$$P(\text{a head and an even number}) = \frac{3}{12} = \frac{1}{4}$$

# STP Maths 9

A head from one toss of the coin and an even number from one throw of the dice are independent events, where

$$P(\text{a head}) = \frac{1}{2} \quad \text{and} \quad P(\text{an even number}) = \frac{3}{6} = \frac{1}{2}$$

But

$$P(\text{a head and an even number}) = \frac{1}{4} = \frac{1}{2} \times \frac{1}{2}$$
$$= P(\text{a head}) \times P(\text{an even number})$$

This example illustrates that

**if $A$ and $B$ are independent events, then $P(A \text{ and } B) = P(A) \times P(B)$**

### Exercise 3c

1. Two coins are tossed. What is the probability that both show heads?

2. Two dice are tossed. Find the probability of getting a double six.

3. Jack has two tubes of sweets. Each tube contains 10 red sweets and 30 sweets of other colours. Jack takes one sweet, chosen at random, from each tube. Find the probability that
   a. he takes a red sweet from a tube
   b. he takes a sweet that is not red from a tube
   c. both the sweets he takes are not red.

4. The probability that Lauren will win the girls' 100 m race is $\frac{2}{5}$ and the probability that Liam will win the boys' 100 m race is $\frac{3}{5}$. What is the probability that
   a. both of them will win their events
   b. neither of them will win their event?

5. A mother has an equal chance of giving birth to a boy or a girl. Holly plans to have two children.
   a. What is the probability that the first is a girl?
   b. What is the probability that both are boys?
   c. What is the probability that neither is a boy?

6. The probability that Olivia will have to wait before she can cross Westgate Street is $\frac{1}{3}$ and the probability that she will be able to cross High Street without waiting is $\frac{1}{4}$.
   What is the probability that
   a. she does not have to wait to cross Westgate Street
   b. she has to wait to cross High Street
   c. she can cross both streets without waiting?

**7** A bag contains three red sweets and two green sweets. Camilla takes one sweet at random and eats it. She then takes another sweet, also at random.
  **a** Make a possibility table to show the possible combinations of colours of the two sweets and use it to find the probability that both sweets removed are red.
  **b** Explain why, in this case, the multiplication rule does not give the correct answer to part **a**.

### Exercise 3d

In questions **1** to **5**, some of the events described are mutually exclusive and some are independent.

**1** A red dice and a blue dice are rolled. Find the probability of getting
  **a** a 5 or a 6 on the red dice
  **b** a 1 or a 2 on the blue dice
  **c** a 2 on both dice
  **d** an even number on both dice.

**2** A card is drawn at random from an ordinary pack of 52 playing cards.
What is the probability that the card is
  **a** a 2
  **b** a red ace
  **c** a 2 or a red ace?

**3** When Kim goes to the cinema the probability that she returns on foot is $\frac{2}{3}$, by bus $\frac{1}{6}$ and in a friend's car $\frac{1}{6}$.
What is the probability that she returns
  **a** by bus or in a friend's car
  **b** on foot or by bus?

**4** The probability that Sam will complete the 5000 km race is 0.9 and the probability that Aaron will complete it is 0.6. What is the probability that both Sam and Aaron will complete the 5000 km race?

**5** A pack of cards is cut, reshuffled and cut again. What is the probability that
  **a** the first card cut is an ace or a king
  **b** the second card cut is an ace or a king
  **c** both cards cut are aces?

# Tree diagrams

When two coins are tossed, one possible outcome is a head and a tail. This outcome involves two events but they do not fit neatly into the 'either … or' category or the 'both … and' category. This is because a head and a tail can be obtained by getting

**either** a head on the first coin **and** a tail on the second,
**or** a tail on the first coin **and** a head on the second.

So getting a head and a tail when two coins are tossed involves a mixture of independent and mutually exclusive events, and we need an organised approach to deal with such a combination. One such approach is to draw up a table showing all the equally likely outcomes, but this method cannot be used if all the possible outcomes are not equally likely, such as the possible outcomes when two people take a driving test.

Now suppose that three coins are tossed and we want the probability of getting two heads and a tail. Three events are involved here, so we cannot use a table to list all the outcomes because a table can only cope with two events.

These examples show that we need a different way of listing outcomes and finding probabilities. We will illustrate this approach with a simple example.

Suppose that we have two discs, a red one marked A on one side and B on the other, and a blue one marked E on one side and F on the other.

Tossing the red disc, the probability that we get A is $\frac{1}{2}$ and the probability that we get B is also $\frac{1}{2}$. This information can be shown in the adjacent diagram.

Suppose that the red disc shows A and we go on to toss the blue disc. The probability of getting E is $\frac{1}{2}$ and the probability of getting F is $\frac{1}{2}$. We can add this information to the diagram.

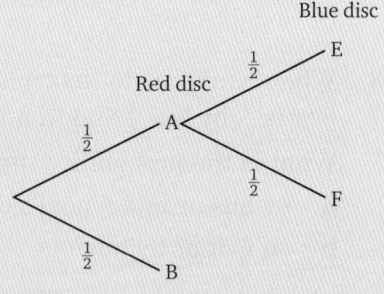

We complete the diagram by considering what the probabilities are if the red disc shows a B before we toss the blue disc.

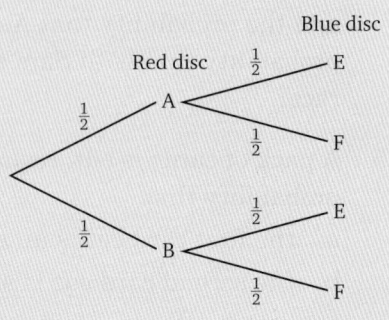

# 3 Probability

Diagrams like this are called **tree diagrams** or **probability trees**. To use the tree diagram to find the probability that we get first an A and then an E, follow the path from left to right for an A on the first branch and an E on the second. The two probabilities we find there are $\frac{1}{2}$ and $\frac{1}{2}$. The blue disc showing E is independent of the letter obtained on the red disc so we multiply the probabilities together to get $\frac{1}{4}$.

To find the probability that we get a B on the red disc and an F on the blue one, follow the B and F path and multiply the probabilities

$$P(B \text{ and } F) = \frac{1}{2} \times \frac{1}{2}$$
$$= \frac{1}{4}$$

**In general, we multiply probabilities when we follow a path along branches.**

## Exercise 3e

### Worked example

→ A coin is tossed and a dice is thrown. Find the probability that
  a  the coin lands head up and the dice does not show a six
  b  the coin lands tail up and the dice shows a six.

 There are only two possible outcomes when the coin is tossed, so we need two 'branches' to show these. There are six possible outcomes when the dice is thrown, but we only need to consider these in two groups, throwing a six or not throwing a six, and we need only two branches to show these.

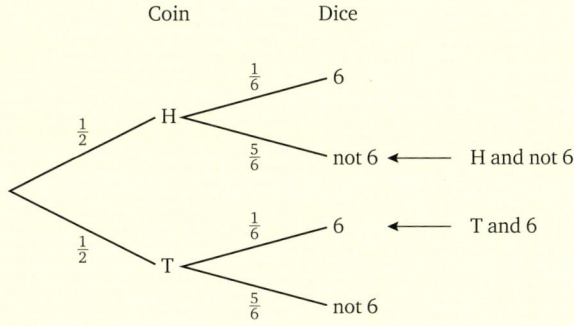

a  $P(H \text{ and not } 6) = \frac{1}{2} \times \frac{5}{6} = \frac{5}{12}$

b  $P(T \text{ and } 6) = \frac{1}{2} \times \frac{1}{6} = \frac{1}{12}$

1. The probability that Mark gets to work on time is $\frac{7}{8}$ and the probability that he leaves work on time is $\frac{3}{5}$.

   a. Find the probability that he does not leave work on time.

   b. Copy and complete the tree diagram.

   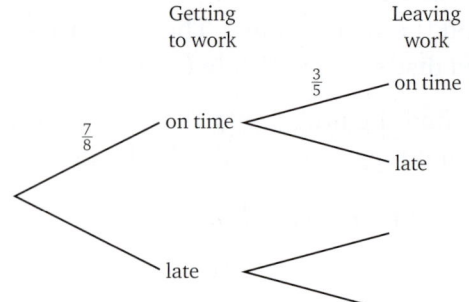

   What is the probability that

   c. Mark gets to work on time but does not leave on time

   d. Mark is late for work but leaves on time?

2. When a drawing pin falls to the ground the probability that it lands point up is 0.2.

   a. Find the probability that a pin does not land point up.

   Two drawing pins fall one after the other.

   b. Copy and complete the tree diagram.

   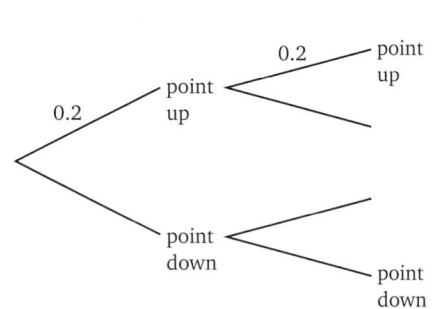

   Find the probability that

   c. both drawing pins land point up

   d. both drawing pins land point down.

3. The first of two boxes of tennis balls contains one white and two yellow balls. The second box contains three yellow and two lime green balls. A ball is taken at random from each box.

   a. Copy and complete the tree diagram.

   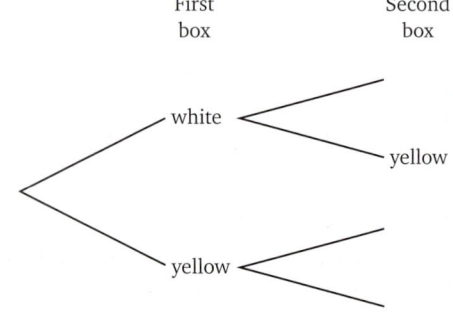

   Find the probability that

   b. both balls are yellow

   c. one is white and one is lime green.

4   Two archers fire arrows at a target. The probability that Becker hits the target is 0.5 and the probability that Crossley does not hit the target is 0.3. Becker fires at the target first, then Crossley fires.

Draw a tree diagram to show the possibilities and use it to find the probability that
   a   both Becker and Crossley hit the target
   b   neither hits the target
   c   Becker hits the target but Crossley misses.
   d   Crossley hits the target but Becker misses.

## Worked example

→ Two coins are tossed. Find the probability that they land showing a head and a tail.

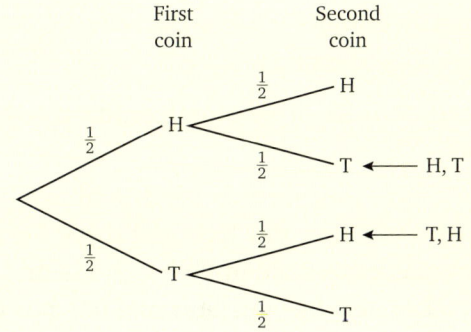

We start by drawing a tree diagram.

 We can see that there are two paths through the tree that give a head and a tail, where

$$P(H \text{ on first coin and } T \text{ on the second}) = P(H, T) = \frac{1}{2} \times \frac{1}{2} = \frac{1}{4}$$

$$P(T \text{ on first coin and } H \text{ on the second}) = P(T, H) = \frac{1}{2} \times \frac{1}{2} = \frac{1}{4}$$

Now the coins can land showing either H, T or T, H so these are mutually exclusive. Therefore we can find P(H, T or T, H) by adding the probabilities at the ends of the two paths:

$$P(H \text{ and } T) = \frac{1}{4} + \frac{1}{4} = \frac{1}{2}.$$

$$P(H \text{ and } T) = \left(\frac{1}{2} \times \frac{1}{2}\right) + \left(\frac{1}{2} \times \frac{1}{2}\right)$$

$$= \frac{1}{4} + \frac{1}{4} = \frac{1}{2}$$

The worked example illustrates the general rule that

> we *multiply* the probabilities when we follow a path along the branches and *add* the results of following different paths.

**5** The probability that Ryan's bus has to wait at the traffic lights in the morning on the way to school is $\frac{1}{5}$.

Draw a probability tree to show the possibilities that the bus has to wait or can drive through the traffic lights on two consecutive mornings. Find the probability that, on two consecutive mornings, the bus
  a  has to wait at the lights on both occasions
  b  does not have to wait on either morning
  c  has to wait on just one morning.

**6  a**  If a dice is rolled, what is the probability of getting
  i  a six    ii  a number other than six?
  **b**  Two dice, one red and the other blue, are rolled. Draw a tree diagram to show the possibilities of getting a 6 or not getting a 6 on each dice. Find the probability that
    i   both dice show a 6
    ii  the red dice shows a 6 but the blue dice does not
    iii the blue dice shows a 6 but the red dice does not
    iv  just one dice shows a 6.

For each of questions **7** to **13**, draw a probability tree to illustrate the given information.

**7** In a group of six girls, four have fair hair and two have dark hair. In a group of five boys, two have fair hair and three have dark hair. One boy and one girl are picked at random. What is the probability that, of the two students picked, one has fair hair and one has dark hair?

**8** In a class of 20, four students are left-handed. In a second class of 24, six students are left-handed. One student is chosen at random from each class. What is the probability that one of the students is left-handed and one is not?

**9** Derek and Alexis keep changing their minds about whether to send Christmas cards to each other. In any one year, the probability that Derek sends a card is $\frac{3}{4}$ and that Alexis sends one is $\frac{5}{6}$.
Find the probability that next year
  a  they both send cards
  b  only one of them sends a card
  c  neither sends a card.
What should the three answers add up to and why?

**10** Copy the tree diagram in the worked example on the previous page and add branches to the right to show the following information.
Three unbiased coins are tossed, one after the other.
Find the probability that
  a  three heads appear
  b  three tails appear
  c  two heads and one tail appear in any order.

**11** The weather forecast gives the probability that it will rain on Saturday as 0.07 and the probability that it will not rain on Sunday as 0.89.
   a  On which of these two days is it more likely to rain and why?
   b  Copy and, complete this tree diagram.

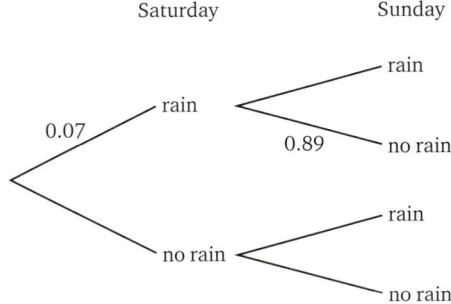

   c  Use your tree diagram to find the probability that it will rain on
      i  both days    ii  just one of the days.

The probability that it will rain on Monday is 0.3. Add more branches to your tree diagram to include Monday.
   d  Use your new tree to find the probability that it will rain on
      i  none of the three days    ii  at least one of the three days.

**12** A coin is tossed three times. Use your tree diagram from question **10** to find the probability of getting
   a  a head and two tails          c  at least one head
   b  exactly one tail              d  at least two heads.

**13** In a group of 120 girls, 24 have blue eyes, 48 have hazel eyes, 36 have green eyes and the remainder have brown eyes. All the girls have either long hair or short hair and the probability that a given girl has long hair is 0.25. The probability that a girl has freckles is 0.65. Assume that each attribute is independent of the others. What is the probability that a girl chosen at random from this group has
   a  brown eyes, freckles and short hair
   b  long hair, no freckles and either blue or green eyes?

## Venn diagrams

**Venn diagrams** can help find the number of times that two events can occur when those events are not mutually exclusive.

Consider a class of 30 students.
Mr Edwards asked them if they had a calculator with them. 20 students put up a hand.
Mr Edwards then asked them if they had a protractor with them. 16 students put up a hand.
There were 6 students who had neither a calculator nor a protractor.

Before we can answer questions such as 'What is probability that a student chosen at random from the class has both a calculator and a protractor?' we need to find how many students have both.

Having a calculator or having a protractor are not mutually exclusive because some students will have both. This is the number of students in the intersection of the sets {students with a calculator} and {students with a protractor}.

We do not know how many students have both, so we will use $x$ as that number.

We can now use a Venn diagram to illustrate this information. We cannot list the students because we do not know their names, so we will use the number of students in each set.

We know that the number in area common to both the circles, representing {students with a calculator} and {students with a protractor}, is $x$.

We also know that the number in the set {students with a calculator} is 20.

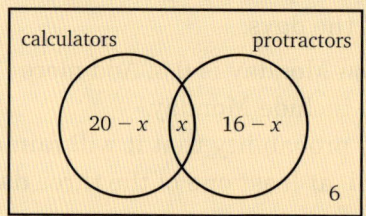

Therefore, the number in the part of the circle representing {calculators but not protractors} is $20 - x$.

Similarly, the number in the part of the circle representing {protractors but not calculators} is $16 - x$.

Outside the two circles is the number who have neither a calculator nor protractor.

We know that there are 30 students in the class so we can form the equation

$$(20 - x) + x + (16 - x) + 6 = 30$$

Solving this equation gives
$$42 - x = 30$$

giving
$$x = 12$$

Now we can give the probability that a student has both a calculator and a protractor as $\frac{12}{30} = \frac{2}{5}$

### Exercise 3f

1. Use the Venn diagram above to find the probability that one student chosen at random from the class

   a. has a calculator but not a protractor

   b. has a calculator and/or a protractor.

2  The Venn diagram shows how many students in a class of 30 own a mobile phone and/or a tablet.

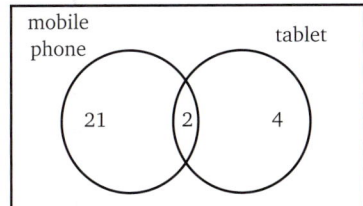

a  How many students do not own either a mobile phone or a tablet?

What is the probability that one student chosen at random from the class

b  owns a tablet but not a mobile phone
c  owns a mobile phone?

3  100 adults were asked how they paid for goods bought in a shop. Some said they used a credit card, some said they paid cash and some said they used both. Some adults used other means to pay for their goods.
Some of these results are shown in the Venn diagram.

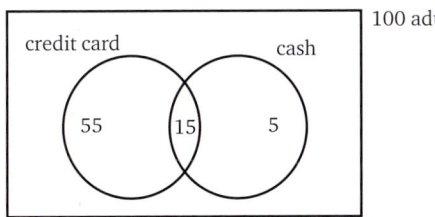

a  Copy and complete the Venn diagram.
b  How many adults paid for goods without using a credit card or cash?
c  What is the probability that one of these adults chosen at random only used cash?
d  What is the probability that one of these adults chosen at random didn't use cash?

4  In a squad of 35 cricketers, 20 said that they could bat and 8 said that they could bat and bowl. Show this information on a Venn diagram. How many more were willing to bowl than to bat?

5  In a group of 24 children, each child had a dog or a cat or both. 18 had a dog and 5 of these also had a cat.
Show this information on a Venn diagram and hence find the probability that one of these children chosen at random had
a  a cat
b  only a dog
c  just one of these as a pet.

6  A group of 50 television addicts were asked if they watched sports programmes and nature programmes. The replies revealed that they all watched one or other or both: 21 watched both sports and nature programmes and 9 watched only nature programmes. Show this information on a Venn diagram and use it to find the probability that one of these people chosen at random
   a  watched sports programmes
   b  did not watch nature programmes
   c  watched either sports or nature programmes but not both.

7  In a youth club, 35 teenagers said that they went to football matches, discos or both. Of the 22 who said they went to football matches, 12 said they also went to discos. A further 10 teenagers said they did not go to either. Show this information on a Venn diagram.
   a  How many teenagers went to football matches or discos, but not to both?
   b  One of this group of teenagers is chosen at random. What is the probability that the teenager went to discos but not to football matches?

8  There are 28 students in a class, all of whom take history or geography or both. 14 take history, 5 of whom also take geography.
   a  Show this information on a Venn diagram.

   One student is chosen at random. What is the probability that the student takes
   b  geography
   c  history but not geography
   d  just one of these subjects?

9  The Venn diagram shows some information about how many students in a class of 32 had goldfish (*G*), budgerigars (*B*) or both.

   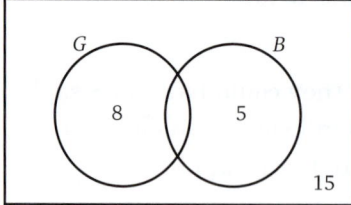

   a  Use the Venn diagram to find the number of students who had both goldfish and budgerigars.

   One student is chosen at random. What is the probability that the student
   b  did not have a budgerigar
   c  had at least one of these pets?

10 The passengers on a coach were questioned about the newspapers and weekly magazines they bought.

3 bought both a daily newspaper and a weekly magazine.
15 bought a daily newspaper. 8 bought a weekly magazine.
8 did not buy either a daily paper or a weekly magazine.
   a  Show this information on a Venn diagram.
   b  How many passengers were there on the coach?
   c  What is the probability that one of the passengers, chosen at random, bought a daily newspaper, a weekly magazine or both?

11 One evening all 78 members of a youth club were asked whether they liked swimming ($S$) and/or dancing ($D$). It was found that 34 liked swimming, 41 liked dancing and 8 liked neither.
   a  Show this information on a Venn diagram.
   b  Use the diagram to find how many liked both swimming and dancing.

What is the probability that one of the members, chosen at random, likes
   c  swimming but not dancing
   d  dancing or swimming but not both?

12 During April, 36 cars were taken to a testing station for a road worthiness certificate. The results showed that 17 cars passed the test, 10 had defective brakes and 13 had defective lights.
   a  Show this information on a Venn diagram.

One of these cars is chosen at random. What is the probability that it
   b  failed the test
   c  had both defects
   d  had just one defect?

13 a  Write down the members of the set of all possible outcomes when an ordinary unbiased six sided dice is rolled.
   b  Write down the members of the set of outcomes that are
      i  prime numbers      ii  even numbers.
   c  The dice is rolled once. What is probability that it scores a number that is neither prime nor even?

14 a  List the members of the set, $S$, of whole numbers from 1 to 16 inclusive.
   b  List the set, $A$, of numbers that are factors of 12 and the set, $B$, of numbers that are factors of 16.
   c  Show the members of all three sets in a Venn diagram.
   d  One number is chosen at random from the set $S$. What is the probability that the number is
      i  a factor of both 12 and 16      ii  a factor of neither 12 nor 16?

**15** The Venn diagram shows the number of students taking geography (G), history (H) and art (A) in a class of 43. Every student takes at least one of these subjects

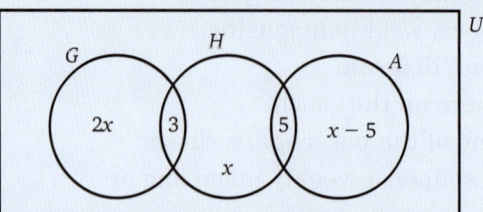

a Write down an expression, in terms of $x$, for the number of students who take history.

b Write down an equation, in terms of $x$, which shows the information given.

c Find the probability that one of these students, chosen at random
  i takes geography only     ii takes art.

## Mixed exercise

The next exercise contains mixed problems on probability. You can answer some of the questions directly from the basic definition of probability and you can answer some using the sum and product rules. Draw a tree diagram or a possibility table or a Venn diagram only when you think it is needed.

### Exercise 3g

 **1** A letter is picked at random from the word CATASTROPHE. Find the probability that
  a the letter is a vowel          b the letter is A or T.

**2** A knitting wool sample card has 1 green, 1 black, 4 blue and 2 red samples. If one sample is picked at random, what is the probability that it is
  a yellow                         b black, green, red or blue?

 **3** The scores on a four-sided spinner are 1, 2, 3 or 4. On a second four-sided spinner the scores are 5, 6, 7 or 8. If the two are spun, find the probability that
  a the score on both spinners is odd
  b the score on both spinners is even
  c the score on neither spinner is prime.

4 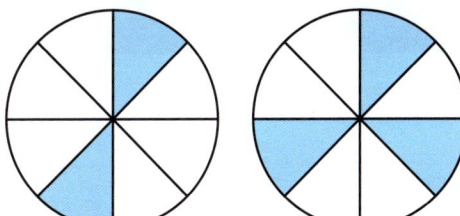 A sector is chosen at random from each circle. What is the probability that
   a   both sectors picked are blue
   b   both sectors picked are white
   c   one is blue and the other is not?

5  There are two bags. The first contains 2 white and 3 black marbles and the second contains 1 red and 2 blue marbles. Two marbles are taken at random, one from each bag. Find the probability that
   a   a white and a blue marble are taken
   b   a black and a red marble are taken
   c   neither a white marble nor a red marble is taken.

6  In a game of skittles the probability that Ted scores more than 5 is $\frac{2}{7}$ and the probability that George scores more than 5 is $\frac{2}{9}$. Ted goes first followed by George. Use a tree diagram to find the probability that
   a   both Ted and George score more than 5
   b   Ted scores more than 5 but George does not
   c   both score 5 or less
   d   one scores more than 5 but the other does not.

7  Mr Aziz sells vegetables from a market stall. One morning he makes a note of the sales of carrots and yams to the first 50 customers.
   25 bought carrots
   36 bought yams
   12 bought neither carrots nor yams.
   a   Find the number of customers who bought carrots and yams.
   b   If one customer from the list is chosen at random, what is the probability that the customer bought carrots but not yams?

## Consider again

Max could win on the fruit machine if cherries appear on all three drums, or lemons appear on all three drums. Before he plays he would like to assess the risk of losing money.

To do this, he needs to know how to find the probability of combined events such as
a   getting either a cherry or a lemon on the first drum
b   getting a cherry on both the first and the second drum
c   getting a cherry on all three drums

There are 10 different types of fruit on each drum.

Now can you give Max the answers he needs?

## Consider also

An ordinary unbiased dice is rolled. What is the probability of throwing a prime number or an even number?

Now can you answer this question?

# 4 Percentages

## Consider

Ben is a buyer in the men's department of a large department store. He buys a batch of shirts from a manufacturer for £11.50 each. His normal practice is to add a mark-up of 50%, then he has to add value added tax (VAT) at 20% to give the selling price.

What does this give as the selling price?

He prefers to sell the shirt at just under £20, so decides to sell it at £19.95 including VAT.

Ben needs to work out how much of the £19.95 is VAT, because he will have to pay this amount to the tax authority. How much VAT will he have to pay?

*You should be able to solve this problem after you have worked through this chapter.*

## Class discussion

Discuss what you need to be able to do to solve the following problems.

1. The total number of a certain species of bird is estimated to be 5000. It is believed that the number is decreasing at the rate of 5% a year. The Society for the Protection of Birds would like to find out how many years it will be before the population drops below 2000.

2. The cost of a unit of electricity is 8.65 p including VAT at 8%. It is rumoured that, if a particular party wins power at the next general election, the rate of VAT that applies to domestic fuel will rise from 8% to $17\frac{1}{2}$%, whereas if the other main party wins they will abolish VAT on domestic fuel altogether. Assuming that the basic cost of electricity stays the same, Matthew would like to know how much one unit of electricity will cost depending on which of these two parties forms the next government. He would also like to work out the percentage difference between the two.

Discussion of the above situations shows that when dealing with a variety of problems involving percentages, a clear understanding of *what the percentage is of* is very important.

## Percentage increase and percentage decrease

Percentage increase or decrease arises in many different situations. We may read that:
- certain workers are to receive an increase in their wages of 8%
- value added tax (VAT) may be increased from 20% to 22%
- the basic rate of income tax should be reduced from 20% to 18%
- all the items in a sale are offered at a discount of 20%.

Changes are expressed in percentage terms as *a percentage of the quantity before any changes are made*. This is because this makes it easier to calculate the actual change in a particular case and to compare one change with another.

If a wage of £100 per week is increased by 8%, then the new wage is

$$108\% \text{ of } £100 = 1.08 \times £100$$
$$= £108$$

If an article costs £55 plus sales tax at $17\frac{1}{2}\%$, then the full cost is

$$117\frac{1}{2}\% \text{ of } £55 = 1.175 \times £55$$
$$= £64.63 \text{ (correct to the nearest penny)}$$

If a woman earns £550 and has to pay tax on it at the rate of 23%, she actually receives $(100 - 23)\%$ of £550, which is 77% of £550.

$$77\% \text{ of } £550 = 0.77 \times £550$$
$$= £423.50$$

If a discount of 25% is offered in a sale, a piece of furniture, originally marked at £760, will cost

$$75\% \text{ of } £760 = 0.75 \times £760$$
$$= £570$$

Retailers buy in goods, which are usually sold at an increased price. The increase is often called the mark-up and is normally given as a percentage of the buying-in price. Occasionally goods are sold at a decreased price, which means they are sold at a loss. The loss is also given as a percentage of the buying-in price.

If a store puts a *mark-up* of 50% on an article it buys for £100, its mark-up is

$$50\% \text{ of } £100 = 0.5 \times £100$$
$$= £50$$

and the selling price is

$$150\% \text{ of } £100 = 1.5 \times £100$$
$$= £150$$

## Exercise 4a

### Worked example

→ A second-hand car dealer bought a car for £3500 and sold it for £4340. Find his percentage mark-up.

Mark-up = selling price − buying-in price

= £4340 − £3500

= £840

% mark-up = $\dfrac{\text{mark-up}}{\text{buying-in price}} \times 100$

= $\dfrac{£840}{£3500} \times 100 = 24$

Therefore the mark-up is 24%.

 Remember, percentage mark-up is the mark-up expressed as percentage of the buying-in price.

In questions **1** to **4**, find the percentage mark-up.

1. Buying-in price £12, mark-up £3
2. Buying-in price £28, mark-up £8.40
3. Buying-in price £16, mark-up £4
4. Buying-in price £55, mark-up £5.50

### Worked example

→ A retailer bought a leather chair for £375 and sold it for £285. Find his percentage loss.

loss = buying-in price − selling price

= £375 − £285

= £90

% loss = $\dfrac{\text{loss}}{\text{buying-in price}} \times 100$

= $\dfrac{£90}{£375} \times 100 = 24$

Therefore the loss is 24%.

 % loss is the loss as a percentage of the buying-in price.

In questions **5** to **8**, find the percentage loss.

5. Buying-in price £20, loss £4
6. Buying-in price £125, loss £25
7. Buying-in price £64, loss £9.60
8. Buying-in price £160, loss £38.40

## Worked example

→ An article costing £30 is sold at a gain of 25%.
Find the selling price.

The selling price is 100% of £30 + 25% of £30, which is

125% of £30

Selling price = 125% of £30
= 1.25 × £30
= £37.50

Therefore the selling price is £37.50.

 Alternatively, we can find the gain, which is 25% of the original cost, and add this to the original cost.

Gain = 25% of £30
= 0.25 × £30
= £7.50

Selling price = £30 + £7.50

Therefore the selling price is £37.50.

In questions **9** to **14**, find the selling price.

9 Cost £50, gain 12%

10 Cost £64, gain 122%

11 Cost £29, gain 110%

12 Cost £36, loss 50%

13 Cost £75, loss 64%

14 Cost £128, loss $37\frac{1}{2}$%

In questions **15** to **19**, find the weekly cash increase for each employee.

15 Ian earns £320 per week and receives a rise of 1%.

16 Nairn earns £280 per week and receives a rise of 2%.

17 Sitara earns £225 per week and receives a rise of 3%.

18 Lyn earns £370 per week and receives a rise of 5%.

19 Joe earns £400 per week and receives a rise of $2\frac{1}{2}$%.

For questions **20** to **22,** which is the better cash pay rise, and by how much?

20 a 6% on a weekly wage of £100, or
   b 4% on a weekly wage of £250

21 a $3\frac{1}{2}$% on a weekly wage of £180, or
   b 2% on a weekly wage of £400

22 a 4% on a weekly wage of £300, or
   b 3% on a weekly wage of £400

In questions **23** to **28**, find the purchase price of the given item.

**23** An electric heater marked £90 plus VAT at 20%

**24** A food mixer marked £64 plus VAT at 18%

**25** A calculator marked £8.40 plus VAT at $17\frac{1}{2}\%$

**26** A car tyre costing £114 plus VAT at 25%

**27** A bathroom suite costing £1800 plus VAT at 17%

**28** A ticket for a concert costing £20 plus VAT at $18\frac{1}{2}\%$

### Income tax

Income tax is deducted from everybody's taxable income. A person's taxable income is their earnings over and above the value of their allowances. Usually people pay most of their tax at the basic rate (for example, 20% of their earnings), while those with high incomes have to pay a certain amount of higher rate tax.

(Rates of tax and allowances change from year to year.)

### Exercise 4b

**Worked example**

→ Piotr has a taxable income of £20 000 per annum. How much tax will he pay if the basic rate is 20%?

Tax due = 20% of £20 000
= 0.20 × £20 000
= £4000

Assuming that the basic rate of income tax is 20%, find the yearly tax on the following taxable incomes.

**1** £5000  **3** £6500  **5** £6450

**2** £8000  **4** £12 500  **6** £8260

Find the yearly income tax due on a taxable income of

**7** £10 000 if the basic tax rate is 33%

**8** £8000 if the basic tax rate is 25%

**9** £16 000 if the basic tax rate is 25%

**10** £24 000 if the basic tax rate is 32%

## Worked example

→ Edgar Brooks earns £30 000 each year. If his tax free allowances amount to £10 000, how much tax will he pay when the basic rate is 23%?

Taxable income = £30 000 − £10 000
= £20 000

Tax due = 23% of £20 000
= 0.23 × £20 000
= £4600

Use the following details to find the income tax due in each case.

|    | Name        | Gross income | Allowances | Basic tax rate |
|----|-------------|--------------|------------|----------------|
| 11 | Miss Deats  | £8000        | £2000      | 20%            |
| 12 | Mr Evans    | £10 000      | £3000      | 23%            |
| 13 | Mrs Khan    | £15 000      | £3200      | 15%            |
| 14 | Mr Ames     | £9000        | £2600      | 33%            |
| 15 | Miss Piatek | £20 000      | £4750      | 28%            |

## Worked example

→ A property developer has an annual income of £240 000 and can claim allowances of £30 270. The basic rate of income tax is 20%, which is payable on the first £32 010 of his taxable income. A higher rate of 40% is due on the next £117 990 of his taxable income and a rate of 45% on any remaining taxable income. Find the total tax payable.

Taxable income is £240 000 − £30 270 = £209 730

Tax due on £32 010 at 20% = 0.20 × £32 010
= £6402

Higher rate tax on next £117 990 at 40% = 0.40 × £117 990
= £47 196

Tax at the rate of 45% is due on £209 730 − £32 010 − £117 990
= £59 730

Tax at 45% on £59 730 = 0.45 × £59 730
= £26 878.50

Total tax payable = £6402 + £47 196 + £26 878.50
= £80 476.50

**16** For each case in the table below, calculate the income tax payable if a lower rate of 20% is charged on the first £30 950 of taxable income and the remainder of the taxable income is charged at the rate of 40%.

|   | Gross income | Allowances |
|---|---|---|
| a | £45 500 | £11 250 |
| b | £48 800 | £14 336 |
| c | £53 660 | £12 660 |

**17** Repeat question **16** if the lower rate of tax is decreased from 20% to 18%.

**18** For each case in the table below, calculate the income tax payable if a lower rate of 10% is charged on the first £2950 of taxable income, the basic rate is charged on the next £22 600 and the remainder of the taxable income is taxed at the higher rate.

|   | Yearly income | Allowances | Basic rate tax | Higher rate tax |
|---|---|---|---|---|
| a | £45 400 | £10 600 | 20% | 40% |
| b | £39 700 | £11 740 | 24% | 40% |
| c | £42 350 | £10 800 | 21% | 60% |
| d | £253 000 | £20 280 | 25% | 50% |

## Sale reductions

Stores regularly encourage customers to make purchases by reducing prices. The reduced price of an article is often called the discounted price. A price reduction or *discount* is frequently expressed as a percentage of the price before the reduction is made.

### Exercise 4c

#### Worked examples

→ During the January sales, a department store offers a discount of 10% off marked prices. What is the discounted price of a dinner service marked £84.50?

If the discount is 10%, the discounted price is 90% of the marked selling price, i.e. the purchase price of the dinner service is 90% of £84.50

$= 0.9 \times £84.50$

$= £76.05$

Alternatively, if the discount is 10% on the marked price
Discount $= 0.1 \times £84.50$
$= £8.45$
∴ purchase price $= £84.50 - £8.45$
$= £76.05$

→ What is the discounted price of a pair of jeans marked £16.30?

Similarly, the discounted price of the jeans is 90% of £16.30

   = 0.9 × £16.30
   = £14.67

 Alternatively, discount of 10% on £16.30 = 0.1 × £16.30
   = £1.63
∴ purchase price = £16.30 − £1.63
   = £14.67

In a sale, a shop offers a discount of 20%. What would be the discounted price for each of the following items?

1  A dress marked £35
2  A lawn mower marked £115
3  A pair of shoes marked £62
4  A set of garden tools marked £72.50
5  Light fittings marked £82 each

In a sale, a department store offers a discount of 50% on the following items. Find their discounted price.

6  A pair of curtains marked £76.50
7  A leather football marked £32.30
8  A boy's jacket marked £28.60
9  A girl's coat marked £64.50
10 In order to clear a large quantity of woollen goods, a shopkeeper puts them on sale at a discount of 33%. Find the discounted price of
   a  a jumper marked £18.30
   b  a skirt marked £22.20

11 A shopkeeper buys in T-shirts at £12.50 each and marks them up by 50%. Some are later sold in a sale at a discount of 35%. Does the shopkeeper gain or lose on these T-shirts and by how much?

## Finding the original quantity

Sometimes we are given an increased or decreased quantity and we want to find the original quantity. For example, if the cost of a chair including VAT at 20% is £176.25, we might need to find the price of the chair before the tax was added.

## Exercise 4d

### Worked example

 An item is sold for £252. If this includes a mark-up of 5%, find the buying-in price.

There is a mark-up of 5%
Selling price = 105% of the buying-in price
= 1.05 × buying-in price

*Remember that the mark-up of 5% is 5% of the buying-in price. We do not know the buying-in price, so we call it £x.*

If the buying-in price is £x, then
$$252 = 1.05 \times x$$
$$\frac{252}{1.05} = x \quad \text{giving} \quad 240 = x$$

Therefore the buying-in price is £240.

In questions **1** to **18**, selling price is abbreviated to SP.

Find the buying-in price.

**1** SP £98, mark-up 40%

**2** SP £64, mark-up 60%

**3** SP £28, mark-up 75%

**4** SP £12, mark-up 100%

**5** SP £40, mark-up 25%

**6** SP £920, mark-up 15%

**7** SP £1008, mark-up 125%

**8** SP £21.50, mark-up $7\frac{1}{2}$%

### Worked example

 A book is sold for £6.30 at a loss of 30%. Find the buying-in price.

Selling price = 70% of the buying-in price
Then if the buying-in price is x pence
$$630 = 0.7 \times x$$
$$630 \div 0.7 = x \quad \text{i.e.} \quad x = 900$$

*The loss of 30% is 30% of the buying-in price, so the selling price is (100% − 30%), i.e. 70% of the buying-in price.*

The buying price of the book is 900 p, or £9.

Find the buying-in price.

**9** SP £30, loss 25%

**10** SP £56, loss 30%

**11** SP £70, loss 65%

**12** SP £12, loss 33%

**13** SP £8.16, loss 40%

**14** SP £45, loss 10%

**15** SP £120, loss 25%

**16** SP £8.50, loss 50%

**17** SP £64, loss 60%

**18** SP £1200, loss 40%

## Worked example

→ After a pay rise of 5% Joshua's weekly pay is £126. How much did he earn before the rise?

If Joshua's original pay was £$x$

then  $126 = 105\%$ of $x$

$126 = 1.05 \times x$

i.e.  $x = \dfrac{126}{1.05}$

$= 120$

> The 5% pay rise is 5% of Joshua's original pay, so his new pay will be (100% + 5%), i.e. 105% of his original pay.

Joshua's original weekly pay was £120.

The table below shows the weekly wage of a number of employees after percentage increases as shown. Find the original weekly wage of each employee, to the nearest whole penny where necessary.

|    | Name | % increase in pay | Weekly wage after increase |
|----|------|-------------------|----------------------------|
| 19 | George Black | 10% | £232 |
| 20 | Kasia Gorski | 8% | £335 |
| 21 | John Rowlands | 15% | £499 |
| 22 | Bianca Lewis | 7% | £396.88 |
| 23 | Priti Kadam | 4% | £185.68 |

## Worked example

→ The purchase price of a watch is £70.50. This includes sales tax at $17\tfrac{1}{2}\%$. Find the price before sales tax was added.

If the price of the watch before tax is added is £$C$

then  $1.175 \times C = 70.5$

i.e.  $C = 70.5 \div 1.175$

$= 60$

> Sales tax is a percentage of the price before the tax has been added, i.e. purchase price = (100% + $17\tfrac{1}{2}\%$) of the price before tax is added.

The price before sales tax was added was £60.

24  The purchase price of a hairdryer is £13.80. If this includes sales tax at 15%, find the price before tax was added.

25  I paid £763.75 for a dining table and four chairs. If the price includes VAT at $17\tfrac{1}{2}\%$, find the price before VAT was added.

26  John's income last week was £336 after income tax at 30% had been deducted. Calculate his pay before the tax was deducted.

27 Water increases in volume by 4% when it is frozen. How much water is required to make 884 cm³ of ice?

28 The stretched length of an elastic string is 31 cm. If this is 24% more than its unstretched length, find its unstretched length.

29 Because of a change in the exchange rate, the cost of the Stones' family holiday was increased by $5\frac{1}{2}$% to £1631.03. What was the cost of the holiday before the increase?

30 Sara's take-home pay last week was £226.46. This was after deductions that amounted to 32.4% of her gross pay.
Find   a  her gross pay   b  her deductions.

## Mixed problems involving percentage increase and decrease

Remember that a percentage increase or decrease is always calculated as a percentage of the quantity before the change. In questions that ask you to find the original quantity, for example question 4, you can check your answer by working through the given information using your answer.

### Exercise 4e

1 A house is bought for £108 000 and then sold at a profit of 14%. Find the selling price.

2 Carpets that had been bought for £18.50 per square metre were sold at a loss of 26%. Find the selling price per square metre.

3 Potatoes bought at £18 per 50 kg bag are sold at 48 p per kg. Find the percentage profit.

4 An art dealer sold a picture for £1980, making a profit of 65%. What did she pay for it?

5 Rachel's present average weekly grocery bill is £113.40, which is 8% more than she paid, on average, for the same goods each week last year. What was Rachel's average weekly grocery bill last year?

6 What is Fred's gross weekly wage if, after paying deductions of 35%, he is left with £223.60?

7 Between two elections the size of the electorate in a constituency fell by 16%. For the second election, 37 191 people were entitled to vote. How many were entitled to vote at the first election?

8 When the rate of sales tax is $17\frac{1}{2}\%$, a CD costs £7.05. What will it cost if the rate of sales tax is
   a  increased to 20%
   b  decreased to 15%?

9 If Nathan begins his journey after 9 a.m. he is allowed a discount of 30% on the cost of his rail ticket. He pays £22.75 for a ticket to London, leaving on the 9.15 a.m.
   a  Express the discounted price as a percentage of the full price.
   b  Calculate the cost of the ticket before deducting the discount.

10 A camera bought for £300 loses 40% of its value in the first year.
   a  What is it worth when it is one year old?
   b  Express its value after one year as a percentage of the amount by which it has depreciated (decreased in value).

11 Andrew Bullen received £2055 pay last month. This sum was made up of a fixed basic wage of £545 plus commission at 2% of the value of the goods he had sold in the preceding month. Find the value of the sales Andrew made last month.

12 In a sale at a department store, a tea set is sold at a discount of 15%. The sale price is £59.49.
   Find    a  the pre-sale price    b  the discount.

13 In Speake's electrical store, a Blu-Ray player is priced £167.50.
   a  Harri pays cash and so is given a discount of 8%. How much does the Blu-Ray player cost him?
   b  The ticket price of the Blu-Ray player now is 3.5% more than it was this time last year. What was the ticket price a year ago?

14 Piedro has a faulty washing machine. He calls out the service engineer, who repairs it.
   Copy and complete his bill, which is given below. Give any value that is not exact correct to the nearest penny.

| | £ |
|---|---|
| Fixed call-out charge | 38.50 |
| Labour: $1\frac{1}{2}$ hours at £42/h | _____ |
| Parts | _____ |
| Total before VAT | _____ |
| VAT at 20% | _____ |
| Total due | 178.80 |

**15** A jeweller buys in watches from a wholesaler. To the price she pays for each watch, she adds a mark-up of 50% to give the retail price. She then adds value added tax at $17\frac{1}{2}\%$ to the retail price and this gives the price to the customer, which is rounded up to the nearest penny. A customer buys a watch for £79.99. What did the shopkeeper pay the wholesaler for it?

**16** Holidays Abroad charge $1\frac{1}{2}\%$ commission when selling foreign currency. How much, in pounds sterling, will Kerry have to pay for 800 euros if the exchange rate is £1 ≡ €1.21, that is, £1 buys €1.21.

**17** Ravinder is going to Canada. Her bank will exchange 1.55 Canadian dollars for each £1, and only gives a whole number of dollars. It also charges 1% commission for the transaction. Ravinder wants to spend at most £250 for her dollars.

    **a** What is the maximum whole number of dollars she can buy?

    **b** How much commission is she charged?

    **c** How much, if any, of her £250 is left over?

**18** The purchase price of a diamond ring increases by £20 when the sales tax is increased from 12% to 20%. Find the original purchase price of the ring.

**19** The table shows the original price and sale price of several items in a clothes shop.

| Original price (£P) | 24 | 56 | 65 | 88 |
|---|---|---|---|---|
| Sale price (£S) | 19.20 | 44.80 | 52 | 70.40 |

Plot these values on a graph using 2 cm ≡ £10 on both axes. Scale the *P*-axis from 0 to 100 and the *S*-axis from 0 to 80. Draw a straight line to pass through the four points, and use your graph to find

    **a** the sale price of a garment originally marked
        **i** £36     **ii** £75

    **b** the original price of a dress whose sale price is
        **i** £44     **ii** £66.

Find the gradient of the line and give a meaning to its value. Express this gradient as a percentage and use its value to find

    **c** the sale price as a percentage of the original price

    **d** the percentage discount.

## 4 Percentages

### Interest

Many organisations such as banks and building societies offer savings accounts. If you put money into such an account, the bank uses your money for other purposes and pays you for that use. The amount that the bank pays for the use of your money is called **interest**.

The time may come when you wish to use someone else's money to buy an expensive item such as a car or even a house. You will normally have to pay for the use of borrowed money, that is, you will have to repay more than you borrow and the extra you repay is also called interest.

Interest on money borrowed (or lent) is usually a percentage of the sum borrowed (or lent). This percentage is often given as a charge per year (per annum, or p.a.) and it is then called the *interest rate*. For example, if £100 is put into a building society account with an interest rate of 2% p.a., then after one year, the society pays

  2% of £100,  i.e.  £2

### Exercise 4f

1. Jade is given £500 for her 18th birthday. She puts the money in a savings account with an interest rate of 2.5% p.a. How much interest is added to her account after one year?

2. Ann Peters is given a loan of £650 from the bank which she agrees to repay after one year. How much does she have to repay if the interest rate is $12\frac{1}{2}$% p.a.?

Find the interest payable after one year on each of the following sums of money invested (i.e. put in a savings account) at the given interest rate.

3. £352 at 1.5% p.a.

4. £10 000 at 4.25% p.a.

5. £2600 at 3.3% p.a.

6. £5840 at 6.4% p.a.

7. What annual rate of interest is necessary to give interest of £238 after one year on an investment of £2800?

8. What is the original size of a loan that costs £45 when repaid after one year at an interest rate of 9% p.a.?

9. Find the original sum borrowed if £287.50 has to be repaid after one year when the interest rate is 15% p.a.

10 Mr and Mrs Surefoot invest a sum of money in a deposit account with the Highway Building Society. The society quotes a gross rate of 4% but add the net interest to the account after deducting income tax at 20%. One year after the investment was made the statement shows that net interest of £800 was added. Find
   a   the gross interest paid by the building society
   b   the amount of money invested.

## Compound percentage problems

There are many occasions when a percentage increase or decrease happens more than once. Suppose that a house is bought for £220 000 and increases in value (appreciates) by 10% of its value each year.

After one year, its value will be 110% of its initial value,

i.e.     110% of £220 000 = 1.1 × £220 000 = £242 000

The next year it will increase by 10% of the £242 000 it was worth at the beginning of the year, so its value after two years will be

110% of £242 000 = 1.1 × £242 000 = £266 200

While some things increase in value year after year, many things decrease in value (depreciate) each year. If you buy a car or a motorcycle, it will probably depreciate in value more quickly than anything else you buy.

If you invest money in a savings account and do not spend the interest, your money will increase by larger amounts each year if the interest rate stays the same.

This kind of interest is called **compound interest**. The sum on which the interest is calculated is called the **principal** and changes each year.

### Exercise 4g

#### Worked example

→ Find the compound interest on £260.60 invested for 2 years at 8% p.a.

Interest for first year at 8% is 8% of the original principal.

New principal at end of first year

   = 100% of original principal + 8% of original principal
   = 108% of the original principal
   = 1.08 × original principal

∴ principal at end of first year

   = 1.08 × £260.60
   = £281.448

 Use all available figures when an intermediate calculation does not work out exactly.

96

Similarly, new principal at end of second year
= 108% of the principal at the beginning of the second year
= 1.08 × £281.448

∴ principal at end of second year
= £303.963...

So the compound interest on £260.60 for 2 years
= principal at end of second year − original principal
= £303.963... − £260.60 = £43.363... = £43.36 correct to the nearest penny

In questions **1** to **7**, give all answers that are not exact correct to the nearest penny. But remember to use all the available figures when your intermediate calculations to not work out exactly. Find the compound interest on

1. £200 for 2 years at 10% p.a.
2. £300 for 2 years at 2% p.a.
3. £400 for 3 years at 8% p.a.
4. £650 for 3 years at 3% p.a.
5. £520 for 2 years at 1% p.a.
6. £690 for 2 years at 4% p.a.
7. £624 for 3 years at 1.2% p.a.
8. A house is bought for £110 000 and appreciates at 8% a year. What will it be worth in 2 years' time?
9. A rare postage stamp increases in value by 15% each year. If it is bought for £50, what will it be worth in 3 years' time?
10. A motorcycle bought for £3500 depreciates in value by 20% each year. Find its value after 3 years.

## Worked examples

→ An antique silver teapot is valued at £750 and appreciates by 12% a year. Find its value after 3 years.

Value of teapot
after 1 year = 112% × £750
= 1.12 × £750
= £840

after 2 years = 1.12 × £840
= £940.80

after 3 years = 1.12 × £940.80
= £1053.70 (to the nearest penny)

→ By contrast, a hi-fi system costing £750 depreciates by 12% a year. Find its value after 3 years.

Value of hi-fi system

after 1 year = 88% × £750
= 0.88 × £750 = £660

after 2 years = 0.88 × £660
= £580.80

after 3 years = 0.88 × £580.80
= £511.10 (to the nearest penny)

→ Express the value of the hi-fi system after 3 years as a percentage of the value of the teapot after 3 years.

Value of hi-fi system after 3 years as a percentage of the value of the tea pot after 3 years

$= \dfrac{\text{Value of hi-fi}}{\text{Value of teapot}} \times 100\%$

$= \dfrac{£511.10}{£1053.70} \times 100\%$

= 48.5% (correct to 3 s.f.)

Remember to use all the available figures when an intermediate calculation does not work out exactly.

**11** Three years ago David and Charles each invested £30 000. David put his money into shares in a pharmaceutical company while Charles invested his money in a really spectacular car. The value of the car depreciated by 20% a year, while shares in the pharmaceutical company appreciated by the same percentage. Find the value of each investment now.

**12** A new car costing £25 000 depreciated in value each year by 18% of its value at the beginning of that year.
  **a** Find its value
    **i** after 1 year   **ii** after 4 years.
  **b** Calculate the percentage decrease in the value of the car over 4 years.
  (Give your answer correct to the nearest tenth of a per cent.)

**13** A new motorbike costing £8000 depreciates each year by 18% of its value at the beginning of the year. Find
  **a** its value   **i** after 1 year   **ii** after 3 years
  **b** the percentage decrease in value after 3 years.
  (Give your answer correct to the nearest tenth of a per cent.)

14 The present toll for a car crossing the Midford suspension bridge is £3.80. This charge is set to rise by 8% this year and by a further 6% for each of the following 5 years. Each increase is rounded down to the nearest 5 p. How much will it cost to cross Midford bridge
   a  in 2 years' time
   b  in 4 years' time?

15 In the state of Necka the current amount that a person can earn without paying any income tax is £3500. This amount, rounded up to the nearest £10, is set to increase in line with inflation, the projected rates of which are given in the table.

| Number of years from now | 1 | 2 | 3 | 4 | 5 |
|---|---|---|---|---|---|
| Expected rate of inflation for that year | 3% | 2.5% | 4.5% | 6.8% | 8.3% |

Use these values to find the tax-free amount a single person can earn in
   a  3 years' time
   b  5 years' time.

16 The rabbit population of Ditcher's Heath has increased by 20% a year for the last 3 years. The estimated population is now 400 rabbits.
   a  Estimate, to the nearest 10, the number of rabbits on the heath
      i  1 year ago
      ii  2 years ago
      iii  3 years ago.
   b  State whether each of the following statements is true or false?
      A  The rabbit population of Ditcher's Heath has increased by the same number each year.
      B  Every year the increase in the number of rabbits is more than the increase was the year before.
      C  If the rate of increase stays the same, there will be 420 rabbits next year.
      D  At the present rate of increase, there will be more than twice as many rabbits on the heath within 10 years.

17 The population of Roxley has increased by 5.5% a year for the last 5 years and is expected to grow at the same rate for the foreseeable future. The present population of Roxley is 20 000.
   a  Find the expected population
      i  next year     ii  in 2 years' time.
   b  Find the population of Roxley
      i  1 year ago     ii  2 years ago.

## Worked example

William buys a fixed rate bond for £5000 at the start of 2014.

Interest is added at the end of each year at 4% of the value of the bond at the start of the year. For how many years will William need to keep the bond before its value is at least £6000?

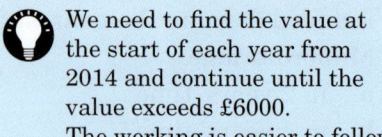

We need to find the value at the start of each year from 2014 and continue until the value exceeds £6000.
The working is easier to follow when it is laid out in a table.

| Date | Elapsed time (years) | Value |
|---|---|---|
| 2014 | 0 | £5000 |
| 2015 | 1 | 104% of £5000 = 1.04 × £5000 = £5200 |
| 2016 | 2 | 104% of £5200 = 1.04 × £5200 = £5408 |
| 2017 | 3 | 1.04 × £5408 = £5624.32 |
| 2018 | 4 | 1.04 × £5624.32 = £5849.2928 |
| 2019 | 5 | 1.04 × £5849.2928 = £6083.26… |

William will need to keep the bond for 5 years.

**18** Mr James grows prize marrows. When growing conditions are ideal, his marrows increase in weight by 10% each day. How many days will it take a marrow whose weight is now 700 g to

    **a** increase in weight to at least 800 g

    **b** increase in weight to at least 1000 g

    **c** at least double in weight?

**19** A new town of 2000 houses is to be built in the year 2020 and is then planned to increase by 20% each year. At the start of which year will the number of houses be at least doubled?

**20** A pest eradication scheme for a railway system aims to decrease the number of rats by 30% a month. At the start of July it is estimated that there are 10 000 rats. After how many months will the number of rats be below

    **a** 5000                       **b** 500?

**21** A classic car is bought for £$P$. Its value appreciates by 8% each year. Find, in terms of $P$, an expression for the value of the car after

    **a** 1 year                 **c** 6 years

    **b** 2 years              **d** $n$ years.

Use the formula you found in part **d** to find the value after 10 years of a classic car bought for £500.

## Mixed exercise

### Exercise 4h

1. Tim earns £310 per week and has just received a rise of 4%. What is his new weekly wage?

2. In a sale a store gives a reduction of 20% off the ticket price. A winter coat is marked £99. How much is the discounted price?

3. Justine May earned £28 500 last year. No income tax was due on the first £3900 of her income but she had to pay tax at 20% on the next £3750 and 23% on the remainder. How much income tax did she pay altogether?

4. A shopkeeper buys 80 items for £200 and sells them for £3.50 each. Find his percentage profit.

5. Engine modifications were made to a particular model of car. As a result the number of kilometres it travels on one litre of petrol increased by 8%. If the new petrol consumption is 16.2 km/litre, what was the previous value?

6. At the end of 2013 the number of a rare species of animal in North America was 6000. It is predicted that the number will decrease by 14% each year.
   a. How many animals will be left at the end of 2016?
   b. By the end of which year will the number first be less than 3000?

## Consider again

Ben buys a batch of shirts from a manufacturer for £11.50 each. His normal practice is to add a mark-up of 50%, then he has to add VAT at 20% to give the selling price.

What does this give as the selling price?

He prefers to sell the shirt at just under £20, so decides to sell it at £19.95 including VAT.

Ben needs to work out how much of the £19.95 is VAT, because he will have to pay this amount to the tax authority. How much VAT will he have to pay?

Now can you answer these questions?

# 5 Ratio and proportion

## Consider

Mina often drives from London to Southampton.

When she drives at an average speed of 80 km/h she uses 10 litres of petrol and the journey takes $2\frac{1}{2}$ hours.

When Mina increases her average speed to 100 km/h, how much petrol does she use and how long does the journey take her?

Describe the relationship between Mina's speed and the amount of petrol used, and also between Mina's speed and the time taken for this journey. State any assumptions you have made.

*You should be able to solve these problems after you have worked through this chapter.*

## Direct proportion and inverse proportion

Richard has an old scooter.

The instructions state that the tank must be filled with fuel mixed from oil and petrol in the ratio 1 : 50. If Richard puts 20 ml of oil in the tank, the ratio 1 : 50 tells him he needs to add 50 times as much petrol, which is 1000 ml or 1 litre. The ratio also tells him that 100 ml of oil has to be mixed with 100 × 50 ml of petrol, and so on. The relationship between the quantity of oil and the quantity of petrol is always the same, that is, 1 : 50.

- Any two quantities that vary so that they are always in the same ratio behave in the same way, so if one quantity is doubled, the other is too; if one quantity is increased by a factor of 4, the other is too, and so on. Quantities that are related in this way are said to be in **direct proportion**.

Richard uses his scooter to travel to work. The distance is 10 miles. If he leaves home at 8 a.m. the journey takes him 40 minutes. Using average speed = $\frac{\text{distance}}{\text{time}}$ gives his average speed as 15 mph.

If he leaves at 7.30 a.m. he avoids the rush hour and the journey only takes 20 minutes. In this case his average speed is 30 mph.

- As the distance is fixed, there is a relationship between the average speed and the time the journey takes. In this case, however, when the speed doubles, the time halves. So the relationship between the speed and the time is not one where they are in the same ratio. Quantities that are related in this way (for example, when one increases by becoming 4 times larger, the other decreases to $\frac{1}{4}$ of its original size) are said to be in **inverse proportion**.

In this chapter we are going to work with these two forms of relationship. First, however, you need to be able to recognise when quantities are related in one of these two ways, and equally importantly, when they are not.

# 5 Ratio and proportion

## Class discussion

In each question discuss how the quantities are related. Some may be directly proportional, some may be inversely proportional. Some may be related in a different way from either of these and others may not be related in any way.

1. Emma's pay for working a seven-hour day and her pay for working a 35-hour week, assuming the rate of pay per hour is constant.
2. Simon's pay for working an eight-hour day and his pay for working a 45-hour week, when this includes overtime pay at a higher rate than the standard rate of pay per hour.
3. The age of a woman and her weight.
4. The number of £1 coins in a pile and the height of the pile.
5. The time it takes to fill a swimming pool and the number of hoses used to fill it, assuming that the rate of flow of water from each hose is the same.
6. The number of sweets that can be bought for £1 and the cost per kilogram of those sweets.
7. The number of towels in a washing machine and the time it takes to wash them.
8. The cost of a telephone bill and the number of calls made using the telephone.
9. The cost of an electricity bill and the number of units used.
10. The area of a square and the length of its side.
11. The size of an interior angle of a polygon and the number of sides.

## Ratios

Ratio and direct proportion are introduced in Book 8. You can use Revision exercise 2, page 17, at the front of this book to revise this work. The next exercise also revises ratios but in different contexts.

## Exercise 5a

### Worked example

→ In a jar of 357 mixed raisins and peanuts there are 153 raisins. Find the ratio of the number of raisins to the number of peanuts.

The number of peanuts is 357 − 153 = 204
Ratio of raisins to peanuts (by number)

= 153 : 204
= 3 : 4

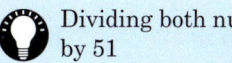 Dividing both numbers by 51

103

1 In a school of 1029 students, 504 are girls. What is the ratio of the number of boys to the number of girls?

2 I spend £7.20 on groceries and £4.80 on vegetables.
 What is the ratio of the cost of
 a  groceries to vegetables
 b  vegetables to groceries
 c  groceries to the total?

3 One rectangle has a length of 6 cm and a width of 4.5 cm.
 A second rectangle has a length of 9 cm and a width of 2.5 cm.
 Find the ratios of
 a  their lengths      c  their perimeters
 b  their widths       d  their areas.

4 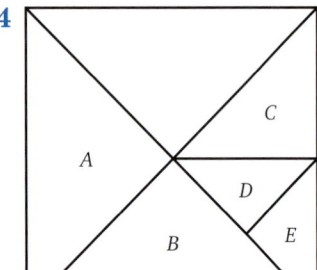 Use the diagram to find the ratios of the following areas.
 a  $B:A$
 b  $C:B$
 c  $E:A+B$
 d  $E:D$
 e  $E:C+D$
 f  $C$:whole square

5 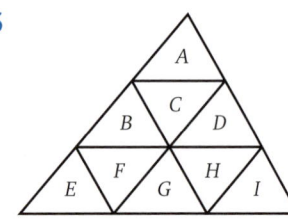 In the diagram, the areas of the small triangles are equal. Find the ratios of the following areas.
 a  $A$:whole figure
 b  $A:A+B+C+D$
 c  $B+E+F+G$:whole figure

### Worked example

→ Express the ratio $5:7$ in the form $1:n$

$5:7 = 1:\frac{7}{5}$
$\phantom{5:7} = 1:1.4$

> Remember that a ratio is unaltered when each number is multiplied (or divided) by the same amount. In this case we divide both numbers by 5 to give 1 as the first number in the ratio.

In questions **6** to **17**, express the following ratios in the form $1:n$, giving $n$ correct to 3 significant figures where necessary.

6   $2:3$  
7   $5:12$  
8   $7:6$  
9   $11:30$  
10  $5:3$  
11  $8:21$  
12  $3:4$  
13  $4:3$  
14  $7:10$  
15  $\frac{1}{3}:\frac{1}{4}$  
16  $0.75:0.25$  
17  $1\frac{1}{2}:2\frac{2}{3}$

## Worked example

→ For one brand of emulsion paint, the information given is that 5 litres covers 45 m². Another brand of emulsion paint has an average coverage of 20 m² for 2 litres. Compare the ratios of the volume of paint to area covered for the two brands.

First brand:  volume of paint : area covered = 5 : 45
Second brand: volume of paint : area covered = 2 : 20

The ratios are easier to compare if they are both expressed in the form $1 : n$

First brand:  ratio = 1 : 9
Second brand: ratio = 1 : 10

Hence the second brand of paint covers a greater area per litre than the first.

In questions **18** to **22**, compare the ratio of quantity to price and hence state which is cheaper.

**18** Gravel at 4 p per kilogram or at £38 per tonne.

**19** Eggs at 18 p each or £2.40 per dozen.

**20** Gold-coloured chain at £16.20 per metre or 15 p per centimetre.

**21** Screws at 72 p for 20 or 4 p each.

**22** A 500 ml bottle of liquid fertiliser that covers 90 m² and costs £2.56 or a 5 litre bag of granular fertiliser that covers 300 m² and costs £28.40.

## Division in a given ratio

The tank on Richard's scooter holds 20 litres of fuel. To fill the tank Richard needs to work out how much oil to put in. He can then fill the tank with petrol from the pump.

The fuel has to be 1 part oil to 50 parts petrol, therefore the 20 litres of fuel to fill the tank has to comprise 51 parts.

Hence the quantity of oil required is $\frac{1}{51}$ of 20 litres

$= \frac{1}{51} \times 20\,000$ ml

$= 392$ ml (correct to 3 s.f.)

 We have given 392 to the nearest millilitre. This is probably more accurate than necessary in this context for several reasons, one of which is that it is not easy or necessary to measure a quantity of oil as accurately as this.

# STP Maths 9

## Exercise 5b

Give your answer to each question as accurately as you consider to be appropriate in the context of the problem.

1. Concentrated orange juice has to be diluted with water in the ratio 2:5 by volume. How many millilitres of concentrated juice are needed to make up 2 litres of juice to drink?

2. The tank on a chemical spray holds 5 litres. For removing moss on hard surfaces, the instructions on a bottle of moss killer recommend dilution in the ratio 3:50.
   a. If the tank is to be filled, how much moss killer should be put in?
   b. If only 3 litres of spray is to be made up, how much moss killer is required?

3. Bronze is a metal alloy which, for one purpose, contains copper and tin in the ratio 3 : 22 by mass. What mass of copper is needed to make 5 kg of this bronze?

4. Mr Brown, Mrs Smith and Mr Shah work for AB Engineering plc. Their salaries are £24 000, £30 000 and £28 000 respectively. A bonus of £6000 is to be divided between these three employees in the ratio of their salaries.
   a. In what ratio are their salaries?
   b. What bonus is paid to each employee?

5. Two students were presented with a telephone bill for £70.22 which they had to pay between them. They decided to share the cost in the ratio of the number of calls they had each made.
   a. James had made 42 calls and Sarah had made 75 calls. How much should they each pay?
   b. After they had paid the bill, Sarah remembered that she had in fact made 6 more calls. How much should she pay James?

## Simple direct proportion

If we know the cost of one article, we can easily find the cost of 10 similar articles. In the same way, if we know what someone is paid for 1 hour's work, we can find what their pay is for 5 hours.

### Exercise 5c

**Worked examples**

→ If 1 cm³ of lead weighs 11.3 g, what is the weight of 6 cm³?

1 cm³ weighs 11.3 g
6 cm³ weighs 11.3 × 6 g = 67.8 g

→ If 1 cm³ of lead weighs 11.3 g, what is the weight of 0.8 cm³?

1 cm³ weighs 11.3 g
0.8 cm³ weighs 11.3 × 0.8 g = 9.04 g

**1** The cost of 1 kg of sugar is 90 p. What is the cost of
    **a** 3 kg      **b** 12 kg?

**2** In one hour an electric heater uses $1\frac{1}{2}$ units of electricity. Find how much it uses in
    **a** 4 hours      **b** $\frac{1}{2}$ hour.

**3** A car uses one litre of petrol to travel 18 km. At the same rate, how far does the car travel on
    **a** 4 litres      **b** 6.6 litres?

**4** The cost of 1 kg of mushrooms is £3.30. Find the cost of
    **a** 500 g      **b** 2.4 kg.

We can reverse the process and, for example, find the cost of one article if we know the cost of three similar articles.

**Worked example**

→ If 18 cm³ of copper weighs 162 g. What is the weight of 1 cm³?

18 cm³ weighs 162 g
1 cm³ weighs $\frac{162}{18}$ g = 9 g

**5** Six pens cost £7.20. What is the cost of one pen?

**6** A car uses 8 litres of petrol to travel 124 km. At the same rate, how far can it travel on 1 litre?

**7** Nick walks steadily for 3 hours and covers 13 km. How far does he walk in 1 hour?

**8** A carpet costs £235.20. Its area is 12 m². What is the cost of 1 m²?

We can use the same process even if the quantities given are not whole numbers of units.

# STP Maths 9

### Worked example

→ The mass of $0.6\,\text{cm}^3$ of a metal is $3\,\text{g}$. What is the mass of $1\,\text{cm}^3$?

The mass of $0.6\,\text{cm}^3$ is $3\,\text{g}$

The mass of $1\,\text{cm}^3$ is $\frac{3}{0.6}\,\text{g} = 5\,\text{g}$

**9** $8.6\,\text{m}^2$ of carpet cost £142.76. What is the cost of $1\,\text{m}^2$?

**10** The cost of running a refrigerator for 3.2 hours is 4.8 p. What is the cost of running a refrigerator for one hour?

**11** A bricklayer takes 0.8 hours to build a wall 1.2 m high. How high a wall (of the same length) could he build in 1 hour?

**12** A piece of fabric is 12.4 cm long and its area is $68.2\,\text{cm}^2$. What is the area of a piece of this fabric that is 1 cm long?

## Direct proportion

If two varying quantities are always in the same ratio, they are said to be *directly proportional* to one another (or sometimes simply *proportional*).

For example, when buying pads of paper that each cost the same amount, the total cost is proportional to the number of pads. The ratio of the cost of 11 pads to the cost of 14 pads is $11:14$, and if we know the cost of 11 pads, we can find the cost of 14 pads.

One method for solving problems involving direct proportion uses ratio. Another uses the ideas in the **Exercise 5c** and is called the *unitary method* because it makes use of the cost of one item or the time taken by one person to complete a piece of work, and so on.

## Exercise 5d

### Worked example

→ The mass of $16\,\text{cm}^3$ of a metal alloy is $24\,\text{g}$. What is the mass of $20\,\text{cm}^3$ of the same alloy?

**First method** (using ratios)

Let the mass of $20\,\text{cm}^3$ be $x$ grams.

> 💡 The ratio of mass to volume stays the same. Write it starting with $x$.

Then $x : 20 = 24 : 16$

i.e $\quad \dfrac{x}{20} = \dfrac{24}{16}$

$$20 \times \dfrac{x}{20} = \overset{5}{\cancel{20}} \times \dfrac{\overset{6}{\cancel{24}}}{\underset{1}{\cancel{16}}}$$

so $\quad x = 30$

The mass of $20\,\text{cm}^3$ is 30 grams.

**Second method** (unitary method)

16 cm³ has a mass of 24 g

 Rewrite the first sentence so that it ends with the quantity you want, i.e. the mass.

∴  1 cm³ has a mass of $\frac{24}{16}$ g

 There is no need to work out the value of $\frac{24}{16}$ yet.

so   20 cm³ has a mass of $\overset{5}{\cancel{20}} \times \frac{\overset{6}{\cancel{24}}}{\underset{1}{\cancel{16}}} = 30$ g

1  At a steady speed a car uses 4 litres of petrol to travel 75 km. At the same speed, how much petrol is needed to travel 60 km?

2  A hiker walked steadily for 4 hours, covering 16 km. How long did she take to cover 12 km?

3  An electric heater uses $7\frac{1}{2}$ units of electricity in 3 hours.
   a  How many units does it use in 5 hours?
   b  How long does the heater take to use 9 units?

4  A rail journey of 300 miles costs £168. At the same rate per mile
   a  what would be the cost of travelling 250 miles
   b  how far could you travel for £63?

5  It costs £362 to turf a lawn of area 63 m². How much would it cost to turf a lawn of area 56 m²?

6  A machine in a soft drinks factory fills 840 bottles in 6 hours. How many could it fill in 5 hours?

7  A 6 kg bag of sprouts costs £6.96. At the same rate, what would an 8 kg bag cost?

8  The instructions for setting the tension on a knitting machine are 55 rows to measure 10 cm. How many rows should be knitted to give 12 cm?

9  A scale model of a ship is such that the mast is 9 cm high and the mast of the original ship is 12 m high. The length of the original ship is 27 m. How long is the model ship?

Either method will work, whether the numbers are complicated or simple. Even if the problem is about something unfamiliar, it is sufficient to know that the quantities are proportional.

> **Worked example**
>
> In a spring balance the extension in the spring is proportional to the load. If the extension is 2.5 cm when the load is 8 newtons, what is the extension when the load is 3.6 newtons?
>
> **Ratio method**
> Let the extension be $x$ cm.
>
> $$x : 3.6 = 2.5 : 8$$
>
> i.e. $\dfrac{x}{3.6} = \dfrac{2.5}{8}$
>
> so $3.6 \times \dfrac{x}{3.6} = 3.6 \times \dfrac{2.5}{8}$ giving $x = 1.125$
>
> The extension is 1.125 cm.
>
> **Unitary method**
> If a load of 8 newtons gives an extension of 2.5 cm,
>
> then a load of 1 newton gives an extension of $\dfrac{2.5}{8}$ cm,
>
> so a load of 3.6 newtons gives an extension of
>
> $$3.6 \times \dfrac{2.5}{8} \text{ cm} = 1.125 \text{ cm}$$

**10** The rates of currency exchange published in the newspapers on a certain day showed that 14 kroner could be exchanged for 28 pesos. How many pesos could be obtained for 32 kroner?

**11** At a steady speed, a car uses 15 litres of petrol to travel 164 km. At the same speed, what distance could it travel on 6 litres?

**12** If a 2 kg bag of sugar contains $9 \times 10^6$ sugar crystals, how many sugar crystals are there in
    **a** 5 kg     **b** 8 kg     **c** 0.03 kg?

**13** The current flowing through a lamp is proportional to the voltage across the lamp. If the voltage across the lamp is 10 volts, the current is 0.6 amps. What voltage is required to make a current of 0.9 amps flow?

**14** The amount of energy carried by an electric current is proportional to the number of coulombs. If 5 coulombs carry 19 joules of energy, how many joules are carried by 6.5 coulombs?

**15** Two varying quantities, $x$ and $y$, are directly proportional. Copy and complete this table.

| $x$ | 2  | 4 | 6 | 8 |
|-----|----|---|---|---|
| $y$ | 10 |   |   |   |

**16** A recipe for date squares lists the following quantities:

| Ingredients | Costs |
|---|---|
| 125 g of brown sugar | 500 g cost £1.35 |
| 75 g of oats | 750 g cost £2.04 |
| 75 g of flour | 1.5 kg cost £1.10 |
| 100 g of margarine | 250 g cost 72 p |
| 100 g of dates | 250 g cost £2.08 |
| Pinch of bicarbonate of soda | – |
| Squeeze of lemon juice | 5 p |

Find the cost of making these date squares as accurately as possible, then give your answer correct to the nearest penny.

## Inverse proportion

At the start of this chapter we found that some quantities are not directly proportional to one another, although there is a connection between them. When we considered the relationship between the average speed of Richard's scooter and the time taken to travel 10 miles, we saw that when the average speed doubled, the time halved. This means that the reciprocal, or inverse, of the time is proportional to the average speed.

Further, when the speed changes from 20 mph to 40 mph,

the time changes from 30 minutes to 15 minutes.

If we multiply the speed and the corresponding time together, the result is 600 in both cases.

As another example, suppose that a fixed amount of food is available for several days. If each person eats the same amount each day, the more people there are, the shorter the time is that the food will last. In fact, if the food will last 2 people for 6 days, it will last 6 people for 2 days. So when we treble the number of people, the number of days the food will last decreases to a third. Hence the number of days the food will last is inversely proportional to the number of people eating it.

Again, we see that the product of the number of days and the number of people the food will feed is 12 in both cases.

These examples illustrate the fact that

**when two quantities are inversely proportional, their product remains constant.**

### Exercise 5e

In this exercise, assume that the rates are constant.

> **Worked example**
>
> → Four bricklayers can build a certain wall in 10 days. How long would it take five bricklayers to build it?
>
> **Constant product method**
> Suppose it takes five bricklayers $x$ days to build the wall,
> then $\quad 5 \times x = 4 \times 10$
> i.e $\quad 5x = 40 \quad$ giving $\quad x = 8$
> It would take them 8 days.
>
> **Unitary method**
> Four bricklayers take 10 days.
> One bricklayer would take 40 days.
> Five bricklayers would take $\frac{40}{5}$ days = 8 days.

1. Eleven taps fill a tank in 3 hours. How long would it take to fill the tank if only six taps are working?

2. The length of an essay is 174 lines with an average of 14 words per line. If it is rewritten with an average of 12 words per line, how many lines will be needed?

3. Nine children share out the sweets in a jar equally and get eight each. If there were only six children, how many would each get?

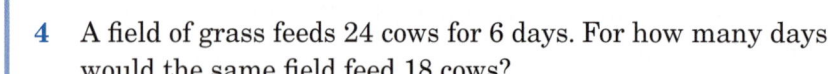

4. A field of grass feeds 24 cows for 6 days. For how many days would the same field feed 18 cows?

5. The dimensions of a block of stamps are 30 cm wide by 20 cm high. The same number of stamps could also have been arranged in a block 24 cm wide. How high would this second block be?

6. A batch of bottles were packed in 25 boxes taking 12 bottles each. If the same batch had been packed in boxes taking 15 each, how many boxes would be filled?

7. When used to knit a scarf 48 stitches wide, one ball of wool produces a length of 18 cm. If there had been 54 stitches instead, how long a piece would the same ball produce?

8  In a school, 33 classrooms are required if each class has 32 students. How many classrooms would be required if the class size was reduced to 22?

9  A factory requires 42 machines to produce a given number of articles in 63 days. How many machines would be required to produce the same number of articles in 54 days?

10  Two quantities $p$ and $q$, which can vary in value, are inversely proportional.
Copy and complete the following table.

| $p$ | 20  | 5 | 0.5 | 0.01 |
|---|---|---|---|---|
| $q$ | 0.5 |   |     |      |

### Exercise 5f

This exercise contains a mixture of problems, some of which cannot be answered because the quantities are in neither direct nor inverse proportion. In these cases give a reason why there is no answer.
For those problems that can be solved, give answers correct to three significant figures where necessary.

1  The list of exchange rates states that £1 = 1.6 US dollars and £1 = 1.4 euros, so that 1.6 US dollars = 1.4 euros.
a  How many euros can 54 dollars be exchanged for?
b  How many euros can be exchanged for 100 dollars?

2  A man earned £80.60 for an eight-hour day. How much would he earn at the same rate for a 38-hour week?

3  A secretary types 3690 words in $4\frac{1}{2}$ hours. How long would it take him to type 2870 words at the same rate?

4  At the age of twelve, a boy is 1.6 m tall. How tall will he be at the age of eighteen?

5  A ream of paper (500 sheets) is 6.2 cm thick. How thick is a pile of 360 sheets of the same paper?

6  If I buy balloons at 30 p each, I can buy 63 of them. If the price of a balloon increases to 40 p, how many can I buy for the same amount of money?

7  A boy's mark for a test is 18 out of a total of 30 marks. If the test had been marked out of 40 what would the boy's mark have been?

8  Twenty-four identical mathematics text books occupy 60 cm of shelf space. How many books will fit into 85 cm?

**9** A lamp post 4 m tall has a shadow 3.2 m long cast by the Sun. A man 1.8 m tall is standing by the lamp post. At the same moment, what is the length of his shadow?

**10** A contractor decides that he can build a barn in nine weeks using four men. If he employs two more men, how long will the job take? Assume that all the men work at the same rate.

**11** A twelve-year-old girl gained 27 marks in a competition. How many marks did her six-year-old sister gain?

**12** For a given voltage, the current flowing is inversely proportional to the resistance. When the current flowing is 2.5 amps the resistance is 0.9 ohms. What is the current when the resistance is 1.5 ohms?

**13** The tables give some corresponding values of two variables. Decide whether the variables are directly proportional, inversely proportional or neither.

a
| $x$ | 2 | 4 | 7 | 8 | 9 | 12 |
|---|---|---|---|---|---|---|
| $y$ | 6 | 12 | 21 | 24 | 27 | 36 |

b
| $p$ | 2 | 3 | 6 | 7 | 9 | 10 |
|---|---|---|---|---|---|---|
| $q$ | 20 | 10 | 5 | 6 | 6.5 | 10 |

c
| $w$ | 1 | 2 | 4 | 8 | 10 | 20 |
|---|---|---|---|---|---|---|
| $p$ | 18 | 9 | 4.5 | 2.25 | 1.8 | 0.9 |

## Mixed exercise

### Exercise 5g

**1** Complete the ratio  ☐ : 9 = 2 : 5

**2** A car uses 7 litres of petrol for a 100 km journey. At the same rate, how far could it go on 8 litres?

**3** Eight typists together could complete a task in 5 hours. If all the typists work at the same rate, how long would six typists take?

**4** The ratio of zinc to copper in one type of brass is 3 : 7. What is the weight of zinc in 10 kg of this brass?

**5** **a** Express the ratio 3 : 5 in the form $n : 1$
  **b** Give the ratio 5 : 6 in the form $1 : n$

**6** The ratio of two sums of money is 4 : 5. The first sum is £6. What is the second?

7  A typist charges £90 for work which took her 6 hours. How much would she charge for 9 hours' work at the same rate?

8  James Bond takes $4\frac{1}{2}$ minutes to complete a Grand Prix circuit when driving at an average speed of 97 mph. How long would it take to complete one circuit if he increases his average speed to 112 mph?

### Consider again

Mina often drives from London to Southampton.

When she drives at an average speed of 80 km/h she uses 10 litres of petrol and the journey takes $2\frac{1}{2}$ hours.

When Mina increases her average speed to 100 km/h, how much petrol does she use and how long does the journey take her?

Describe the relationship between Mina's speed and the amount of petrol used, and also between Mina's speed and the time taken for this journey. State any assumptions you have made.

Now can you answer these questions?

###  Investigation

Sweets at a 'Pick and mix' counter are sold by weight at £0.95 per 100 grams.

a  If *x* grams cost *y* pence, copy and complete this table giving values of *y* corresponding to values of *x*.

| x | 20 | 50 | 100 | 200 | 500 | 1000 |
|---|----|----|-----|-----|-----|------|
| y |    |    |     |     |     |      |

b  Use a scale of 1 cm for 50 units on the *x*-axis and a scale of 2 cm for 50 units on the *y*-axis to plot these points on a graph.

c  What do you notice about these points? Can you use your graph to find the cost of 162 grams?

d  The cost and weight of these sweets are directly proportional. Investigate the graphical relationship between other quantities that are directly proportional. What do you notice? Is this always true?

e  Extend your work to investigate the graphical relationship between two quantities that are inversely proportional. Use some of the questions in **Exercise 5e** as examples of quantities that are inversely proportional.

# Summary 1

## Distance, speed and time

The relationship between distance, speed and time is given by

$$\text{Distance} = \text{speed} \times \text{time}$$

which can also be expressed as $\quad \text{Speed} = \dfrac{\text{distance}}{\text{time}}$

or as $\quad \text{Time} = \dfrac{\text{distance}}{\text{speed}}$

A useful way to remember these relationships is to use the triangle. (Cover up the one you want to find.)

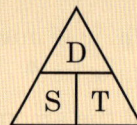

$$Average\ speed \text{ for a journey} = \dfrac{\text{total distance covered}}{\text{total time taken}}$$

## Numbers

### Rounded numbers

When a number has been rounded, its true value lies within a range that can be shown on a number line.

For example, if a nail is 23.5 mm long correct to 1 decimal place, then the length, $x$ mm, is in the range from 23.45 mm up to, but not including, 23.55 mm. This range is shown on the number line below.

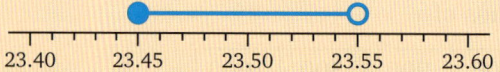

$23.45 \leqslant x < 23.55$

23.45 is the *lower bound* and 23.55 is the *upper bound* of $x$.

### Reciprocals

The reciprocal of a number is 1 divided by that number,

for example, the reciprocal of 4 is $\tfrac{1}{4} = 0.25$

and the reciprocal of 0.8 is $1 \div 0.8 = 1.25$

### Recurring decimals

When some fractions are changed to decimals, the result is a recurring pattern of digits that repeats indefinitely. To save space, we place a dot over the first and last numbers in the group of digits that recur.

For example, $\quad \tfrac{1}{11} = 1 \div 11 = 0.090\,909... = 0.\dot{0}\dot{9}$

### Division in a given ratio

To divide £200 into three amounts of money in the ratio $2:5:3$, we have to divide it into $2 + 5 + 3 = 10$ equal parts;
so the first amount is 2 of these parts, that is, $\frac{2}{10}$ of £200,
the second amount is $\frac{5}{10}$ of £200 and the third amount is $\frac{3}{10}$ of £200.

### Direct proportion

When two quantities are related so that when one of them trebles, say, the other also trebles, the quantities are directly proportional (that is, they are always in the same ratio).

### Inverse proportion

When two quantities are related so that when one of them trebles, say, the other becomes a third of its original size, the quantities are inversely proportional and their product is constant.

## Percentage

### Percentage change

A percentage change in a quantity is expressed as a percentage of that quantity before any changes were made.

For example, if a shirt is sold for £22.50 after a discount of 10% has been made, the discount is 10% of the price before it is reduced,

i.e. £22.50 = original price − 10% of original price
= 90% × original price

### Interest

When a sum of money is borrowed, interest is usually charged on a yearly basis and is given as a percentage of the sum borrowed, for example, 3% p.a. When money is lent, interest is paid using the same principle.

### Compound percentage change

Compound percentage change is an accumulating change. If, for example, the value of a house increases by 5% of its value at the start of each year; its value after one year is 105% of its initial value; after another year, its value is 105% of its value at the start of that year, that is, 105% of its increased value, and so on.

## Probability

### Adding probabilities

We add probabilities when we want the probability that one or other of two (or more) events will happen, provided that only one of the events can happen at a time.

Events such that only one of them can happen on any one occasion are called *mutually exclusive*.

For example, when one dice is rolled,

P(scoring 5 or 6) = P(scoring 5) + P(scoring 6)

## Multiplying probabilities

We multiply probabilities when we want the probability that two (or more) events both happen, provided that each event has no influence on whether or not the other occurs.

Events such that each has no influence on whether the other occurs are called *independent events*.

For example, when two dice, A and B, are rolled,

P(scoring 6 on A and B) = P(scoring 6 on A) × P(scoring 6 on B)

## Tree diagrams

Tree diagrams can be used to illustrate the outcomes when two or more events occur.

For example, this tree diagram shows the possible outcomes when two coins are tossed.

We *multiply* the probabilities when we follow a path along the branches and *add* the results of following several paths.

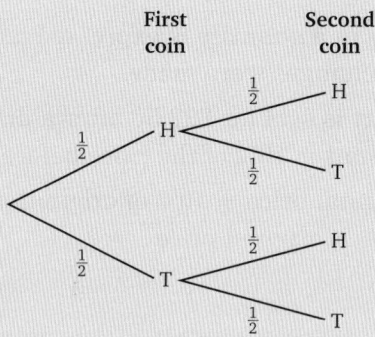

## Venn diagrams

Venn diagrams can be used to show outcomes when two or more events are not mutually exclusive. When there are several outcomes in each category, we use the number of outcomes rather than list each outcome.

This Venn diagram shows the numbers of students in Year 9 of a school who belong to the sports club and who belong to the chess club.

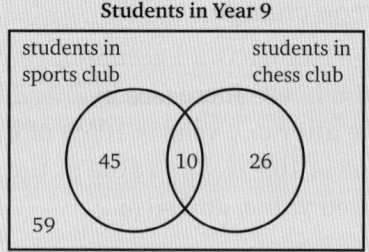

This shows that 45 students are in the sports club but not in the chess club, 10 students are in both clubs, 26 are in the chess club but not in the sports club and 59 students are in neither club.

# REVISION EXERCISE 1.1 (Chapters 1 to 3)

**1** The graph shows the journey of an athlete in a race.

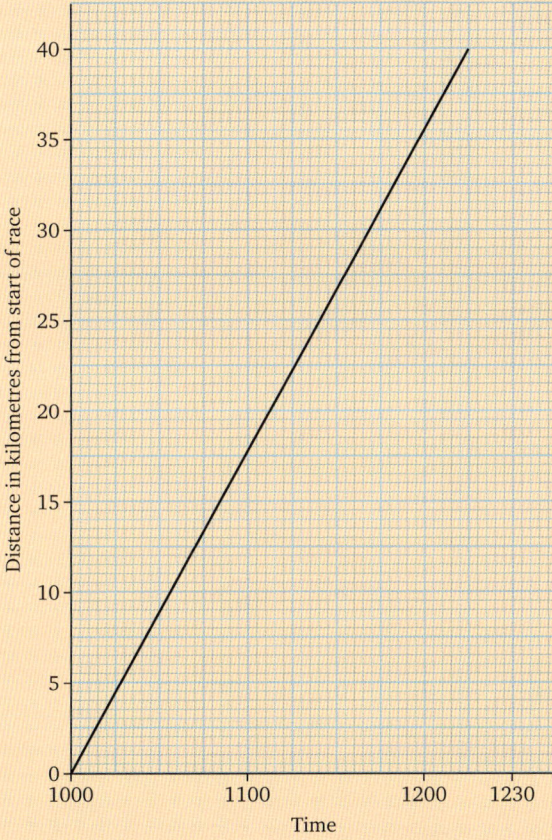

**a** What was the length of the race?

**b** How long did the athlete take?

**c** What was his average speed for the whole journey?

**d** How far did he travel in the first $1\frac{1}{4}$ hours?

**e** Did the athlete stop at any time during the race?

**f** Did the athlete travel at more than one speed?

**2** The graph shows the journey of a car from Amberley to Brickworth and on to Coldham.

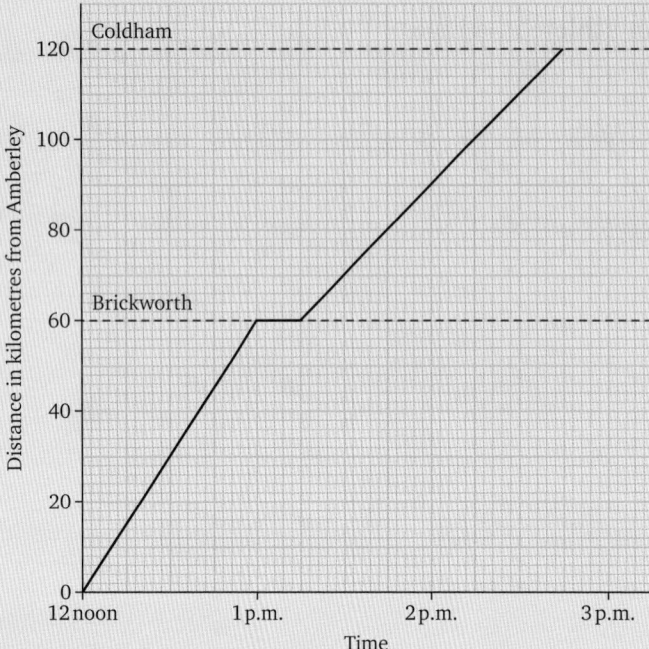

  **a**  Where was the car at
     **i**  12.30
     **ii**  2.15?

  **b**  What was the average speed of the car between
     **i**  Amberley and Brickworth
     **ii**  Brickworth and Coldham?

  **c**  For how long did the car stop at Brickworth?

  **d**  How long did the journey take, including the stop?

  **e**  What was the average speed of the car for the whole journey? Give your answer correct to the nearest whole number.

**3 a**  A car travels at 72 km/h. How far will it travel in
     **i**  $1\tfrac{1}{2}$ hours
     **ii**  20 minutes
     **iii**  1 hour 40 minutes?

  **b**  How long will Rob, walking at 6 km/h, take to walk
     **i**  12 km
     **ii**  20 km?

  **c**  Pete cycles at 12 mph. How long will it take him to cycle
     **i**  18 miles
     **ii**  30 miles?

**4** The graph represents the bicycle journeys of three school friends, Ali, Bianca and Craig, from the village in which they live to Buckwell, the nearest main town, which is 30 km away.
Use the graph to find

  **a** their order of arrival at Buckwell

  **b** Ali's average speed for the journey

  **c** Bianca's average speed for the journey

  **d** Craig's average speed for the journey

  **e** where and when Craig passes Ali

  **f** how far each is from Buckwell at 2 p.m.

  **g** how far Bianca is ahead of Craig at 2.15 p.m.

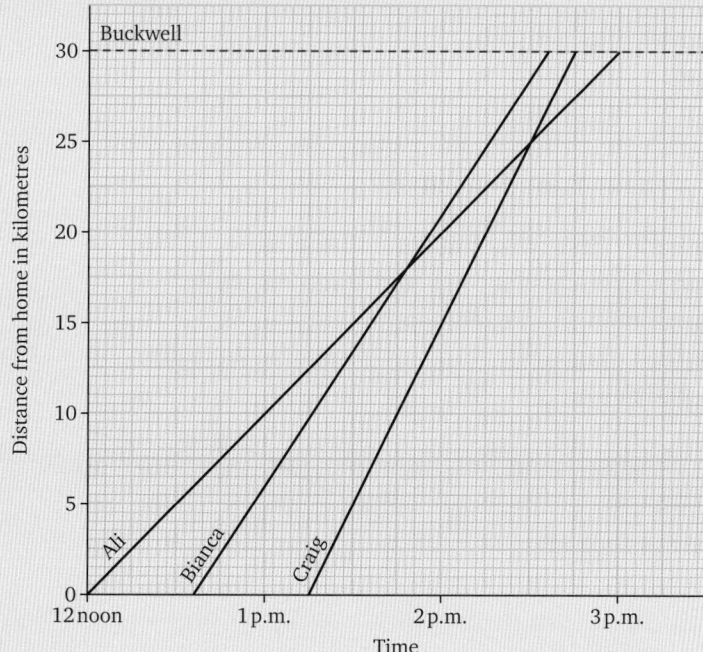

**5** Find    **a** $4 \div \frac{2}{3}$    **b** $2\frac{1}{3} \div 1\frac{5}{9}$    **c** $3\frac{2}{3} \div \left(\frac{2}{3} + \frac{4}{5}\right)$

**6 a** Express each fraction as a decimal.
    **i** $\frac{7}{20}$    **ii** $\frac{5}{16}$    **iii** $\frac{1}{400}$    **iv** $\frac{15}{32}$

   **b** Express each decimal as a fraction in its lowest terms,
    **i** 0.55    **ii** 0.875    **iii** 0.015    **iv** 2.208

**7** Use dot notation to write the following fractions as decimals.
   **a** $\frac{1}{15}$    **b** $\frac{4}{7}$    **c** $\frac{11}{12}$

**8** Write down the reciprocal of    **a** $\frac{1}{5}$    **b** 1.8    **c** $\frac{2}{9}$

**9 a** It is stated that the number of screws in a box is 200 to the nearest 10. Find the range in which the actual number of screws lies.

   **b** A number is given as 2.47 correct to 2 decimal places. Illustrate on a number line the range in which this number lies.

**10** A card is selected at random from an ordinary pack of 52. A second card is selected at random from another ordinary pack of 52 cards. What is the probability that
  **a** the first card is a red 2
  **b** the second card is a black queen
  **c** the first card is a red 2 and the second card is a black queen?

**11** The probability that Samuel completes his maths homework is $\frac{7}{8}$ and that Hanna completes her maths homework is $\frac{3}{4}$.
What is the probability that
  **a** Samuel does not complete his maths homework
  **b** both Samuel and Hanna complete their maths homework
  **c** neither Samuel nor Hanna complete their maths homework?

**12** Kim rolls an ordinary dice and tosses a coin.
What is the probability that
  **a** the dice shows a six
  **b** the coin lands head up
  **c** the dice shows a six and the coin lands head up?

**13** When the post arrived Val had 3 first-class letters and 4 second-class letters, while Dirk had 5 first-class letters and 2 second-class letters. A letter is taken at random from each batch.
  **a** Copy and complete the tree diagram by writing in the probabilities on each branch.
  Find the probability that
  **b** both letters are first class
  **c** one is first class and one is second class.

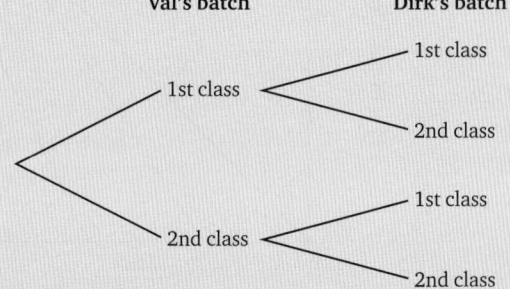

**14** Vicki has a box of 36 biscuits, 12 of which are square biscuits and the remainder of which are round biscuits. Joe also has a box of 36 biscuits but 20 of his are square and the remainder are round. A biscuit is taken at random from each box.
  **a** Copy and complete the tree diagram by writing in the probabilities on each branch.
  Find the probability that
  **b** both biscuits are round
  **c** one biscuit is square and the other is round.

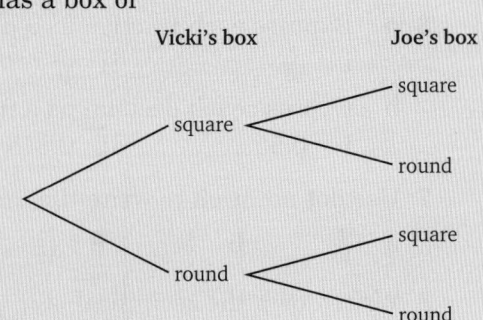

**15 a** Show in a Venn diagram, $A$ = {the different letters in the word SQUASH} and $B$ = {the different letters in the word SCHOOL}.
  **b** One letter is chosen at random from the alphabet. What is the probability that it is not a letter in either SQUASH or SCHOOL?

## REVISION EXERCISE 1.2 (Chapters 4 and 5)

1. A chair costing £85 is sold at a gain of 60%. Find the selling price.

2. Find the cost of an electric heater marked £34 + VAT at $17\frac{1}{2}\%$.

3. The selling price, excluding VAT, of a CD player is £127.50. If this includes a mark-up of 50% find the buying-in price.

4. Mrs Denham has a gross annual income of £12 750 and allowances of £3925. She pays tax at 23% on her taxable income.
   a. How much tax must she pay?
   b. Express the tax as a percentage of her gross pay. Give your answer correct to the nearest whole number.

5. Find the compound interest on £800 invested for 2 years at 5%.

6. a. Express in the form $1:n$
      i. $5:9$   ii. $\frac{1}{4}:\frac{1}{5}$   iii. $3:7$
      (giving $n$ correct to 3 significant figures when necessary)
   b. Express in the form $n:1$
      i. $7:4$   ii. $\frac{1}{4}:2$   iii. $11:3$
      (giving $n$ correct to 3 significant figures when necessary)

7. Kyle spends £1.20 on a pen and 90 p on a pencil. What is the ratio of the cost of
   a. the pen to the pencil
   b. the pencil to the pen
   c. the pencil to the total?

8. A bus journey of 120 miles costs £15.60. At the same rate per mile
   a. what would be the cost of travelling 150 miles
   b. how far you can travel for £26?

9. If it takes 21 men 4 days to mark out the field in a stadium for an athletics meeting, how long would it take 12 men to do the same job?

10. For a youth camp the organisers take enough supplies to support 42 campers for 14 days. In the event 49 campers attend. How long will the supplies last?

## REVISION EXERCISE 1.3 (Chapters 1 to 5)

1. Luiz sets out on his bike on a journey of 12 km. He has cycled 10 km at 15 km/h when his bicycle suffers a puncture. As a result, he pushes his bike the rest of the distance at 5 km/h.
   a. For how long does he cycle before the puncture?
   b. How far does he have to push the bike and for how long?
   c. Find the total time for his journey.
   d. Find his average speed for the whole journey.

**2 a** Give 34.678 cm correct to the nearest tenth of a centimetre.

**b** The acceptable thickness of a metal plate is 4 mm, to the nearest mm, Find the range in which the actual thickness lies.

**3** Find    **a** $\frac{3}{4} \div \frac{1}{2}$    **b** $4\frac{1}{7} \div 2\frac{5}{12}$    **c** $\frac{2}{5} \times \left(\frac{5}{6} - \frac{5}{12}\right) + \frac{1}{2}$

**4** Write down

**a** the reciprocal of $\frac{3}{7}$

**b** $\frac{19}{11}$ as a decimal using dot notation

**c** 0.275 as a fraction in its lowest terms.

**5** When Mrs Hussain needs to buy bread, the probability that she buys it at the supermarket is $\frac{4}{9}$, while the probability that she buys it at the local bakery is $\frac{1}{3}$. What is the probability that

**a** she buys bread either at the supermarket or at the local bakery

**b** she buys bread somewhere else?

**6 a** Sam Nolan has a taxable income of £7600 a year. How much tax must he pay if the rate of tax is 23%?

**b** In a sale, a shirt is marked at £16.50, which includes a reduction of 25%. What was the pre-sale price?

**7** Because they are emigrating, Mr and Mrs Thomson are anxious to sell their house quickly. They put the house on the market at £100 000 but reduce the price every week by 4% of its price at the beginning of that week. It is sold the first week the price drops below £90 000.

**a** How many weeks did it take to sell?

**b** What was the selling price? (Give your answer correct to the nearest £1000.)

**8** A car bought for £12 000 depreciates in value by 20% each year. Find its value after 3 years.

**9** A hotel charges £259 per person per seven-day week. What would be the charge for 16 days at the same rate?

**10** Under normal conditions a school boiler consumes 0.75 t of fuel a day and the stock of fuel is sufficient to last for 12 days. A cold spell causes the consumption rate to rise to 0.9 t per day. How long will the fuel last?

## REVISION EXERCISE 1.4 (Chapters 1 to 5)

**1 a** Express $\frac{9}{11}$ as a decimal correct to 3 decimal places.

**b** Use your answer to part **a** to find $\frac{9}{11}$ of £50. Then use fractions to find $\frac{9}{11}$ of £50 correct to the nearest penny. Explain your two answers.

2 Express as a fraction   a  $0.\dot{6}$   b  $0.0\dot{6}$   c  $0.60\dot{6}$

3 Find   a  $\frac{2}{3} \div 1\frac{2}{7}$   b  $\left(3 - 1\frac{7}{8}\right) \times 1\frac{1}{3}$   c  $\left(1\frac{2}{3} + 2\frac{1}{5}\right) \div 5\frac{4}{5}$

4 Carl takes one disc at random from a bag containing 2 red discs and 3 blue discs. He then takes another disc from a second bag containing 3 red discs and 5 blue discs. What is the probability that
   a  both discs are red
   b  both discs are blue
   c  either both discs are red or both discs are blue?

5 Courtney has two bags. Bag A contains 4 sweet apples and 5 sour apples, while bag B contains 6 sweet apples and 4 sour apples. Jamie takes one apple from each bag.
   a  Copy and complete the tree diagram by filling in the probabilities on the branches.
   Find the probability that
   b  both apples are sweet
   c  one apple is sweet and the other is sour.

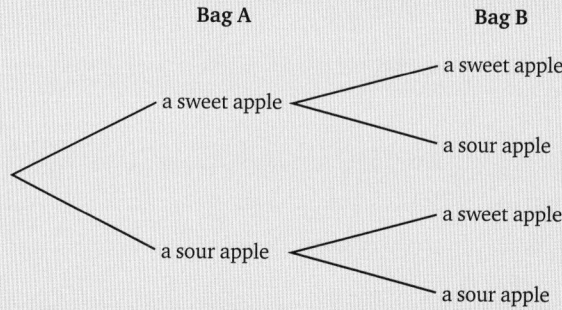

6 A dress costing £55 is sold at a loss of 20%. Find the sale price.

7 When a fireside rug is sold for £39 the retailer suffers a loss of 40%. Find the buying-in price.

8 a  Find the interest payable after one year if £8500 is invested at 5.25% a year.
   b  If the original sum, plus the interest, remains invested at the same rate, find the total value of the investment at the end of the second year.

9 A bookshelf will hold 360 books of thickness 2 cm. How many copies of *National Geographic* magazine will it hold if each copy is 6 mm thick?

10 $U$ = {whole numbers from 1 to 14 inclusive}
   $A$ = {even numbers from 3 to 13}
   $B$ = {multiples of 3 between 1 and 14}
   One number is chosen at random from the members of $U$. What is probability that it is not a member of $A$ or of $B$?

## Mental arithmetic practice 1

1. Find $2.1 \div 0.7$
2. What is the value of $(-2)^4$?
3. Find $1\frac{1}{2} - \frac{3}{5}$
4. Find $0.15 \times 300$
5. What is the second significant figure in 0.0203?
6. Find $2 - 0.05$
7. Which is larger, 0.7 or $\frac{5}{8}$?
8. Give 1.25 as a percentage.
9. What has to be added to 37 to give 100?
10. Express 2.15 as a fraction.
11. Express 25 cm as a fraction of 1 metre.
12. Divide 35 kg into two parts in the ratio $3:4$
13. A discount of 20% is given in a sale. What is the sale price of a table marked £120?
14. What is the value of $5^{-3}$?
15. Helen weighs 64 kg to the nearest kilogram. What is the upper bound of her weight?
16. Give 0.3 recurring as a fraction.
17. What is the reciprocal of 0.5?
18. The probability that it will rain tomorrow is 0.15. What is the probability that it will not rain tomorrow?
19. Express $\frac{1}{6}$ as a decimal.
20. Express 12.5% as a fraction.
21. Find $1.4^2$
22. Find $3 - 0.08$
23. Find $0.5^3$
24. Write 36 000 in standard form.
25. Simplify $a^7 \div a^5$
26. Express 1.45 as a percentage.
27. Find $\frac{1}{4} \div \frac{1}{8}$
28. Which is larger: $10^{-5}$ or $10^{-6}$?
29. Find the value of 6 minus $-7$.
30. Express $\frac{23}{25}$ as a percentage.

## Summary 1

31  Ben's height is 160 cm correct to the nearest cm. What is his least possible height?

32  Write $\frac{3}{11}$ as a recurring decimal.

33  What number is halfway between 1.2 and 1.7?

34  What is the reciprocal of 0.2?

35  Find the square root of 0.16

36  Find $1 \div 0.2^2$

37  A book costing £9 is sold at a profit of 80%. Find the selling price.

38  Find $\frac{1}{4} \div \frac{1}{3}$

39  Estimate $38.4 \times 5.31$

40  Which number is nearer to 0: $10^2$ or $10^{-2}$?

41  Find $\frac{3}{5} - \frac{1}{2}$

42  Find $2\frac{1}{2}\%$ of £60.

43  Kate pays tax at 22% on £7000 of her income. How much tax does she pay?

44  What is the reciprocal of $\frac{3}{4}$?

45  A cuboid is 3 cm deep, 4 cm wide and 5 cm high. What is its volume?

46  Express $3\frac{1}{4}\%$ as a decimal.

47  Increase £70 by 5%.

48  Express 48 cm as a percentage of 2 metres.

49  Give 64% as a fraction in its lowest terms.

50  Express 3.75 as a percentage.

51  Write 0.000 006 in standard form.

52  What is the simple interest on £750 invested for 1 year at 8%?

53  Increase 55 cm by 10%.

54  Find $\frac{3}{8} \div \frac{3}{4}$

55  What name is given to a polygon with 8 sides?

56  Find $3a \times 4a$

57  Find $12a \div 3a$

58  The lengths of the two shorter sides of a right-angled triangle are 6 cm and 8 cm. How long is the third side?

59  What is the lowest common multiple of 9 and 12?

60  Decrease 60 cm$^2$ by 15%.

# 6 Algebraic products

## Consider

Andy asked how he could find an expression for the area of this rectangle.

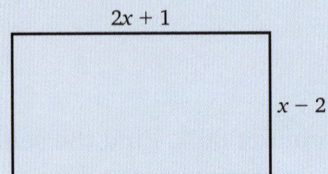

Anna said, 'It is $2x^2 - x - 2$'

Is Anna right? Give a reason for your answer.

You should be able to answer this question after you have worked through this chapter.

## Brackets

We will start with a reminder about multiplying out simpler brackets.

Remember that $\quad 5(x + 1) = 5x + 5$

and that $\quad 4x(y + z) = 4xy + 4xz$

### Exercise 6a

Expand, i.e. multiply out

1. $2(x + 1)$
2. $3(x - 1)$
3. $4(x + 3)$
4. $-3(x + 2)$
5. $5x(3y + z)$
6. $4y(4x + 3z)$
7. $2n(3p - 5q)$
8. $-3b(a - c)$
9. $8r(2t - s)$
10. $3a(b - 5c)$
11. $4x(3y + 2z)$
12. $-3x(y - 2z)$

## The product of two brackets

Frequently, we need to find the product of two brackets, each of which contains two terms, for example, $(a + b)(c + d)$.

Each term in the second brackets has to be multiplied by each term in the first bracket

so $\qquad (a + b)(c + d) \quad$ means $\quad a \times (c + d) + b \times (c + d)$

Now $\qquad a \times (c + d) + b \times (c + d) = ac + ad + bc + bd$

Therefore, $\qquad (a + b)(c + d) = ac + ad + bc + bd$

In the same way, $\quad (p + q)(2r - 3s) = p(2r - 3s) + q(2r - 3s)$

$\qquad\qquad\qquad\qquad\qquad = 2pr - 3ps + 2qr - 3qs$

## 6 Algebraic products

We can miss out the middle step and we can do the multiplication in any order. However, it is easier if we always follow the same pattern to make sure we don't miss anything out.

Always multiply the brackets together in the following order:

1. the first terms in the brackets
2. the outside terms
3. the inside terms
4. the second terms in the brackets.

Thus

$$(a + b)(c + d) = ac + ad + bc + bd$$

### Exercise 6b

**Worked example**

 Expand $(x + 2y)(2y - z)$

$(x + 2y)(2y - z) = 2xy - xz + 4y^2 - 2yz$

> Remember that the product of two numbers with the same sign gives a positive result and the product of two numbers with different signs gives a negative result.

Expand

1. $(a + b)(c + d)$
2. $(p + q)(s + t)$
3. $(2a + b)(c + 2d)$
4. $(5x + 2y)(z + 3)$
5. $(x + y)(z - 4)$
6. $(a - b)(c + d)$
7. $(x + y)(w + z)$
8. $(2a + b)(3c + d)$
9. $(5x + 4y)(z + 2)$
10. $(3x - 2y)(5 - z)$
11. $(p + q)(2s - 3t)$
12. $(a - 2b)(c - d)$
13. $(6u - 5v)(w - 5r)$
14. $(3a + 4b)(2c - 3d)$
15. $(3x + 2y)(3z + 2)$
16. $(3p - q)(4r - 3s)$
17. $(3a - 4b)(3c + 4d)$
18. $(7x - 2y)(3 - 2z)$
19. $(2a + b)(5c - 2)$
20. $(5a - 4b)(3 - 2d)$

129

We get a slightly simpler form when we find the product of two brackets that contain the same letter together with a number, such as $(x + 2)$ and $(x + 4)$.

For example, using the order we chose earlier

$$(x + 2)(x + 4) = x^2 + 4x + 2x + 8$$
$$= x^2 + 6x + 8$$

 Because $4x$ and $2x$ are like terms, they can be collected together.

i.e. $(x + 2)(x + 4) = x^2 + 6x + 8$

## Exercise 6c

Expand

**1** $(x + 3)(x + 4)$

**2** $(x + 4)(x + 2)$

**3** $(x + 1)(x + 6)$

**4** $(x + 5)(x + 2)$

**5** $(x + 8)(x + 3)$

**6** $(a + 4)(a + 5)$

**7** $(b + 2)(b + 7)$

**8** $(c + 4)(c + 6)$

**9** $(p + 3)(p + 12)$

**10** $(q + 7)(q + 10)$

### Worked example

→ Expand $(x - 4)(x - 6)$

$$(x - 4)(x - 6) = x^2 - 6x - 4x + 24$$
$$= x^2 - 10x + 24$$

 Remember that $(-4) \times (-6) = +24$.

Expand

**11** $(x - 2)(x - 3)$

**12** $(x - 5)(x - 7)$

**13** $(a - 2)(a - 8)$

**14** $(x - 10)(x - 3)$

**15** $(b - 5)(b - 5)$

**16** $(x - 3)(x - 4)$

**17** $(x - 4)(x - 8)$

**18** $(b - 4)(b - 2)$

**19** $(a - 4)(a - 4)$

**20** $(p - 7)(p - 8)$

# 6 Algebraic products

## Worked example

→ Expand $(x + 3)(x - 6)$

$(x + 3)(x - 6) = x^2 - 6x + 3x - 18$

$\qquad\qquad\qquad = x^2 - 3x - 18$

Expand

21  $(x + 3)(x - 2)$

22  $(x - 4)(x + 5)$

23  $(x - 7)(x + 4)$

24  $(a + 3)(a - 10)$

25  $(p + 5)(p - 5)$

26  $(x + 7)(x - 2)$

27  $(x - 5)(x + 6)$

28  $(x + 10)(x - 1)$

29  $(b - 8)(b + 7)$

30  $(z + 1)(z - 12)$

## Finding the pattern

You may have noticed in **Exercise 6c** that when you expanded the brackets and simplified the answers, there was a definite pattern.

e.g. $\qquad (x + 5)(x + 9) = x^2 + 9x + 5x + 45$

$\qquad\qquad\qquad\qquad = x^2 + 14x + 45$

We could have written this as

$\qquad (x + 5)(x + 9) = x^2 + (9 + 5)x + (5) \times (9)$

$\qquad\qquad\qquad\qquad = x^2 + 14x + 45$

Similarly $\qquad (x + 4)(x - 7) = x^2 + (-7 + 4)x + (4) \times (-7)$

$\qquad\qquad\qquad\qquad = x^2 - 3x - 28$

and $\qquad (x - 3)(x - 8) = x^2 + (-8 - 3)x + (-3) \times (-8)$

$\qquad\qquad\qquad\qquad = x^2 - 11x + 24$

In each case there is a pattern:

**the *product* of the two numbers in the brackets gives the number term in the expansion, while *collecting* them gives the number of *x*s.**

# Exercise 6d

Use the pattern given on the previous page to expand the following products.

1. $(x + 4)(x + 5)$
2. $(a + 2)(a + 5)$
3. $(x - 4)(x - 5)$
4. $(a - 2)(a - 5)$
5. $(x + 8)(x + 6)$
6. $(a + 10)(a + 7)$
7. $(x - 8)(x - 6)$
8. $(a - 10)(a - 7)$
9. $(a + 2)(a - 5)$
10. $(y - 6)(y + 3)$
11. $(z + 4)(z - 10)$
12. $(p + 5)(p - 8)$
13. $(a - 10)(a + 7)$
14. $(y + 10)(y - 2)$
15. $(z - 12)(z + 1)$
16. $(p + 2)(p - 13)$

The pattern is similar when the brackets are slightly more complicated.

# Exercise 6e

**Worked example**

→ Expand $(2x + 3)(x + 2)$

$$(2x + 3)(x + 2) = 2x^2 + 4x + 3x + 6$$
$$= 2x^2 + 7x + 6$$

Expand the following products.

1. $(2x + 1)(x + 1)$
2. $(x + 2)(5x + 2)$
3. $(5x + 2)(x + 3)$
4. $(3x + 4)(x + 5)$
5. $(3x + 2)(x + 1)$
6. $(x + 3)(3x + 2)$
7. $(4x + 3)(x + 1)$
8. $(7x + 2)(x + 3)$

**Worked example**

→ Expand $(3x - 2)(2x + 5)$

$$(3x - 2)(2x + 5) = 6x^2 + 15x - 4x - 10$$
$$= 6x^2 + 11x - 10$$

Expand

9  $(3x + 2)(2x + 3)$
10 $(4x - 3)(3x - 4)$
11 $(5x + 6)(2x - 3)$
12 $(7a - 3)(3a - 7)$
13 $(5x + 3)(2x + 5)$
14 $(7x - 2)(3x - 2)$
15 $(3x - 2)(4x + 1)$
16 $(3b + 5)(2b - 5)$

17 $(2a + 3)(2a - 3)$
18 $(3b - 7)(3b + 7)$
19 $(7y - 5)(7y + 5)$
20 $(5a + 4)(4a - 3)$
21 $(4x + 3)(4x - 3)$
22 $(5y - 2)(5y + 2)$
23 $(3x - 1)(3x + 1)$
24 $(4x - 7)(4x + 5)$

## Worked example

→ Expand $(4 + x)(2 - x)$

$(4 + x)(2 - x) = 8 - 4x + 2x - x^2$
$\phantom{(4 + x)(2 - x)} = 8 - 2x - x^2$

Expand

25 $(2 - x)(5 + x)$
26 $(4 + 3x)(2 - x)$
27 $(x - 1)(1 - x)$
28 $(5 - y)(4 + y)$

29 $(7 + x)(3 - x)$
30 $(1 + 4x)(2 - x)$
31 $(x - 3)(2 - x)$
32 $(4 - 2p)(5 + 2p)$

## Worked example

→ Expand $(3x - 2)(5 - 2x)$

$(3x - 2)(5 - 2x) = 15x - 6x^2 - 10 + 4x$
$\phantom{(3x - 2)(5 - 2x)} = 19x - 6x^2 - 10$
$\phantom{(3x - 2)(5 - 2x)} = -6x^2 + 19x - 10$

 We have rearranged the terms so that the powers of $x$ are in descending order of size. This is the conventional way of writing expressions such as this one. It is also helpful to have the terms in this order for future work.

Expand

33 $(2x + 1)(1 + 3x)$
34 $(5x + 2)(2 - x)$
35 $(6x - 1)(3 - x)$
36 $(5a - 2)(3 - 7a)$
37 $(3x + 2)(4 - x)$

38 $(5x + 2)(4 + 3x)$
39 $(7x + 4)(3 - 2x)$
40 $(4x - 3)(3 - 5x)$
41 $(3 - p)(4 + p)$
42 $(x - 5)(2 + x)$

## Important products

Three very important products are:

$$(x + a)^2 = (x + a)(x + a)$$
$$= x^2 + xa + ax + a^2$$
$$= x^2 + 2ax + a^2 \text{ (since } xa \text{ is the same as } ax)$$

i.e.  $(x + a)^2 = x^2 + 2ax + a^2$

so  $(x + 3)^2 = x^2 + 6x + 9$

$$(x - a)^2 = (x - a)(x - a)$$
$$= x^2 - xa - ax + a^2$$

i.e.  $(x - a)^2 = x^2 - 2ax + a^2$

so  $(x - 4)^2 = x^2 - 8x + 16$

$$(x + a)(x - a) = x^2 - xa + ax - a^2$$
$$= x^2 - a^2$$

i.e.  $(x + a)(x - a) = x^2 - a^2$

and  $(x - a)(x + a) = x^2 - a^2$

so  $(x + 5)(x - 5) = x^2 - 25$

and  $(x - 3)(x + 3) = x^2 - 9$

You should learn these three results thoroughly, as they will appear time and time again. Given the left-hand side, you should know the right-hand side and vice versa.

## Exercise 6f

### Worked example

→ Expand $(x + 5)^2$

💡 Comparing $(x + 5)^2$ with $(x + a)^2$ tells us that $a = 5$ in this case.
So $x^2 + 2ax + a^2$ becomes $x^2 + 2(5)x + (5)^2$

$(x + 5)^2 = x^2 + 10x + 25$

💡 Alternatively, write $(x + 5)^2$ as $(x + 5)(x + 5)$ and multiply out in the usual way.

Expand, by comparing with $(x + a)^2$

1. $(x + 1)^2$
2. $(x + 2)^2$
3. $(a + 3)^2$
4. $(b + 4)^2$
5. $(x + z)^2$
6. $(y + x)^2$
7. $(c + d)^2$
8. $(m + n)^2$
9. $(a + 9)^2$

Expand by writing as the product of two brackets

10  $(t + 10)^2$
11  $(x + 12)^2$
12  $(x + 8)^2$
13  $(p + 7)^2$
14  $(p + q)^2$
15  $(a + b)^2$
16  $(e + f)^2$
17  $(u + v)^2$
18  $(M + m)^2$

**Worked example**

→ Expand $(2x + 3)^2$

$(2x + 3)^2 = (2x)^2 + 2(2x)(3) + (3)^2$
i.e. $(2x + 3)^2 = 4x^2 + 12x + 9$

 Use $(x + a)^2$ and replace $x$ by $2x$ and $a$ by 3.

Expand

19  $(2x + 1)^2$
20  $(4b + 1)^2$
21  $(5x + 2)^2$
22  $(6c + 1)^2$
23  $(3a + 1)^2$
24  $(2x + 5)^2$
25  $(3a + 4)^2$
26  $(4y + 3)^2$
27  $(3W + 2)^2$

**Worked example**

→ Expand $(2x + 3y)^2$

$(2x + 3y)^2 = 4x^2 + 12xy + 9y^2$

 $(2x + 3y)^2 = (2x)^2 + 2(2x)(3y) + (3y)^2$

Expand

28  $(x + 2y)^2$
29  $(3x + y)^2$
30  $(2x + 5y)^2$
31  $(3a + 2b)^2$
32  $(3a + b)^2$
33  $(p + 4q)^2$
34  $(7x + 2y)^2$
35  $(3s + 4t)^2$
36  $(3s + t)^2$

**Worked example**

→ Expand $(x - 5)^2$

$(x - 5)^2 = x^2 - 10x + 25$

Expand

37  $(x - 2)^2$
38  $(x - 6)^2$
39  $(a - 10)^2$
40  $(x - y)^2$
41  $(x - 3)^2$
42  $(x - 7)^2$
43  $(a - b)^2$
44  $(u - v)^2$

## Worked example

→ Expand $(2x - 7)^2$

$(2x - 7)^2 = 4x^2 - 28x + 49$

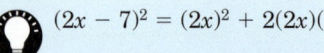
$(2x - 7)^2 = (2x)^2 + 2(2x)(-7) + (-7)^2$

Expand

**45** $(3x - 1)^2$
**46** $(5z - 1)^2$
**47** $(10a - 9)^2$
**48** $(4x - 3)^2$

**49** $(2a - 1)^2$
**50** $(4y - 1)^2$
**51** $(7b - 2)^2$
**52** $(5x - 3)^2$

## Worked example

→ Expand $(7a - 4b)^2$

$(7a - 4b)^2 = 49a^2 - 56ab + 16b^2$

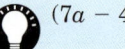
$(7a - 4b)^2 = (7a)^2 + 2(7a)(-4b) + (-4b)^2$

Expand

**53** $(2y - x)^2$
**54** $(5x - y)^2$
**55** $(3m - 2n)^2$
**56** $(7x - 3y)^2$

**57** $(a - 3b)^2$
**58** $(m - 8n)^2$
**59** $(5a - 2b)^2$
**60** $(3p - 5q)^2$

## Exercise 6g

### Worked example

→ Expand  **a** $(a + 2)(a - 2)$  **b** $(2x + 3)(2x - 3)$

**a** $(a + 2)(a - 2) = a^2 - 4$
**b** $(2x + 3)(2x - 3) = 4x^2 - 9$

Expand

**1** $(x + 4)(x - 4)$
**2** $(b + 6)(b - 6)$

**3** $(c - 3)(c + 3)$
**4** $(x + 12)(x - 12)$

# 6 Algebraic products

**5** $(x + 5)(x - 5)$
**6** $(a - 7)(a + 7)$
**7** $(q + 10)(q - 10)$
**8** $(x - 8)(x + 8)$
**9** $(2x - 1)(2x + 1)$
**10** $(3x + 1)(3x - 1)$

**11** $(7a + 2)(7a - 2)$
**12** $(5a - 4)(5a + 4)$
**13** $(5x + 1)(5x - 1)$
**14** $(2a - 3)(2a + 3)$
**15** $(10m - 1)(10m + 1)$
**16** $(6a + 5)(6a - 5)$

### Worked example

→ Expand $(3x + 2y)(3x - 2y)$

$$(3x + 2y)(3x - 2y) = (3x)^2 - (2y)^2$$
$$= 9x^2 - 4y^2$$

Expand

**17** $(3x + 4y)(3x - 4y)$
**18** $(2a - 5b)(2a + 5b)$
**19** $(1 - 2a)(1 + 2a)$
**20** $(7y + 3z)(7y - 3z)$

**21** $(10a - 9b)(10a + 9b)$
**22** $(5a - 4b)(5a + 4b)$
**23** $(1 + 3x)(1 - 3x)$
**24** $(3 - 5x)(3 + 5x)$

The results from this exercise are very important when written the other way around,

i.e. $a^2 - b^2 = (a + b)(a - b)$

We refer to this as **factorising** the **difference between two squares**, and we will deal with it in detail in Chapter 7.

## More complex expansions

### Exercise 6h

### Worked example

→ Simplify $(x + 2)(x + 5) + 2x(x + 7)$

$$(x + 2)(x + 5) + 2x(x + 7) = x^2 + 5x + 2x + 10 + 2x^2 + 14x$$
$$= 3x^2 + 21x + 10$$

Simplify

**1** $(x + 3)(x + 4) + x(x + 2)$
**2** $x(x + 6) + (x + 1)(x + 2)$

**3** $(x + 4)(x + 5) + 6(x + 2)$
**4** $(a - 6)(a - 5) + 2(a + 3)$

**5** $(a-5)(2a+3) - 3(a-4)$

**6** $(x+3)(x+5) + 5(x+2)$

**7** $(x-3)(x+4) - 3(x+3)$

**8** $(x+7)(x-5) - 4(x-3)$

**9** $(2x+1)(3x-4) + (2x+3)(5x-2)$

**10** $(5x-2)(3x+5) - (3x+5)(x+2)$

### Worked example

➤ Expand $(xy - z)^2$

$(xy - z)^2 = x^2y^2 - 2xyz + z^2$

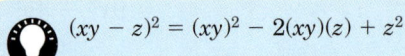 $(xy - z)^2 = (xy)^2 - 2(xy)(z) + z^2$

**11** $(xy - 3)^2$

**12** $(5 - yz)^2$

**13** $(xy + 4)^2$

**14** $(3pq + 8)^2$

**15** $(a - bc)^2$

**16** $(ab - 2)^2$

**17** $(6 - pq)^2$

**18** $(mn + 3)^2$

**19** $(uv - 2w)^2$

### Summary

The following is a summary of the most important types of examples considered in this chapter that will be required in future work.

**1** $2(3x + 4) = 6x + 8$

**2** $(x + 2)(x + 3) = x^2 + 5x + 6$

**3** $(x - 2)(x - 3) = x^2 - 5x + 6$

**4** $(x - 2)(x + 3) = x^2 + x - 6$

**5** $(2x + 1)(3x + 2) = 6x^2 + 7x + 2$

**6** $(2x - 1)(3x - 2) = 6x^2 - 7x + 2$

**7** $(2x + 1)(3x - 2) = 6x^2 - x - 2$

**8** $(2 + x)(3 - x) = 6 + x - x^2$

Note that     **a**   if the signs in the brackets are the same, i.e. both + or −, then the number term is +
(examples **2**, **3**, **5** and **6**)

whereas     **b**   if the signs in the brackets are different, i.e. one + and one −, then the number term is −
(examples **4** and **7**)

           **c**   the middle term is given by collecting the product of the outside terms in the brackets and the product of the inside terms in the brackets,

## 6 Algebraic products

i.e. in **2** the middle term is $3x + 2x$ or $5x$
in **3** the middle term is $-3x - 2x$ or $-5x$
in **4** the middle term is $3x - 2x$ or $x$
in **5** the middle term is $4x + 3x$ or $7x$
in **6** the middle term is $-4x - 3x$ or $-7x$
in **7** the middle term is $-4x + 3x$ or $-x$
in **8** the middle term is $-2x + 3x$ or $x$.

Most important of all, we must remember the general expansions:

$$(x + a)^2 = x^2 + 2ax + a^2$$
$$(x - a)^2 = x^2 - 2ax + a^2$$
$$(x + a)(x - a) = x^2 - a^2$$

## Mixed exercise

### Exercise 6i

Expand

1. $5(x + 2)$
2. $8p(3q - 2r)$
3. $(3a + b)(2a - 5b)$
4. $(4x + 1)(3x - 5)$
5. $(x + 6)(x + 10)$
6. $(x - 8)(x - 12)$
7. $(4y + 3)(4y - 7)$
8. $(4y - 9)(4y + 9)$
9. $(5x + 2)^2$
10. $(2a - 7b)^2$
11. $4(2 - 5x)$
12. $8a(2 - 3a)$
13. $(4a + 3)(3a - 11)$
14. $(x + 11)(9 - x)$
15. $(2x + 5)(1 - 10x)$
16. $(y + 2z)^2$
17. $(6y - z)(6y + 5z)$
18. $(4a + 1)^2$
19. $(5a - 7)^2$
20. $(6z - 13y)^2$
21. $3(2 - a)$
22. $4a(2b + c)$
23. $(5a + 2b)(2c + 5d)$
24. $(x - 7)(x - 12)$
25. $(a + 7)(a + 9)$
26. $(a + 4)(a - 5)$
27. $(3x + 1)(2x + 3)$
28. $(5x - 2)(5x + 2)$
29. $(3x - 7)^2$
30. $(5x + 2y)(5x - 2y)$

# STP Maths 9

## Expressions, identities and equations

$2x$ is an **expression**. When we write $x + x = 2x$, we are showing the equivalence between two forms of the *same* expression. The equality between two forms of the same expression is called an **identity**. In an identity the equality is true for any value of $x$. For example, when $x = 6$, $x + x = 12$, and $2x = 12$, and so on for any value of $x$.

If, on the other hand, we have $x + 2 = 2x$ then, when $x = 6$, $x + 2 = 8$ and $2x = 12$ i.e. $x + 2$ and $2x$ are *not* equal for all values of $x$.

$x + 2 = 2x$ is called an **equation**. We can find the values of $x$ (if any) for which the equality is true by solving the equation.

### Exercise 6j

Determine which of these are expressions, which are identities and which are equations.

1  $2(x + 1) = 2x + 2$
2  $2(x + 1) = 4$
3  $3x + 6 = 3(x + 2)$
4  $x(x + 1) = x^2 + x$
5  $5(x + 2) - 3(x - 4)$
6  $(x - 1)(x + 1) = x^2 - 1$
7  $3x + 4(x - 2)$
8  $(x + 1)(x + 4) = (x + 2)^2$

### Consider again

Andy asked how he could find an expression for the area of this rectangle.

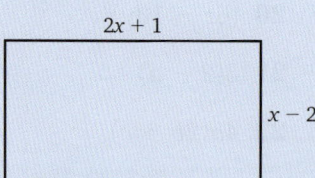

Anna said, 'It is $2x^2 - x - 2$'

Now can you say whether Anna is right? Give a reason for your answer.

# 7 Algebraic factors

> **Consider**
> 
> How can you quickly find the value of $8.8^2 - 1.2^2$ without using a calculator?
> 
> *You should be able to do this after you have worked through this chapter.*

In Chapter 6 we saw that we could expand algebraic expressions involving brackets, for example, $(x + 3)(x + 2)$ can be multiplied out to give $x^2 + 5x + 6$.

Later in this book we meet situations where we need to reverse this process, that is, we need to change from the form $x^2 + 5x + 6$ to the form $(x + 3)(x + 2)$.

In this chapter we build up the skills necessary to do this.

## Finding factors

When we reverse the process of expanding expressions, we are finding the factors of an expression. This is called **factorising**.

### Common factors

In the expression $7a + 14b$ we could write the first term as $7 \times a$ and the second term as $7 \times 2b$,

i.e. $\quad 7a + 14b = 7 \times a + 7 \times 2b$

The 7 is a common factor.

However, we already know that $\quad 7(a + 2b) = 7 \times a + 7 \times 2b$

∴ $\quad 7a + 14b = 7 \times a + 7 \times 2b = 7(a + 2b)$

### Exercise 7a

**Worked example**

→ Factorise $3x - 12$

$3x - 12 = 3 \times x - 3 \times 4$
$\phantom{3x - 12} = 3(x - 4)$

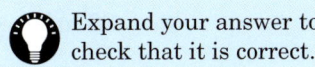 Expand your answer to check that it is correct.

Expand your answer in your head to check that it is correct.

Factorise

1. $4x + 4$
2. $12x - 3$
3. $6a + 2$
4. $5a - 10b$
5. $3t - 9$
6. $10a - 5$
7. $12a + 4$
8. $2a + 4b$
9. $14x - 7$

**Worked example**

→ Factorise $x^2 - 7x$

$x^2 - 7x = x \times x - 7 \times x$
$\phantom{x^2 - 7x} = x(x - 7)$

Factorise

10. $x^2 + 2x$
11. $x^2 - 7x$
12. $a^2 + 6a$
13. $2x^2 + x$
14. $4t - 2t^2$
15. $x^2 + 5x$
16. $x^2 - 4x$
17. $b^2 + 4b$
18. $4a^2 - a$

**Worked example**

→ Factorise $9ab + 12bc$

$9ab + 12bc = 3b \times 3a + 3b \times 4c$
$\phantom{9ab + 12bc} = 3b(3a + 4c)$

Factorise

19. $2x^2 - 6x$
20. $2z^3 + 4z$
21. $25a^2 - 5a$
22. $12x^2 + 16x$
23. $5ab - 10bc$
24. $3y^2 - 27y$
25. $2a^2 - 12a$
26. $6p^2 + 2p$
27. $9y^2 - 6y$

**Worked example**

→ Factorise $ab + 2bc + bd$

$ab + 2bc + bd = b(a + 2c + d)$

Factorise

28. $2x^2 + 4x + 6$
29. $10a^2 - 5a + 20$
30. $ab + 4bc - 3bd$
31. $8x - 4y + 12z$

32  $9ab - 6ac - 3ad$

33  $3x^2 - 6x + 9$

34  $4a^2 + 8a - 4$

35  $5xy + 4xz + 3x$

36  $5ab + 10bc + 5bd$

37  $2xy - 4yz + 8yw$

## Worked example

→ Factorise $8x^3 - 4x^2$

$8x^3 - 4x^2 = 4x^2(2x - 1)$

 Sometimes we do not 'see' all the common factors to begin with. In this case, we may spot that 4 is a common factor and not 'see' the $x^2$, giving
$8x^3 - 4x^2 = 4(2x^3 - x^2)$
A check on the terms inside the bracket shows that there is another common factor, namely $x^2$,
so $\quad 8x^3 - 4x^2 = 4x^2(2x - 1)$

Remember to check that *all* the common factors have been removed from inside the bracket.

Factorise

38  $x^3 + x^2$

39  $x^2 - x^3$

40  $20a^2 - 5a^3$

41  $12x^3 - 16x^2$

42  $4x^4 + 12x^2$

43  $a^2 + a^3$

44  $b^3 - b^2$

45  $4x^3 - 2x^2$

46  $27a^2 - 18a^3$

47  $10x^2 - 15x^4$

48  $12x + 8$

49  $8x^2 + 12x$

50  $9x^2 - 6x + 12$

51  $5x^3 - 10x$

52  $8pq + 4qr$

53  $x^2 - 8x$

54  $12 + 9y^2$

55  $12xy + 16xz + 8x$

56  $4x^3 + 6x$

57  $12abc - 8bcd$

## Worked examples

→ Factorise $2\pi r^2 + 2\pi rh$

$2\pi r^2 + 2\pi rh = 2\pi r(r + h)$

→ Factorise $\tfrac{1}{2}Mu^2 - \tfrac{1}{2}mu^2$

$\tfrac{1}{2}Mu^2 - \tfrac{1}{2}mu^2 = \tfrac{1}{2}u^2(M - m)$

**58** $\frac{1}{2}ah + \frac{1}{2}bh$

**59** $mg - ma$

**60** $\frac{1}{2}mv^2 - \frac{1}{2}mu^2$

**61** $P + \frac{PRT}{100}$

**62** $2\pi r^2 + \pi rh$

**63** $\pi R^2 + \pi r^2$

**64** $2gh_1 - 2gh_2$

**65** $\frac{1}{2}mv^2 - mgh$

**66** $\frac{4}{3}\pi r^3 - \frac{1}{3}\pi r^2 h$

**67** $3\pi r^2 + 2\pi rh$

**68** $\frac{1}{2}mu^2 + \frac{1}{2}mv^2$

**69** $\frac{1}{2}bc - \frac{1}{4}ca$

### Factorising quadratic expressions

The type of expression we are most likely to want to factorise is one such as $x^2 + 7x + 10$.

This type of expression is called a **quadratic expression**. To factorise such an expression, we look for two brackets whose product is the original expression.

When we expanded $(x + 2)(x + 4)$ we had

$(x + 2)(x + 4) = x^2 + 6x + 8$

If we write $x^2 + 6x + 8 = (x + 2)(x + 4)$ we say we have factorised $x^2 + 6x + 8$,

i.e.   just as 10 is $2 \times 5$   so   $x^2 + 6x + 8$ is $(x + 2) \times (x + 4)$.

To factorise an expression of the form $x^2 + 7x + 10$, where all the terms are positive, we need to remind ourselves of the patterns we observed in Chapter 6.

We found when expanding brackets that:
- If the sign in each bracket is +, then the number term in the expansion is +.
- The $x^2$ term comes from $x \times x$.
- The number term in the expansion comes from multiplying the numbers in the brackets together.
- The middle term, or $x$ term in the expansion, comes from collecting the product of the outside terms in the brackets and the product of the inside terms in the brackets.

Using these ideas in reverse order

$x^2 + 7x + 10 = (x +\ \ \ )(x +\ \ \ )$
$\qquad\qquad\quad = (x + 2)(x + 5)$

 Choose two numbers whose product is 10 and whose sum is 7.
The other pair of numbers whose product is 10 is 1 and 10, but the sum of 1 and 10 is 11.

# 7 Algebraic factors

## Exercise 7b

### Worked example

Factorise $x^2 + 8x + 15$

$x^2 + 8x + 15 = (x + 3)(x + 5)$

 The product of 3 and 5 is 15, and their sum is 8. The other possible pair is 1 and 15, but $1 + 15 = 16$ not 8.

 Remember that $2 \times 3$ is the same as $3 \times 2$ so that $(x + 3)(x + 5)$ is the same as $(x + 5)(x + 3)$. The order in which the brackets are written does not matter.

Factorise

1. $x^2 + 3x + 2$
2. $x^2 + 6x + 5$
3. $x^2 + 7x + 12$
4. $x^2 + 8x + 15$
5. $x^2 + 21x + 20$
6. $x^2 + 8x + 7$
7. $x^2 + 8x + 12$
8. $x^2 + 13x + 12$
9. $x^2 + 16x + 15$
10. $x^2 + 12x + 20$
11. $x^2 + 8x + 16$
12. $x^2 + 15x + 36$
13. $x^2 + 19x + 18$
14. $x^2 + 22x + 40$
15. $x^2 + 9x + 8$
16. $x^2 + 6x + 9$
17. $x^2 + 20x + 36$
18. $x^2 + 9x + 18$
19. $x^2 + 11x + 30$
20. $x^2 + 14x + 40$

To factorise an expression of the form $x^2 - 6x + 8$, remember the pattern:
- The numbers in the brackets must multiply to give $+8$, so they must have the same sign. Since the middle term in the expression is $-$, they must both be $-$.
- The $x^2$ term comes from $x \times x$.
- The middle term, or $x$ term, comes from collecting the product of the outside terms and the product of the inside terms.

Thus $x^2 - 6x + 8 = (x - 2)(x - 4)$

since $(-2) \times (-4) = +8$

and $x \times (-4) + (-2) \times x = -4x - 2x = -6x$

145

### Exercise 7c

**Worked example**

→ Factorise $x^2 - 7x + 12$

$x^2 - 7x + 12 = (x - 3)(x - 4)$

 The product of $-3$ and $-4$ is $+12$.
The outside product is $-4x$ and the inside product is $-3x$.
Collecting these gives $-7x$.
Other pairs looked at and discarded are $-2$ and $-6$, and $-1$ and $-12$.

Factorise

1. $x^2 - 9x + 8$
2. $x^2 - 7x + 12$
3. $x^2 - 17x + 30$
4. $x^2 - 11x + 28$
5. $x^2 - 13x + 42$
6. $x^2 - 5x + 6$
7. $x^2 - 16x + 15$
8. $x^2 - 6x + 9$
9. $x^2 - 18x + 32$
10. $x^2 - 16x + 63$

Similarly $\quad x^2 + x - 12 = (x + 4)(x - 3)$

If the number term in the expansion is negative, the signs in the brackets are different.

Thus $\quad (+4) \times (-3) = -12$

Working as before, the product of the outside terms is $-3x$ and the product of the inside term is $+4x$

Therefore the total is $+x$.

Similarly $\quad x^2 + 2x - 15 = (x + 5)(x - 3)$
or $\quad\quad\quad\;\; x^2 + 2x - 15 = (x - 3)(x + 5)$

### Exercise 7d

Factorise

1. $x^2 - x - 6$
2. $x^2 + x - 20$
3. $x^2 - x - 12$
4. $x^2 + 3x - 28$
5. $x^2 + 2x - 15$
6. $x^2 - 2x - 24$
7. $x^2 + 6x - 27$
8. $x^2 - 9x - 22$
9. $x^2 - 2x - 35$
10. $x^2 - 8x - 20$

Most of the values in the previous three exercises have been easy to spot. Should you have difficulty, set out all possible pairs of numbers, as shown below, until you find the pair that gives the original expression when you multiply back.

Factorise

**a** $x^2 - 11x + 24$

| Possible factors | | Sum |
|---|---|---|
| $-1$ | $-24$ | $-25$ |
| $-2$ | $-12$ | $-14$ |
| $-3$ | $-8$ | $-11$ |

(Because the number term is $+$, the two numbers in the brackets must have the same sign.)

$\therefore \quad x^2 - 11x + 24 = (x - 3)(x - 8)$

**b** $x^2 + 5x - 24$

| Possible factors | | Sum |
|---|---|---|
| $-1$ | $+24$ | $+23$ |
| $-2$ | $+12$ | $+10$ |
| $-3$ | $+8$ | $+5$ |

(Because the number term is $-$, the two numbers in the brackets have different signs.)

$\therefore \quad x^2 + 5x - 24 = (x - 3)(x + 8)$

Remember that $+$ before the number term means that the signs in the brackets are the same, whereas $-$ before the number term means that they are different.

## Exercise 7e

### Worked example

Factorise $x^2 + 13x + 36$

$x^2 + 13x + 36 = (x + 4)(x + 9)$

The possible pairs of numbers whose product is 36 are $1 \times 36$, $2 \times 18$, $3 \times 12$, $4 \times 9$ and $6 \times 6$. 4 and 9 is the only pair that gives a sum of 13.

Factorise

**1** $x^2 + 9x + 14$

**2** $x^2 - 10x + 21$

**3** $x^2 + 5x - 14$

**4** $x^2 + x - 30$

**5** $x^2 + 9x + 8$

**6** $x^2 - 10x + 25$

**7** $x^2 + 8x - 9$

**8** $x^2 - 15x + 26$

**9** $x^2 + x - 56$

**10** $x^2 + 32x + 60$

**11** $x^2 - 6x - 27$

**12** $x^2 + 16x - 80$

**13** $x^2 + 14x + 13$

**14** $x^2 + 12x - 28$

**15** $x^2 + 2x - 80$

**16** $x^2 - 11x + 30$

**17** $x^2 + 8x - 48$

**18** $x^2 + 18x + 72$

**19** $x^2 + 17x + 52$

**20** $x^2 - 12x - 28$

**21** $x^2 + 11x + 24$

**22** $x^2 - 11x - 42$

**23** $x^2 - 18x + 32$

**24** $x^2 - 7x - 60$

## Worked example

➔ Factorise $6 + x^2 - 5x$

$6 + x^2 - 5x = x^2 - 5x + 6$
$\phantom{6 + x^2 - 5x} = (x - 2)(x - 3)$

 This needs to be rearranged into the familiar form, i.e. $x^2$ term first, then the $x$ term and finally the number.

 Possible pairs:
1, 6, sum 7; reject
2, 3, sum 5; correct

Factorise

**25** $8 + x^2 + 9x$

**26** $9 + x^2 - 6x$

**27** $11x + 28 + x^2$

**28** $x - 20 + x^2$

**29** $9 + x^2 + 6x$

**30** $8 + x^2 - 9x$

**31** $17x + 30 + x^2$

**32** $6x - 27 + x^2$

**33** $x^2 + 22 + 13x$

**34** $x^2 - 11x - 26$

**35** $7 + x^2 - 8x$

**36** $x + x^2 - 42$

**37** $x^2 - 5x - 24$

**38** $14 + x^2 - 9x$

**39** $28x + 27 + x^2$

**40** $2x - 63 + x^2$

## Worked example

➔ Factorise $x^2 + 6x + 9$

$x^2 + 6x + 9 = (x + 3)(x + 3)$
$\phantom{x^2 + 6x + 9} = (x + 3)^2$

 If you cannot see the numbers required, write down all the pairs whose product is 9.

 $3 \times 3$ or $1 \times 9$

Factorise

**41** $x^2 + 10x + 25$

**42** $x^2 - 10x + 25$

**43** $x^2 + 4x + 4$

**44** $x^2 - 14x + 49$

**45** $x^2 + 12x + 36$

**46** $x^2 - 12x + 36$

**47** $x^2 - 4x + 4$

**48** $x^2 + 16x + 64$

## Exercise 7f

### Worked example

Factorise $6 - 5x - x^2$

When the $x^2$ term is negative, the terms should be arranged: number term, then the $x$ term and finally the $x^2$ term. This means that the $x$ term appears at the end of each bracket.

$6 - 5x - x^2 = (6 + x)(1 - x)$

$2 \times 3$ or $6 \times 1$

Factorise

1. $2 - x - x^2$
2. $6 + x - x^2$
3. $4 - 3x - x^2$
4. $8 + 2x - x^2$
5. $6 - x - x^2$
6. $2 + x - x^2$
7. $8 - 2x - x^2$
8. $5 - 4x - x^2$
9. $10 - 3x - x^2$
10. $12 + 4x - x^2$
11. $5 + 4x - x^2$
12. $14 - 5x - x^2$
13. $6 + 5x - x^2$
14. $20 - x - x^2$
15. $15 - 2x - x^2$
16. $12 + x - x^2$

### The difference between two squares

In Chapter 6, one of the expansions we listed was
$$(x + a)(x - a) = x^2 - a^2$$

If we reverse this, we have
$$x^2 - a^2 = (x + a)(x - a)$$
or
$$x^2 - a^2 = (x - a)(x + a)$$

(The order of multiplication of two brackets makes no difference to the result.)

This result is known as factorising the **difference between two squares** and is very important.

When factorising, do not confuse $x^2 - 4$ with $x^2 - 4x$.
$$x^2 - 4 = (x + 2)(x - 2)$$
whereas $\quad x^2 - 4x = x(x - 4) \quad$ ($4x$ is *not* a square)

# STP Maths 9

## Exercise 7g

**Worked example**

→ Factorise $x^2 - 9$

$x^2 - 9 = x^2 - 3^2$
$\phantom{x^2 - 9} = (x + 3)(x - 3)$ 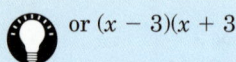 or $(x - 3)(x + 3)$

Factorise

1. $x^2 - 25$
2. $x^2 - 4$
3. $x^2 - 100$
4. $x^2 - 1$
5. $x^2 - 64$
6. $x^2 - 16$
7. $x^2 - 36$
8. $x^2 - 81$
9. $x^2 - 49$

**Worked example**

→ Factorise $4 - x^2$

$4 - x^2 = 2^2 - x^2$
$\phantom{4 - x^2} = (2 + x)(2 - x)$ 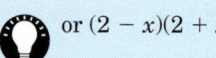 or $(2 - x)(2 + x)$

Factorise

10. $9 - x^2$
11. $36 - x^2$
12. $100 - x^2$
13. $a^2 - b^2$
14. $9y^2 - z^2$
15. $16 - x^2$
16. $25 - x^2$
17. $81 - x^2$
18. $x^2 - y^2$

We began this chapter by considering common factors. The next exercise starts with a reminder of how to extract common factors, followed by factorising expressions such as $2x^2 - 8x - 10$ where 2 is a common factor.

## Exercise 7h

**Worked example**

→ Factorise $12x - 6$

$12x - 6 = 6(2x - 1)$

# 7 Algebraic factors

Factorise

1. $3x + 12$
2. $25x^2 + 10x$
3. $12x^2 - 8$
4. $14x + 21$
5. $4x^2 + 2$
6. $21x - 7$
7. $9x^2 - 18x$
8. $20x + 12$
9. $8x^2 - 4x$

**Worked example**

→ Factorise $2x^2 - 8x - 10$

$2x^2 - 8x - 10 = 2(x^2 - 4x - 5)$
$\qquad\qquad\qquad = 2(x - 5)(x + 1)$

 Now check to see if the quadratic expression factorises.

Factorise

10. $3x^2 + 12x + 9$
11. $5x^2 - 15x - 50$
12. $4x^2 + 8x - 32$
13. $3x^2 - 12$
14. $2x^2 - 18x + 28$
15. $4x^2 - 24x + 20$
16. $3x^2 + 18x + 24$
17. $5x^2 - 45$
18. $3x^2 - 12x - 63$
19. $18 - 3x - 3x^2$

## Calculations using factorising

**Exercise 7i**

**Worked example**

→ Find $1.7^2 + 0.3 \times 1.7$

$1.7^2 + 0.3 \times 1.7 = 1.7(1.7 + 0.3)$
$\qquad\qquad\qquad\quad = 1.7 \times 2$
$\qquad\qquad\qquad\quad = 3.4$

Find, without using a calculator

1. $2.5^2 + 0.5 \times 2.5$
2. $1.3 \times 3.7 + 3.7^2$
3. $5.9^2 - 2.9 \times 5.9$
4. $8.76^2 - 4.76 \times 8.76$
5. $5.2^2 + 0.8 \times 5.2$
6. $2.6 \times 3.4 + 3.4^2$
7. $4.3^2 - 1.3 \times 4.3$
8. $16.27^2 - 5.27 \times 16.27$

## Worked example

→ Find $100^2 - 98^2$

$$100^2 - 98^2 = (100 + 98)(100 - 98)$$
$$= 198 \times 2$$
$$= 396$$

Find, without using a calculator

**9** $55^2 - 45^2$

**10** $20.6^2 - 9.4^2$

**11** $7.82^2 - 2.82^2$

**12** $2.667^2 - 1.333^2$

**13** $10.2^2 - 9.8^2$

**14** $13.5^2 - 6.5^2$

**15** $8.79^2 - 1.21^2$

**16** $0.763^2 - 0.237^2$

## Mixed quadratic expressions

Some quadratic expressions such as $x^2 + 9$ and $x^2 + 3x + 1$ will not factorise. The next exercise in this chapter includes some expressions that will not factorise.

### Exercise 7j

Factorise, where possible

**1** $x^2 + 13x + 40$

**2** $x^2 - 11x + 18$

**3** $x^2 - 36$

**4** $x^2 + 4$

**5** $x^2 - 8x + 12$

**6** $x^2 - 11x - 10$

**7** $x^2 + 6x - 7$

**8** $x^2 + 13x - 30$

**9** $x^2 - 11x + 24$

**10** $x^2 + 11x + 12$

**11** $x^2 + 14x - 15$

**12** $28 - 12x - x^2$

**13** $x^2 + 8x + 12$

**14** $x^2 - x - 30$

**15** $x^2 - 49$

**16** $x^2 - 7x + 2$

**17** $x^2 - 7x - 10$

**18** $x^2 + 13x + 42$

**19** $x^2 - 9$

**20** $x^2 - 10x + 24$

**21** $x^2 + 13x - 68$

**22** $x^2 + 11x - 26$

**23** $a^2 - 16a + 63$

**24** $28 + 3x - x^2$

# Mixed exercises

## Exercise 7k

1. Factorise    **a**   $10a + 20$    **b**   $15p^2 - 10p$
2. Factorise    **a**   $4ab - 8bc$    **b**   $5b^2 + 15b - 5$
3. Factorise    **a**   $a^2 + 9a + 18$    **b**   $x^2 - 7x - 8$
4. Factorise    **a**   $21 + 10x + x^2$    **b**   $10 - 7x + x^2$
5. Factorise    **a**   $a^2 - 36$    **b**   $16 - x^2$
6. Find, without using a calculator
      **a**   $3.7^2 + 1.3 \times 3.7$    **b**   $7.7 \times 2.3 + 2.3^2$

## Exercise 7l

1. Factorise    **a**   $8z^3 - 4z^2$    **b**   $5xy - 20yz$
2. Factorise    **a**   $7a - a^2$    **b**   $x^2 - 6x - 27$
3. Factorise    **a**   $x^2 + 12x + 35$    **b**   $2a^2 - 6a - 8$
4. Factorise    **a**   $100 - x^2$    **b**   $x^2 - 9$
5. Factorise    **a**   $7x - 8 + x^2$    **b**   $a^2 - 14a + 49$
6. Find, without using a calculator
      **a**   $10.3^2 - 9.7^2$    **b**   $0.643^2 - 0.357^2$

## Exercise 7m

1. Factorise    **a**   $12z^2 - 6z$    **b**   $8xy - 12yz$
2. Factorise    **a**   $x^2 + 10x + 25$    **b**   $x^2 - 2x - 24$
3. Factorise    **a**   $a^2 + a - 6$    **b**   $x^2 + 7x - 44$
4. Factorise    **a**   $b^2 - 49$    **b**   $16p^2 - p$
5. Factorise    **a**   $30 - 17x + x^2$    **b**   $12 - x - x^2$
6. Find, without using a calculator
      **a**   $13.2 \times 6.8 + 13.2^2$    **b**   $997^2 - 797^2$

## Consider again

Now do you know how to quickly find the value of $8.8^2 - 1.2^2$ without using a calculator?

# 8 Organising and summarising data

**Consider**

The local health authority wants some information about the heights of five-year-old children in their area. The information is obtained from the first school medical examination, when the height of each child is recorded.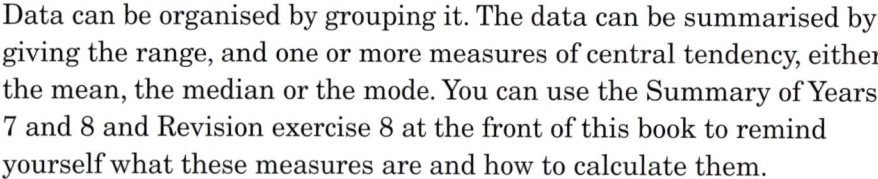

- There are 1256 five-year-olds in primary schools in the area. What does the health authority need to do to the data in order to obtain useful information from it?
- The health authority may want to compare the distribution of heights in its area with those in other health authority areas. What measures can it usefully use to compare different sets of information? Give reasons for your answers.

*You should be able to answer these questions after you have worked through this chapter.*

Data can be organised by grouping it. The data can be summarised by giving the range, and one or more measures of central tendency, either the mean, the median or the mode. You can use the Summary of Years 7 and 8 and Revision exercise 8 at the front of this book to remind yourself what these measures are and how to calculate them.

**Class discussion**

1. In Area A, the height of the shortest child is 92 cm and that of the tallest child is 112 cm. The range of heights is therefore 20 cm.

   In Area B, the range of heights is also 20 cm.

   a Does this mean that the height of the shortest child in Area B is 92 cm?

   b What can you say about the height of the tallest child in Area B?

   c Is the range on its own a good way of describing a set of data?

   d An unusually tall five-year-old joined a primary school in Area B. As a consequence, the range of heights in this area increased to 30 cm. Is it reasonable to use the ranges alone to compare the heights in Area B with those in Area A?

2. The first three students entering a classroom had with them 2 books, 2 books and 8 books respectively. The next three students entering the room had with them 3 books, 4 books and 5 books respectively.

   a For the first group of students, write down the mean, median and the range of the number of books.

**b** Repeat part **a** for the second group of students.

**c** Is the mean number of books, on its own, a satisfactory way of describing either set?

**d** If the mean and the range are used, is this a better way to describe a set of data?

**e** Is there any advantage in using the median instead of the mean?

**f** The numbers of books brought in by the first three students lie in the interval shown on this number line.

```
├──┼──┼──┼──┼──┼──┼──┼──┼──┤
0  1  2  3  4  5  6  7  8  9
```

For a third group of three students who enter the room, the range is 4 books. Try drawing this range on a number line. What problems do you have?

**f** For the third group of students, the median number of books is 4. Does this help to locate the range on a number line?

**3** The health authority wanted to investigate the possible long-term harm that students may suffer by carrying heavy loads to and from school in unsuitable bags. They made a start by gathering information about the weights carried by 800 secondary school students.

The data collected made it possible to ask questions such as

What is the range of weights?
What is the mean weight?
What is the range of the lowest 90% of these weights?

Discuss how you could go about answering these questions and what problems you might have.

## Analysing large sets of information

Your discussion of the questions in the class discussion should show that, to give a reasonable summary of a set of data, we need to use either the mean or the median together with the range. It should also show that the range has disadvantages as a measure of spread; other ways of describing the spread of a distribution are discussed later in this chapter.

In your discussion of question **3**, you may have concluded that it is difficult to carry out the calculations needed to analyse a large set of data unless the data is entered into a spreadsheet, or other statistical computer program.

Consider the following figures, collected on behalf of the local health authority.

**Heights (rounded down to the nearest centimetre) of 90 five-year-olds from one primary school**

| 99  | 107 | 102 | 98  | 115 | 95  | 106 | 110 | 108 | 105 | 118 | 102 | 114 | 108 | 94  |
| --- | --- | --- | --- | --- | --- | --- | --- | --- | --- | --- | --- | --- | --- | --- |
| 104 | 113 | 102 | 105 | 95  | 105 | 110 | 109 | 101 | 106 | 108 | 107 | 107 | 101 | 109 |
| 105 | 116 | 109 | 114 | 110 | 97  | 110 | 113 | 116 | 112 | 101 | 92  | 105 | 104 | 115 |
| 111 | 103 | 110 | 99  | 93  | 104 | 103 | 113 | 107 | 94  | 102 | 117 | 116 | 104 | 99  |
| 114 | 106 | 114 | 98  | 109 | 107 | 114 | 106 | 107 | 109 | 113 | 112 | 100 | 109 | 113 |
| 118 | 104 | 94  | 114 | 107 | 96  | 108 | 103 | 112 | 106 | 115 | 111 | 115 | 101 | 108 |

These figures were written on record cards in the same order as the children came into the medical examination, so the heights are listed in a random order. Disorganised figures like these are called **raw data**. Some form of summary, such as the mean, the median, the mode and the range, are needed to describe these figures.

- If the mean height of five-year-olds in one school is required, it can be found from the raw data. For the 90 heights given, the mean can be calculated by adding up the heights and then dividing by 90; this is a tedious job and, even with the help of a calculator, mistakes are likely. With 1256 heights, it is not sensible to use the raw data unless they have been entered into, say, a spreadsheet that can do the calculations.
- Grouping the data not only helps to give a 'picture' of the distribution of heights, it also reduces the complexity by replacing hundreds of individual figures with a much smaller number of groups of figures. However, it does reduce the amount of detail given by the individual figures.

First we will organise the data into groups. If we use a number line to represent the height of these children, there is no point on the line which could not represent someone's height, that is, heights are continuous data. So the grouping we choose must not have any 'gaps' between the values included in consecutive groups.

Taking $h$ cm to represent the height of any child, a suitable grouping is

$90 \leq h < 95$, $95 \leq h < 100$, $100 \leq h < 105$, $105 \leq h < 110$, $110 \leq h < 115$, $115 \leq h < 120$

and these are used to construct the following **frequency** table.

| Height, $h$ (cm) | Tally | Frequency, $f$ |
| --- | --- | --- |
| $90 \leq h < 95$ | \|\|\|\| | 5 |
| $95 \leq h < 100$ | \|\|\|\| \|\|\|\| | 9 |
| $100 \leq h < 105$ | \|\|\|\| \|\|\|\| \|\|\|\| \|\| | 17 |
| $105 \leq h < 110$ | \|\|\|\| \|\|\|\| \|\|\|\| \|\|\|\| \|\|\|\| \|\|\| | 28 |
| $110 \leq h < 115$ | \|\|\|\| \|\|\|\| \|\|\|\| \|\|\|\| \| | 21 |
| $115 \leq h < 120$ | \|\|\|\| \|\|\|\| | 10 |
| Total | | 90 |

 When you make a frequency table from raw data, work down the columns, making a tally mark in the appropriate row for each value. Do not go through the data looking for values that fit into the first group and the second group and so on.

Now we can see that the modal group is 105 cm to 110 cm. We can estimate the range of height as

(upper bound of last group − lower bound of first group) = (120 − 90) cm = 30 cm

## Stem-and-leaf diagrams

Placing the data in groups loses some of the detail. For example, we can give the modal group but not the mode from the table on the previous page. Another way of organising data is to draw a **stem-and-leaf diagram**. This also groups the data but preserves the detail of individual figures.

This is a list of the number of pages in twenty-five text books.

201, 325, 188, 410, 241, 377, 506, 309, 220, 162, 386, 180, 424, 258, 275, 463, 336, 174, 310, 379, 510, 234, 371, 292, 380

The groups form the stem: the number of pages are all between 100 and 600, so we use the number of hundreds as the numbers for the stem, i.e, 1, 2, 3, 4, and 5. These are written down the left-hand side.

The leaves are the corresponding tens and units, and these are written on the right, next to the appropriate stem. Therefore, 201 is represented by 2 in the stem and 01 in the leaf next to the stem.

Start by working across the rows, marking each number of tens and units in the appropriate place, but do not attempt to order them.

| Stem | Leaves |
|---|---|
| 1 | 88  62  80  74 |
| 2 | 01  41  20  58  75  34  92 |
| 3 | 25  77  09  86  36  10  79  71  80 |
| 4 | 10  24  63 |
| 5 | 06  10 |

1 | 88 means 188, 2 | 20 means 220, 4 | 63 means 463, etc.

Next redraw the diagram with the numbers in order of size and provide a key.

| Number of pages | 1 | 62 means 162 |
|---|---|
| 1 | 62  74  80  88 |
| 2 | 01  20  34  41  58  75  92 |
| 3 | 09  10  25  36  71  77  79  80  86 |
| 4 | 10  24  63 |
| 5 | 06  10 |

Now we can see that there is no mode (all the numbers are different). We can find the median: there are 25 numbers so the median is the 13th number. This is 310, so the median number of pages is 310. We can also find the range: the least number of pages is 162 and the greatest number of pages is 510, so the range is 510 − 162 = 348.

# STP Maths 9

### Exercise 8a

1. Use the stem-and-leaf diagram on the previous page, giving the number of pages in 25 books, to answer these questions.
   a. How many books have more than 250 pages?
   b. How many books have fewer than 350 pages?

2. This stem-and-leaf diagram gives the heights from the list on page 156.

   | Stem | Leaves        9 \| 2 means 92 |
   |------|-------------------------------|
   | 9    | 2 3 4 4 4 5 5 6 7 8 8 9 9 9 |
   | 10   | 0 1 1 1 1 2 2 2 2 3 3 3 3 4 4 4 4 4 5 5 5 5 5 6 6 6 6 6 7 7 7 7 7 7 8 8 8 8 8 9 9 9 9 9 9 |
   | 11   | 0 0 0 0 0 1 1 2 2 2 3 3 3 3 3 4 4 4 4 4 4 5 5 5 5 6 6 6 7 8 8 |

   a. What is the modal height?
   b. What is the median height?
   c. What is the range of heights?
   d. How many of these five-year-olds are less than 103 cm tall?

## Finding the mean of an ungrouped frequency distribution

Next we need to develop methods for finding the mean and the median from grouped data. Firstly, we will remind ourselves how to find the mean of an ungrouped frequency distribution.

In Book 7, we found the mean value of a frequency distribution by multiplying each value by its frequency, adding these products and dividing the result by the total number of values, that is, the sum of the frequencies. This example (from Book 7) summarises the process.

The symbol $\Sigma$ means 'the sum of all items such as'.

**Test marks**

| Mark, $x$ | Frequency, $f$ | Frequency × mark, $fx$ |
|-----------|----------------|------------------------|
| 0 | 1 | 0 |
| 1 | 1 | 1 |
| 2 | 8 | 16 |
| 3 | 11 | 33 |
| 4 | 5 | 20 |
| 5 | 4 | 20 |
|   | Total, $\Sigma f$, = 30 | Total, $\Sigma fx$, = 90 |

The mean mark is given by $\dfrac{\Sigma fx}{\Sigma f} = \dfrac{90}{30} = 3$.

This method can be used to find the mean of any ungrouped frequency distribution, i.e.

**the mean value of a frequency distribution is given by**

$$\dfrac{\Sigma fx}{\Sigma f}$$

where $x$ is the value of an item and $f$ is its frequency.

## Finding the mean of a grouped frequency distribution

From the table below we can see that 5 children had heights, $h$ cm, in the range $90 \leq h < 95$.

| Height, $h$ (cm) | Frequency, $f$ |
|---|---|
| $90 \leq h < 95$ | 5 |
| $95 \leq h < 100$ | 9 |
| $100 \leq h < 105$ | 17 |
| $105 \leq h < 110$ | 28 |
| $110 \leq h < 115$ | 21 |
| $115 \leq h < 120$ | 10 |

 If we assume that the mean height of these five children is halfway between 90 cm and 95 cm, i.e. 92.5 cm, then we can estimate the total heights of the 5 children as $92.5 \times 5$ cm $= 462.5$ cm.

The middle value of a group is called the **midclass value**.

Using the midclass value as an estimate for the mean value in each group, we can find (approximately) the total height of the children in each group and hence the total height of all 90 five-year-olds.

It is easier to keep track of the calculations if we add another two columns to the frequency table.

| Height, $h$ (cm) | Frequency, $f$ | Midclass value, $x$ | $fx$ |
|---|---|---|---|
| $90 \leq h < 95$ | 5 | 92.5 | 462.5 |
| $95 \leq h < 100$ | 9 | 97.5 | 877.5 |
| $100 \leq h < 105$ | 17 | 102.5 | 1742.5 |
| $105 \leq h < 110$ | 28 | 107.5 | 3010 |
| $110 \leq h < 115$ | 21 | 112.5 | 2362.5 |
| $115 \leq h < 120$ | 10 | 117.5 | 1175 |
| Totals | $\Sigma f = 90$ | | $\Sigma fx = 9630$ |

The total height of all 50 children is estimated as 9630 cm, so the mean height is approximately

$$\frac{9630}{90} \text{ cm} = 107 \text{ cm}$$

Remember that this calculation is based on the assumption that the average height in each group is halfway through the group, so what we have found is an estimate for the mean.

This process can be used with any grouped frequency distribution, so

**the estimated mean value of a grouped frequency distribution is given by**

$$\frac{\Sigma fx}{\Sigma f}$$

**where $x$ is the midclass value and $f$ is the frequency of items in the group.**

# STP Maths 9

## Exercise 8b

1. Fifty boxes of peaches were examined and the number of bad peaches in each box was recorded, with the following result. Estimate the mean number of bad peaches per box.

| No. of bad peaches per box | 0–4 | 5–9 | 10–14 | 15–19 |
|---|---|---|---|---|
| Frequency | 34 | 11 | 4 | 1 |

2. Twenty tomato seeds were planted in a seed tray. Four weeks later, the heights of the resulting plants were measured and the following frequency table was made. Estimate the mean height of the seedlings.

| Height, $h$ (cm) | $1 \leqslant h < 4$ | $4 \leqslant h < 7$ | $7 \leqslant h < 10$ | $10 \leqslant h < 13$ |
|---|---|---|---|---|
| Frequency | 2 | 5 | 10 | 3 |

3. The table shows the result of a survey of 100 students on the amount of money each of them spent in the school tuck shop on one particular day. Find an estimate for the mean amount of money spent.

| Amount (pence) | 0–24 | 25–49 | 50–74 | 75–99 |
|---|---|---|---|---|
| Frequency | 26 | 15 | 38 | 21 |

4. The bar chart shows the result of an examination of 20 boxes of screws.

   Make a frequency table and estimate the mean number of defective screws per box.

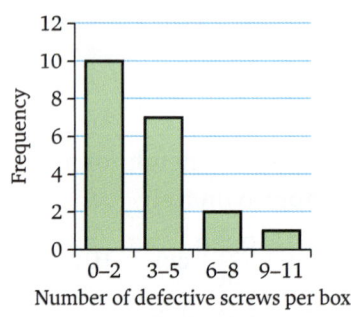

5. The table shows the distribution of heights of 100 adult females, rounded down to the nearest centimetre. Estimate the mean height.

| Height, $h$ (cm) | $145 \leqslant h < 150$ | $150 \leqslant h < 155$ | $155 \leqslant h < 160$ | $160 \leqslant h < 165$ | $165 \leqslant h < 170$ | $170 \leqslant h < 175$ |
|---|---|---|---|---|---|---|
| Frequency | 2 | 6 | 42 | 36 | 10 | 4 |

6  A new income tax form was trialled by asking some people to complete it. The time each person took was recorded and the frequency polygon summarises the results.
   a  How many people were asked to complete the form?
   b  Estimate the range of times taken.
   c  Copy and complete this table.

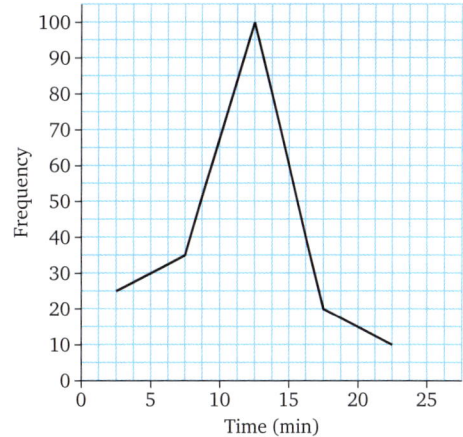

| Midclass value, $t$ (minutes) | 2.5 | | | |
|---|---|---|---|---|
| Frequency | 25 | | | |

   d  Estimate the mean time taken to complete the form.

# The median

The table on page 156 shows the heights of 90 five-year-olds collected from one primary school by a local health authority. These are some of the 1256 heights collected from all primary schools in the area.
- The authority may want to use these figures to find the median height, which is the height matched or exceeded by half the five-year-olds.
- It may also want answers to questions such as 'What proportion of five-year-olds are at least 105 cm tall?' or 'Below what height are the shortest 25% of the children?'

In order to answer these questions, the heights need to be arranged in ascending order. Without the help of a computer, this is a time-consuming task for 90 figures; for 1256 figures it is daunting.

# Running totals

The proportion of the 90 children in one school who are at least 105 cm tall can be found from the grouped frequency table by working out a 'running total'. This means finding the total number of heights below 95 cm, the total number of heights below 100 cm, and so on. We add another column to the table to show the running totals.

| Height, $h$ (cm) | Frequency, $f$ | Running total |
|---|---|---|
| $90 \leq h < 95$ | 5 | 5 heights below 95 cm |
| $95 \leq h < 100$ | 9 | 5 + 9, i.e. 14, heights below 100 cm |
| $100 \leq h < 105$ | 17 | 14 + 17, i.e. 31, heights below 105 cm |
| $105 \leq h < 110$ | 28 | 31 + 28, i.e. 59, heights below 110 cm |
| $110 \leq h < 115$ | 21 | … |
| $115 \leq h < 120$ | 10 | |

The last column shows that 31 out of 90 children are less than 105 cm tall, i.e. $\frac{31}{90}$ = 34.4% are less than 105 cm tall, so 65.6% are taller than 105 cm.

## Exercise 8c

1. The following table shows the separate subject results achieved by a certain student, with the running total given in the fourth column.

| Lesson | Subject | Mark | Running total |
|---|---|---|---|
| 1 | Physics | 54 | 54 |
| 2 | French | 72 | 126 |
| 3 | Biology | 62 | |
| 4 | Chemistry | 45 | |
| 5 | History | 78 | |
| 6 | Mathematics | 64 | |
| 7 | English | 45 | |
| 8 | Geography | 82 | |

Copy and complete the table.

2. The table shows the running totals of students who have school lunch each day during a certain school week. Complete the table to find out how many had lunch on each day.

| Weekday | Number of lunches served each day | Running total of lunches served |
|---|---|---|
| Monday | | 126 |
| Tuesday | | 280 |
| Wednesday | | 424 |
| Thursday | | 599 |
| Friday | | 717 |

3. The mile-posts along the M4 motorway show that the distances, in miles, between various places are as follows.

   Cardiff to Newport 10.
   Newport to Severn Bridge 16.
   Severn Bridge to Leigh Delamere 28.
   Leigh Delamere to Swindon 18.
   Swindon to Reading 39.
   Reading to Heathrow Airport 28.
   Heathrow Airport to Central London 15.

   a  Make a running total of the distances along the motorway from Cardiff to Central London.

   b  Use your table to find the distance from
   i  Swindon to Heathrow Airport
   ii  Newport to Reading.

4   During a week's holiday a family spent the following amounts on snacks.

|  | Amount spent | Running total of expenditure |
|---|---|---|
| Monday | £25 | |
| Tuesday | £48 | |
| Wednesday | £8 | |
| Thursday | £55 | |
| Friday | £34 | |
| Saturday | £15 | |
| Sunday | £5 | |

a   Copy and complete the table.
b   How much in total did the family spend in the week on snacks?
c   By which day had they spent over half of their total expenditure on snacks?

## Cumulative frequency

The running total of frequencies is called the **cumulative frequency**.

A **cumulative frequency table** is constructed by adding each frequency to the sum of all those that have gone before it.

The cumulative frequency table for the 90 heights of five-year-old children can be constructed as follows.

| Height, $h$ (cm) | Frequency, $f$ | Height, $h$ (cm) | Cumulative frequency |
|---|---|---|---|
| $90 \leqslant h < 95$ | 5 | $h < 95$ | 5 |
| $95 \leqslant h < 100$ | 9 | $h < 100$ | (5 + 9 =) 14 |
| $100 \leqslant h < 105$ | 17 | $h < 105$ | (14 + 17 =) 31 |
| $105 \leqslant h < 110$ | 28 | $h < 110$ | (31 + 28 =) 59 |
| $110 \leqslant h < 115$ | 21 | $h < 115$ | (59 + 21 =) 80 |
| $115 \leqslant h < 120$ | 10 | $h < 120$ | (80 + 10 =) 90 |

Even if the second column is omitted, we can find the frequency of heights in any group from the cumulative frequencies.

For example, the number of heights in the group $105 \leqslant h < 110$ is given by the cumulative frequency up to 110 cm minus the cumulative frequency up to 105 cm, that is, $59 - 31 = 28$.

Notice that we can use the last number in the cumulative frequency column as a check on accuracy. It confirms that the total number of heights is 90.

One of the questions the health authority may want to answer is 'What proportion of the children are shorter than 115 cm?'

# STP Maths 9

From the table on page 163, we can see that 80 of the 90 children are less than 115 cm tall, that is, $\frac{80}{90} = \frac{8}{9} = 88\%$ (correct to 2 s.f.) are shorter than 115 cm.

If the health authority needs to know the percentage of children that are taller than 115 cm, then, since 88% are shorter than 115 cm,

(100 − 88), i.e. 12%, are taller than 115 cm.

## Exercise 8d

Keep your tables for questions **1** to **3** because you will need them for the next exercises.

**1** Copy and complete the following table, which shows the distribution of goals scored by the home sides in a football league one Saturday.

| Score | Frequency | Score | Cumulative frequency |
|---|---|---|---|
| 0 | 3 | ⩽ 0 | 3 |
| 1 | 8 | ⩽ 1 | 3 + 8 = 11 |
| 2 | 4 | ⩽ 2 | |
| 3 | 3 | ⩽ 3 | |
| 4 | 5 | ⩽ 4 | |
| 5 | 2 | ⩽ 5 | |
| 6 | 1 | ⩽ 6 | |

**a** How many matches were played?

**b** In how many matches were 3 or more goals scored by the home side?

**2** Copy and complete the following table, which shows the distribution of the marks scored by Year 7 students in their English test.

| Mark | Frequency (no. of students' scores within each range) | Mark | Cumulative frequency |
|---|---|---|---|
| 1–10 | 7 | ⩽ 10 | |
| 11–20 | 14 | ⩽ 20 | |
| 21–30 | 18 | ⩽ 30 | |
| 31–40 | 33 | ⩽ 40 | |
| 41–50 | 36 | ⩽ 50 | |
| 51–60 | 43 | ⩽ 60 | |
| 61–70 | 21 | ⩽ 70 | |
| 71–80 | 15 | ⩽ 80 | |
| 81–90 | 8 | ⩽ 90 | |
| 91–100 | 5 | ⩽ 100 | |

a How many Year 7 students are there?
b How many scored 50 or less?
c How many scored more than 60?
d Can you say how many students scored 75? Explain your answer.
e If you were asked to give the number of students who scored less than 55, what difficulties would you have in providing an answer?

3 The table is based on a cricketer's scores in each innings during one season. Copy and complete the table to show the cumulative frequencies.

| Score | 0–19 | 20–39 | 40–59 | 60–79 | 80–99 | 100–119 | 120–139 |
|---|---|---|---|---|---|---|---|
| Frequency | 8 | 14 | 33 | 6 | 5 | 3 | 1 |
| Score | ⩽ 19 | ⩽ 39 | ⩽ 59 | ⩽ 79 | ⩽ 99 | ⩽ 119 | ⩽ 139 |
| Cumulative frequency | | | | | | | |

a How many innings did he play?
b In how many innings did he score less than 60?
c In how many innings did he score at least 40?

4 A school is organising a Grand Prize Draw to raise money to buy a mini-bus. Tickets are sold at 50 p per book and the school is encouraging students to sell as many books as possible by offering a £20 prize to the student who sells the most books. The table shows the distribution of the numbers of books sold by the students in the school.

| Number of books sold | 0–5 | 6–10 | 11–15 | 16–20 | 21–25 | 26–30 | 31–35 | 36–40 | 41–45 | 46–50 |
|---|---|---|---|---|---|---|---|---|---|---|
| Frequency | 77 | 124 | | | | | 73 | 32 | 22 | 9 |
| Number of books sold | ⩽ 5 | ⩽ 10 | ⩽ 15 | ⩽ 20 | ⩽ 25 | ⩽ 30 | ⩽ 35 | ⩽ 40 | ⩽ 45 | ⩽ 50 |
| Cumulative frequency | | | 383 | 611 | 775 | 867 | | | | |

Copy and complete the table and use it to find
a the number of students who sold more than 30 books
b the number of students who sold fewer than 21 books
c the number of student who sold more than 10 books but fewer than 31 books.
d Was the £20 prize won by one student or could it have been shared? Give a reason for your answer.

# STP Maths 9

5. The table is based on a golfer's scores on the professional circuit one summer.

| Score | 67 | 68 | 69 | 70 | 71 | 72 | 73 | 74 | 75 | 76 | 77 | 78 |
|---|---|---|---|---|---|---|---|---|---|---|---|---|
| Frequency | 2 | 4 | 9 | | | | | | | | | |
| Score | ≤67 | ≤68 | ≤69 | ≤70 | ≤71 | ≤72 | ≤73 | ≤74 | ≤75 | ≤76 | ≤77 | ≤78 |
| Cumulative frequency | 2 | 6 | 15 | 24 | 36 | 51 | 64 | 72 | 77 | 85 | 91 | 95 |

Copy and complete this table and hence find

a  the number of rounds in which she scored 73

b  the number of rounds in which she scored 75 or more.

## Cumulative frequency diagrams

We used this cumulative frequency table for the height of 90 five-year-old children to find the number of children who were less than 105 cm tall.

| Height, $h$ (cm) | Frequency, $f$ | Height, $h$ (cm) | Cumulative frequency |
|---|---|---|---|
| $90 \leq h < 95$ | 5 | $h < 95$ | 5 |
| $95 \leq h < 100$ | 9 | $h < 100$ | 14 |
| $100 \leq h < 105$ | 17 | $h < 105$ | 31 |
| $105 \leq h < 110$ | 28 | $h < 110$ | 59 |
| $110 \leq h < 115$ | 21 | $h < 115$ | 80 |
| $115 \leq h < 120$ | 10 | $h < 120$ | 90 |

It is not possible to find the number of children who are less than 102 cm tall, because 102 cm is in the middle of a group and we do not know the individual heights in this group. They may all be less than 102 cm, or all greater than 102 cm, or any distribution between 100 cm and 105 cm.

We can make an estimate of this number as follows:

102 cm is just under halfway through the group 100 cm to 105 cm,

there are 17 heights in this group so *assume* that just under half of them (8) are less than 102 cm, that is, assume that the heights are evenly spread throughout the group.

This gives 14 + 8 = 22 as an estimate of the total number of heights less than 102 cm.

We can find this estimate more easily from a graph drawn by plotting cumulative frequencies against the upper ends of the groups. When the points are joined with straight lines, the graph is called a **cumulative frequency polygon**. If we draw a smooth curve through the points, the graph is called a **cumulative frequency curve**.

This graph shows the cumulative frequency polygon (blue) and the curve (black) for the distribution of heights given above.

## 8 Organising and summarising data

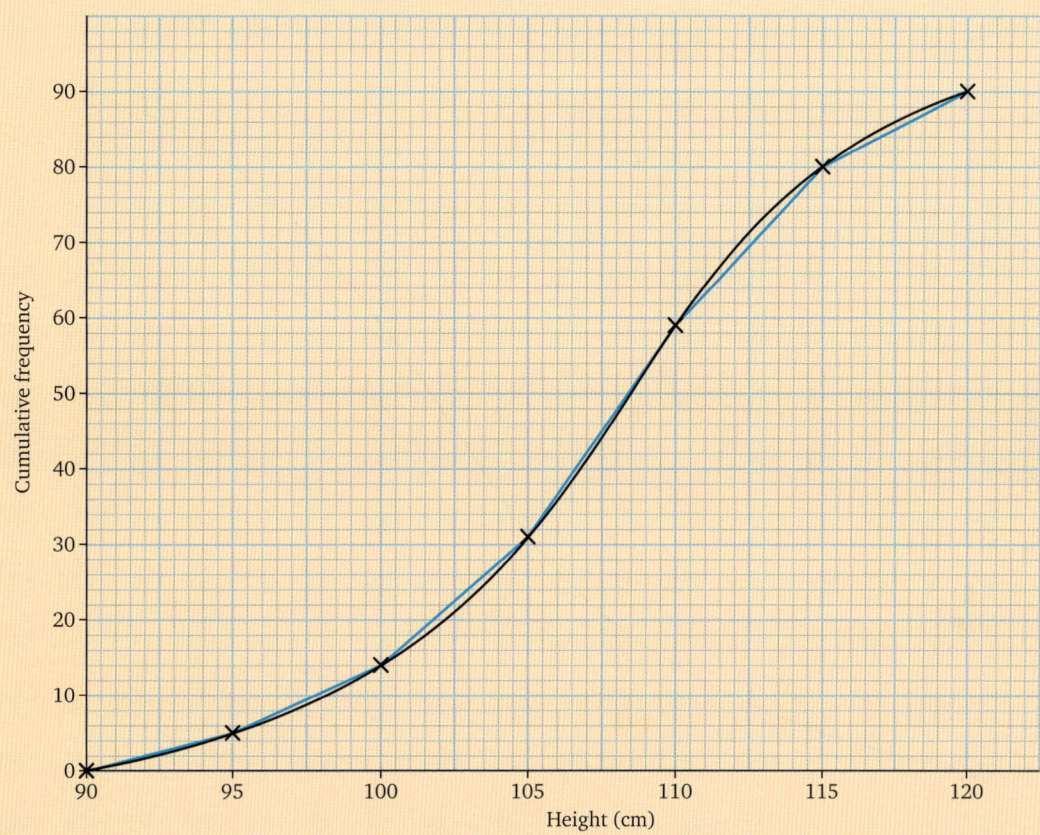

Notice that the graph starts at the point where the cumulative frequency is zero at the lower end of the first group. This is because there are no heights less than 90 cm.

When we join the points with straight lines, we are assuming that the items in any one group are evenly spread throughout that group. For example, we are assuming that the five heights in the group $90 \leq h < 95$ and the nine heights in the group $95 \leq h < 100$ are spread as shown on this number line.

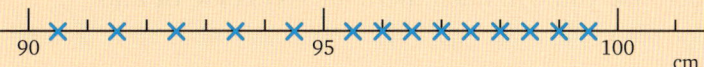

They are more likely to be less evenly spread, for example as shown on this number line.

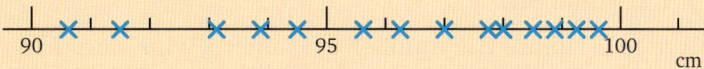

Drawing a smooth curve through the points assumes this more likely distribution of the heights.

We can now use the curve to estimate the number of children whose heights are less than 102 cm.

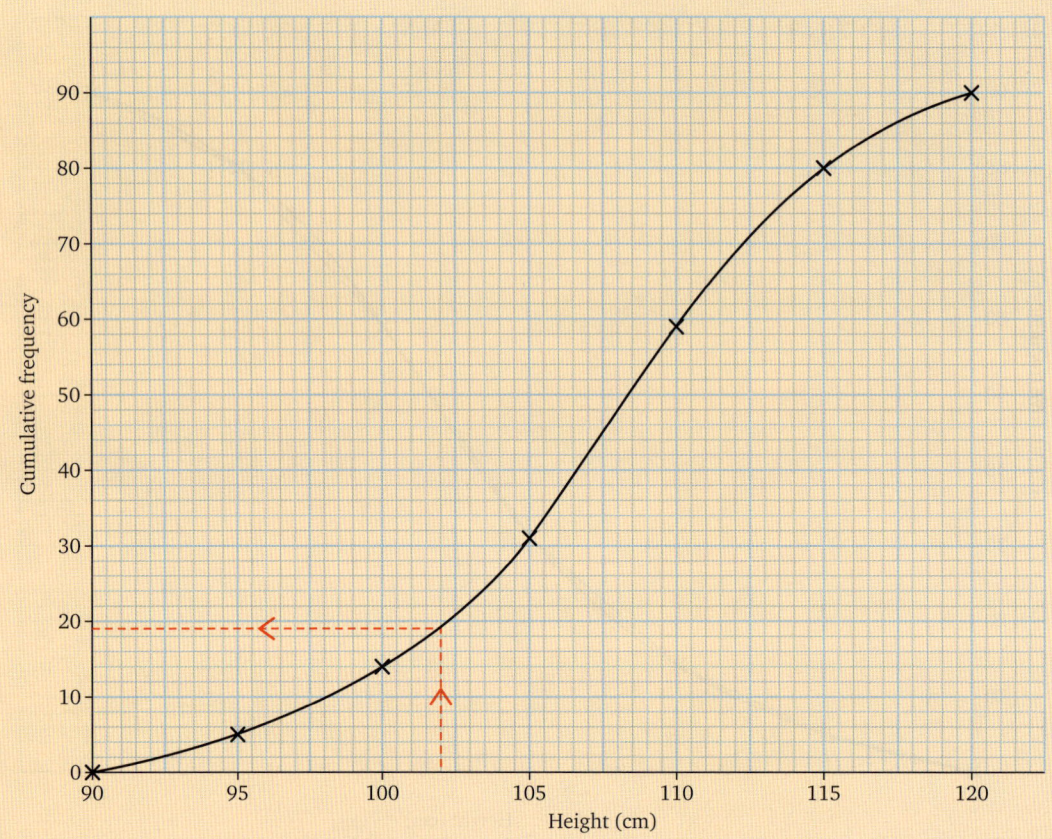

We draw a line up from 102 cm to the curve and then draw a line across to the cumulative frequency axis. The reading is 19, so we estimate that 19 children are shorter than 102 cm.

### Exercise 8e

Keep your curves because you will need them for **Exercise 8f** and **Exercise 8g**.

1. Draw a cumulative frequency curve for question **1** in **Exercise 8d**.

2. **a** Draw a cumulative frequency curve to illustrate the data given in question **2** of **Exercise 8d**.
   **b** Use your curve to estimate the number of students who scored 75 or more.
   **c** The pass mark for the examination was 45%. How many students passed?

3. **a** Draw a cumulative frequency curve for question **3** in **Exercise 8d**.
   **b** Estimate the number of innings in which the batsman scored 90 or more.

## Finding the median of a grouped frequency distribution

For an ungrouped frequency distribution, we can find the median directly from the cumulative frequency table.

For example, this table gives the scores from shots in a shooting competition.

| Score | Cumulative frequency (i.e. number of shots) |
|---|---|
| $\leq 1$ | 3 |
| $\leq 2$ | 7 |
| $\leq 3$ | 25 |
| $\leq 4$ | 41 |
| $\leq 5$ | 50 |

The median of these 50 scores is the $\frac{50+1}{2}$th score, that is, the average of the 25th and 26th score.

From the cumulative frequencies, we see that the 25th score is 3 and the 26th score is 4, so the median score is $\frac{1}{2}(3+4) = 3.5$

Now consider again the grouped distribution of 90 heights.

| Height, $h$ (cm) | Cumulative frequency |
|---|---|
| $h < 95$ | 5 |
| $h < 100$ | 14 |
| $h < 105$ | 31 |
| $h < 110$ | 59 |
| $h < 115$ | 80 |
| $h < 120$ | 90 |

The median of the 90 heights is the average of the 45th and 46th heights when they have been arranged in order of size. The table shows that these both lie between 105 cm and 110 cm.

Therefore, the median is in the interval $105 \leq h < 110$.

Although we cannot locate the median exactly, we can estimate its value from the cumulative frequency curve. When we do this, we read the value from exactly halfway through the total frequency.

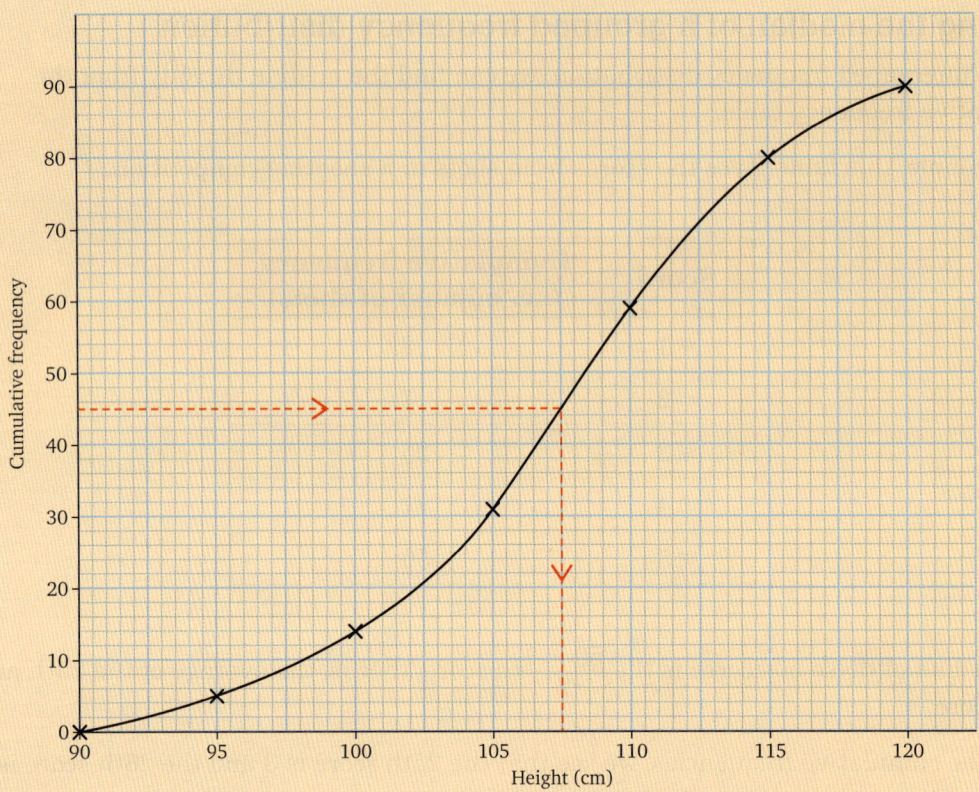

The median is halfway through the 90 heights; 45 is halfway between 0 and 90 on the cumulative frequency axis, so we read across from 45 and then down to the horizontal axis.

The reading here is 107.5 cm, and this is used as an estimate for the median.

We can also use the curve to estimate answers to questions such as 'Below what height are the shortest 25% of the children?'

Because 25% of 90 is 22.5, we read across from 22.5 and then down to the horizontal axis to 103 cm. This gives us the estimate that 25% of the children are shorter than 103 cm.

## Exercise 8f

1. Find the median from each of the graphs you drew for **Exercise 8e**.

2. Use the cumulative frequency table below to draw the cumulative frequency curve for the prices of all the houses advertised in a property magazine one weekend.

| Price (thousands of £s) | ⩽ 70 | ⩽ 80 | ⩽ 90 | ⩽ 100 | ⩽ 110 | ⩽ 120 | ⩽ 130 | ⩽ 140 | ⩽ 150 | ⩽ 160 |
|---|---|---|---|---|---|---|---|---|---|---|
| Cumulative frequency | 10 | 22 | 60 | 128 | 170 | 187 | 197 | 203 | 208 | 210 |

Use your graph to estimate the median advertised price for a house.

3   This is the cumulative frequency curve for the marks awarded in a maths exam.
   a   What is the median mark?
   b   The pass mark will be set so that 25% of candidates fail. What should this mark be?

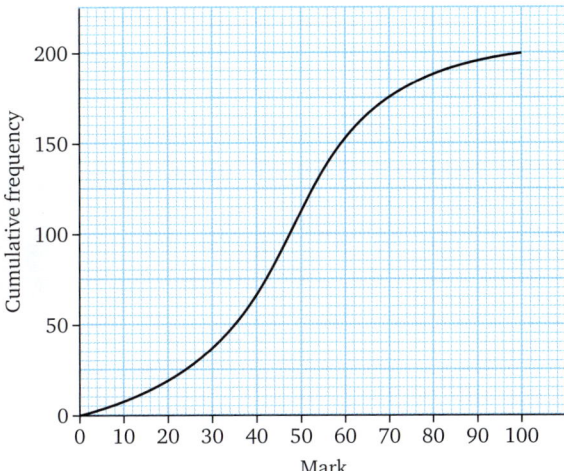

4   The cumulative frequency curve for the marks in an English test is given below. Use the graph to find
   a   the number of students sitting the test
   b   the median mark.

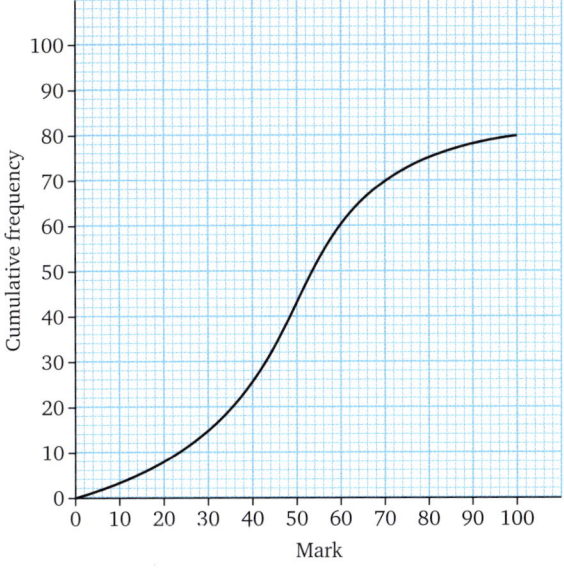

STP Maths 9

5. Use the following cumulative frequency table to draw the corresponding cumulative frequency curve for the marks obtained by candidates in an examination.

| Marks | ⩽ 9 | ⩽ 19 | ⩽ 29 | ⩽ 39 | ⩽ 49 | ⩽ 59 | ⩽ 69 | ⩽ 79 | ⩽ 89 | ⩽ 99 |
|---|---|---|---|---|---|---|---|---|---|---|
| Cumulative frequency | 7 | 16 | 28 | 47 | 80 | 125 | 174 | 202 | 212 | 220 |

Use your graph to estimate the median mark.

6. The number of cars using a ferry on each trip during a particular month was noted and the results are given in the following table.

| Number of cars | 40 | 41 | 42 | 43 | 44 | 45 | 46 | 47 | 48 | 49 | 50 |
|---|---|---|---|---|---|---|---|---|---|---|---|
| Number of trips | 2 | 4 | 6 | 10 | 10 | 12 | 8 | 6 | 2 | 1 | 1 |

Construct the corresponding cumulative frequency table and use it to find the median number of cars making the trip. How many trips did the ferry make during the month?

7. A traffic survey counted the number of cars per hour passing Southwood Post Office each hour of the day from 8 a.m. to 6 p.m. for a week. The results are given in the table.

|  | 8 a.m.–9 a.m. | 9 a.m.–10 a.m. | 10 a.m.–11 a.m. | 11 a.m.–12 noon | 12 noon–1 p.m. | 1 p.m.–2 p.m. | 2 p.m.–3 p.m. | 3 p.m.–4 p.m. | 4 p.m.–5 p.m. | 5 p.m.–6 p.m. |
|---|---|---|---|---|---|---|---|---|---|---|
| Monday | 39 | 37 | 46 | 36 | 41 | 34 | 33 | 32 | 22 | 23 |
| Tuesday | 16 | 31 | 40 | 39 | 42 | 43 | 39 | 37 | 24 | 17 |
| Wednesday | 24 | 39 | 37 | 45 | 44 | 39 | 38 | 36 | 29 | 27 |
| Thursday | 19 | 33 | 32 | 34 | 42 | 38 | 37 | 39 | 25 | 27 |
| Friday | 30 | 37 | 36 | 41 | 48 | 47 | 40 | 43 | 35 | 34 |
| Saturday | 28 | 38 | 46 | 39 | 42 | 48 | 42 | 40 | 31 | 33 |
| Sunday | 3 | 7 | 42 | 14 | 11 | 33 | 36 | 35 | 27 | 26 |

Use groups 0–5, 6–10, 11–15, etc. to make a frequency table and a cumulative frequency table. Draw the cumulative frequency curve and use it to estimate the median number of cars passing Southwood Post Office per hour.

## Interquartile range

Consider this set of marks obtained by a student in the end-of-year exams.

54, 93, 86, 75, 8, 59, 73, 83, 55, 64, 73

The marks range from 8 to 93. All the marks except one, however, are over 50.

Therefore, the range of these marks gives a misleading impression of their spread and it would be better to quote the range of a restricted section of the marks.

The mark of 8 is an example of an **outlier**. An outlier is a value that is much higher or lower than the rest of the values in a set.

Now consider again the heights of five-year-olds collected by a local health authority.

The health authority may want to compare the range of heights with those from other areas. If there are one or two abnormally short or tall children in any one health authority area, comparing the full ranges is not as helpful as comparing the ranges of restricted sections of the heights. Clearly the comparisons are valid only if the *same* sections of each distribution are compared.

The restricted range that we usually use is the middle half, that is, from $\frac{1}{4}$ to $\frac{3}{4}$ of the way through a distribution.

The value that is $\frac{1}{4}$ of the way through a distribution is called the **lower quartile**.

**For $n$ values arranged in order of size,**

**the *lower quartile* is the $\frac{n+1}{4}$th value.**

The value that is $\frac{3}{4}$ of the way through a distribution is called the **upper quartile**.

**For $n$ values arranged in order of size,**

**the *upper quartile* is the $\frac{3(n+1)}{4}$th value.**

We use $Q_1$ to denote the lower quartile and $Q_3$ to denote the upper quartile. ($Q_2$ is used to denote the median.)

For an ungrouped distribution, we can find the lower and upper quartiles exactly.

For a grouped distribution, we use a cumulative frequency curve to estimate

**$Q_1$ as the $\frac{n}{4}$th value, and $Q_3$ as the $\frac{3n}{4}$th value.**

The difference between the quartiles is called the **interquartile range**.

**The interquartile range is $Q_3 - Q_1$.**

This is the cumulative frequency curve of the heights of 90 children.

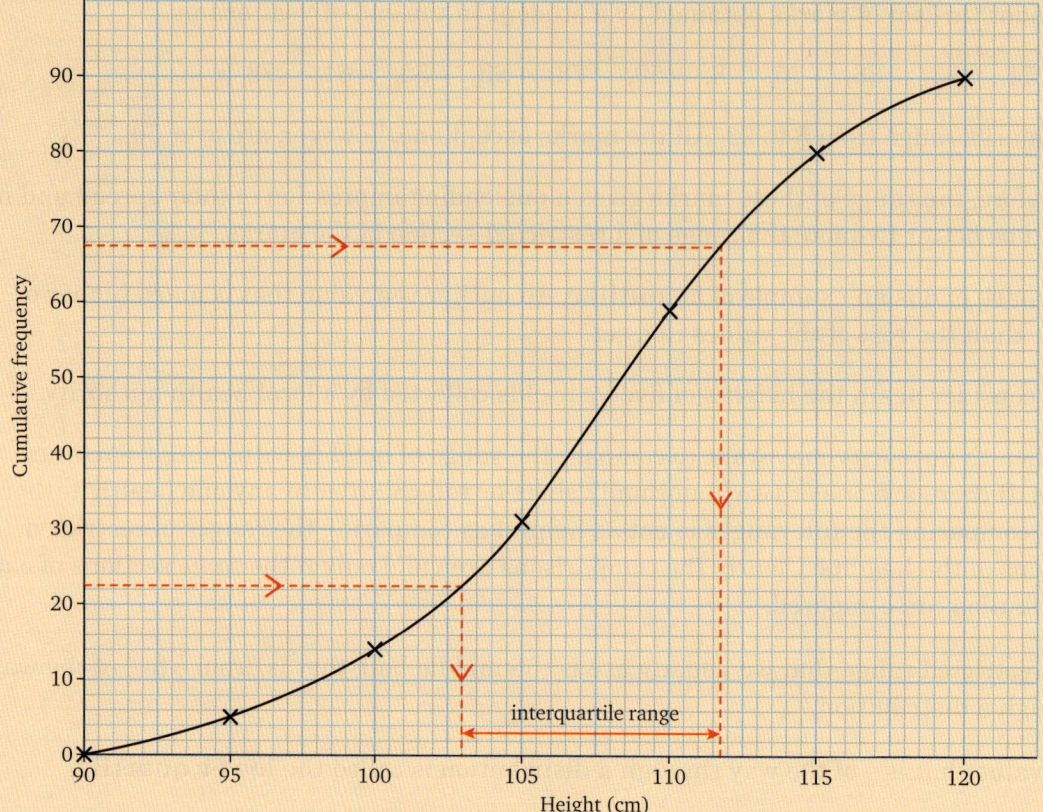

$Q_1$ is the $\frac{90}{4}$th, i.e. the 22.5th, value

and $Q_3$ is the $\frac{(3 \times 90)}{4}$th, i.e. the 67.5th, value.

Reading from the graph, $Q_1 = 103$ cm and $Q_3 = 111.8$ cm

Therefore, the interquartile range is $Q_3 - Q_1 = 111.8 - 103$ cm
$\phantom{Therefore, the interquartile range is Q_3 - Q_1\ } = 8.8$ cm

(Note that readings from graphs can only give estimates for values.)

### Exercise 8g

1. Use the cumulative frequency curves you drew for **Exercise 8e** to find the upper and lower quartiles and the interquartile ranges.

2. The cumulative frequency curve opposite shows the weekly earnings, in pounds, of a group of teenagers.

Use the graph to find
a the median
b the upper and lower quartiles.
c Hence find the interquartile range.

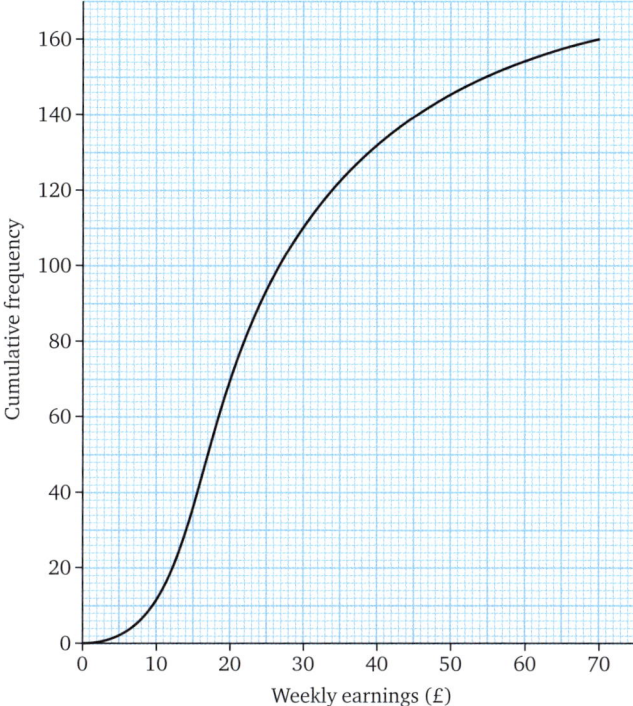

3 The table shows the distribution of the ages of people attending a public concert.

| Age range | 0–19 | 20–39 | 40–59 | 60–79 | 80–99 |
|---|---|---|---|---|---|
| No. of people attending concert | 8 | 26 | 110 | 128 | 56 |

Copy and complete the following cumulative frequency table and use it to draw a cumulative frequency curve.

| Age range | < 20 | < 40 | < 60 | < 80 | < 100 |
|---|---|---|---|---|---|
| No. of people attending concert | | | | | |

Hence find
a the number of people attending the concert
b the median age
c the upper and lower quartile ages, and the interquartile range.

**4** Several darts players were chosen at random and each was asked to throw 50 darts at the bullseye on a dartboard. The number of bullseyes scored by each person was noted and the following frequency table was constructed.

| No. of bullseyes | Frequency | No. of bullseyes | Cumulative frequency |
|---|---|---|---|
| 0–4 | 25 | ≤ 4 | |
| 5–9 | 20 | ≤ 9 | |
| 10–14 | 15 | ≤ 14 | |
| 15–19 | 12 | | |
| 20–24 | 10 | | |
| 25–29 | 3 | | |
| 30–34 | 2 | | |
| 35–39 | 1 | | |
| 40–44 | 1 | | |
| 45– | 0 | | |

Copy the table and complete the third and fourth columns. Draw the corresponding cumulative frequency curve, using a scale of 2 cm to represent 10 bullseyes on one axis and 2 cm to represent a cumulative frequency of 10 on the other axis. Use your graph to estimate
**a** the median
**b** the upper and lower quartiles.

**5** In the first round of a golf tournament the following scores were recorded.

| 70 | 68 | 71 | 67 | 74 | 69 | 69 | 71 | 68 | 70 |
| 71 | 70 | 72 | 69 | 69 | 68 | 71 | 70 | 70 | 72 |
| 72 | 69 | 68 | 70 | 68 | 69 | 67 | 71 | 69 | 70 |
| 68 | 67 | 70 | 70 | 73 | 69 | 71 | 67 | 69 | 68 |

**a** Construct a cumulative frequency table for these scores,
**b** How many rounds of less than 70 were there?
**c** How many rounds of more than 69 were there?
**d** Find the median score.
**e** Explain why you do not need a cumulative frequency curve to find the median.

**6** The marks obtained by the candidates sitting a test are given in the following table.

| Mark | 0–9 | 10–19 | 20–29 | 30–39 | 40–49 | 50–59 | 60–69 | 70–79 |
|---|---|---|---|---|---|---|---|---|
| Frequency | 3 | 13 | 27 | 43 | 28 | 20 | 12 | 8 |

Draw a cumulative frequency curve for these figures. Use 2 cm to represent 10 marks on one axis and a cumulative frequency of 20 on the other axis.

Use your graph to estimate
a   the median
b   the upper and lower quartiles and, hence, the interquartile range
c   the pass mark if 75% of the candidates pass.

## Mixed exercise

### Exercise 8h

1   The following cumulative frequency table gives the percentage marks of 250 students in an English examination.

| Mark | 10 | 20 | 30 | 40 | 50 | 60 | 70 | 80 | 90 | 100 |
|---|---|---|---|---|---|---|---|---|---|---|
| Number of students scoring up to and including this mark | 5 | 15 | 29 | 52 | 89 | 142 | 197 | 223 | 240 | 250 |

a   How many students scored a mark of more than 70?
b   How many students scored a mark from 41 to 60?
c   Plot the values from the table on a graph and draw a smooth curve through your points. (Use a scale of 2 cm to represent 20 marks on one axis and 2 cm to represent a cumulative frequency of 25 on the other axis.)
d   Use your graph to estimate
    i   the median
    ii  the upper and lower quartiles.
e   State the probability that a student chosen at random will have a mark
    i   less than or equal to 50
    ii  greater than 60.

2   The bar chart on the following page illustrates the distribution of the weekly pocket money of the 240 students in Year 6 of a school.
a   How many students received from £6 to £10 inclusive?
b   Half the students received more than £$x$ per week. Estimate the value of $x$.
c   The line AB indicates that the value of the lower quartile of the distribution is £1.50. What does this mean?
d   The value of the interquartile range is £2.70. What is the value of the upper quartile? What information does the interquartile range give us about the weekly pocket money of the group?
e   Estimate the total amount of pocket money received by the Year 6 students.
f   Hence estimate the mean amount of pocket money received by the Year 6 students.

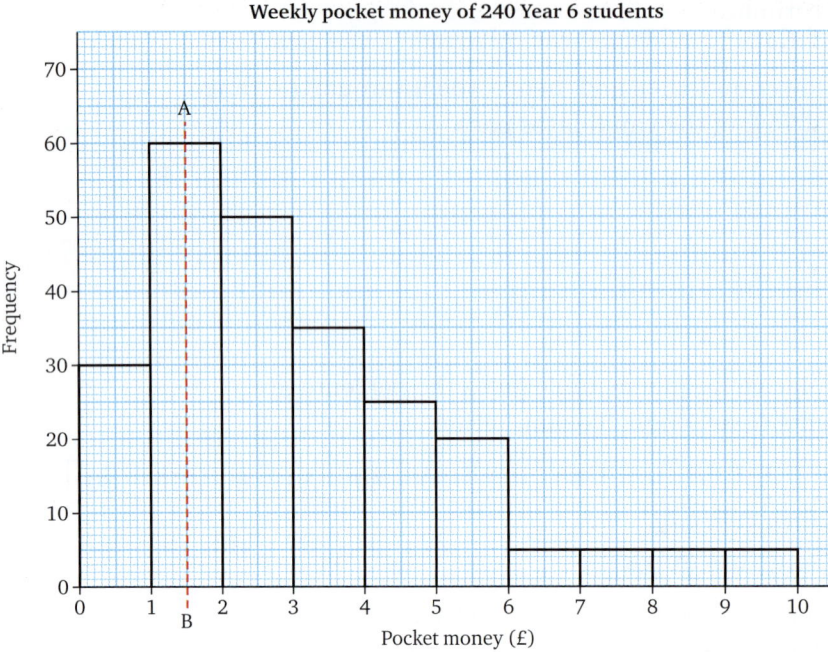

Weekly pocket money of 240 Year 6 students

**3** The following marks were obtained by the 80 candidates in an English test, which was marked out of 60.

```
54  52  31  47  24  36  27  15  44  26   8  20  46  32
27  31  33  57  39  32  43  32  23  33  31  21  38  28
40  19  52  37  38  39   9  30  47  29   8  13  33  35
48  18  36  39  23  58  34  35  16  21  32  38  34  13
27  32  37  23  37  49  25  38  24  27  48  36  45  18
41  34  43  12  47  24   8  29  37  33
```

Copy the table below and use the data to complete it.

| Interval | Tally | Frequency | Mark | Cumulative frequency |
|---|---|---|---|---|
| 0–9 | | | ⩽ 9 | |
| 10–19 | | | ⩽ 19 | |
| 20–29 | | | ⩽ 29 | |
| 30–39 | | | | |
| 40–49 | | | | |
| 50–60 | | | | |

Use the information in your table to draw a cumulative frequency curve and from it estimate
  **a** the median mark
  **b** the upper and lower quartiles
  **c** the number of candidates who passed, if the pass mark was 40
  **d** the pass mark if 70% of the candidates passed
  **e** the probability that a student selected at random scored less than 30.
  **f** Draw a stem-and-leaf diagram of this data. Use the number of 10s as the stem.

# 8 Organising and summarising data

**4** An agricultural researcher wanted to compare the milk yields from two herds of cows. The following evidence was collected.

**Herd A**

| Yield per cow per day, $c$ (litres) | $0 \leq c < 5$ | $5 \leq c < 10$ | $10 \leq c < 15$ | $15 \leq c < 20$ |
|---|---|---|---|---|
| Frequency | 2 | 6 | 24 | 14 |

**Herd B**

| Yield per cow per day, $c$ (litres) | $0 \leq c < 5$ | $5 \leq c < 10$ | $10 \leq c < 15$ | $15 \leq c < 20$ |
|---|---|---|---|---|
| Frequency | 3 | 14 | 18 | 5 |

a  Find the median, the range and the interquartile range for the yields from each herd.

b  Use your results from part **a** to compare the yields from these two herds.

**5** A group of Year 6 students went on a school outing. The diagram shows the distribution of spending money that the students had at the beginning of the trip in the form of two bar charts drawn back to back.

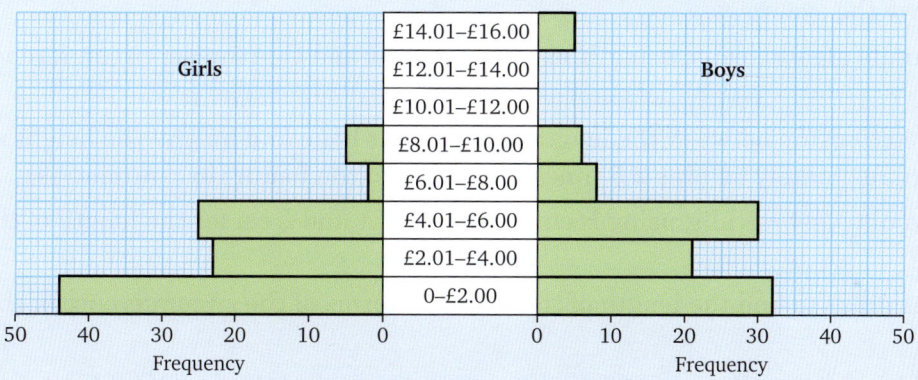

a  How many  **i** girls  **ii** boys  went on the trip?
b  What fraction of the  **i** boys  **ii** girls  had more than £10?
c  Draw a cumulative frequency table for both sets of data.
d  Find the median and interquartile range for each set of data.
e  Give two differences between the girls' and the boys' spending money.

## Consider again

The local health authority wants some information about the heights of five-year-old children in their area.

- There are 1256 five-year-olds in primary schools in the area. What does the health authority need to do to the data in order to obtain useful information from it?
- The health authority may want to compare the distribution of heights in its area with those in other health authority areas. What measures can it usefully use to compare different sets of information? Give reasons for your answers.

Now can you answer these questions?

# 9 Formulas

**Consider**

A newspaper headline reports:

**ANCIENT SKULL FOUND IN PEAT BOG**

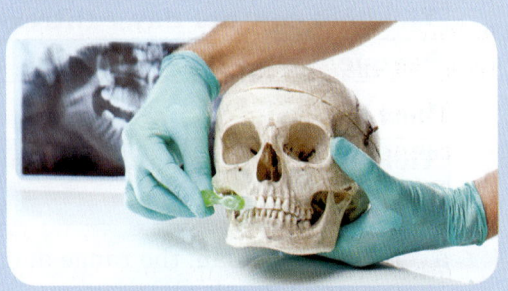

From this single part of the skeleton it is possible to draw several conclusions about the person to whom the skull belonged. For example, the scientists say that it belonged to a man who was between 1.78 m and 1.80 m tall. They worked this out by using data that shows that an adult's height is just over three times the distance measured round their head immediately above the eyebrows.

Express this formula algebraically, using $H$ m for the height and $C$ m for the circumference of the head.

There is also a relationship between the length of the femur bone and height: the height in centimetres is approximately 54 plus 2.5 times the length of the femur (in centimetres).

Express this formula algebraically, using $H$ cm for the height and $L$ cm for the length of the femur.

Use these two formulas to find the length of the femur in terms of the circumference of the head.

*You should be able to answer these questions after you have worked through this chapter.*

When a definite relationship between two quantities has been established, we can find either one when we are given the other. In this case, we can calculate the height if we know the circumference of a person's skull.

The relationships between the quantities referred to in the Consider section is not exact, but in other cases it may be. For example, if grapefruit cost 33 p each, the **formula** $C = 33n$ enables us to find the exact cost, $C$ pence, when we buy $n$ grapefruit.

In this chapter we look at some formulas that are more complicated than the ones we studied in Books 7 and 8.

## Class discussion

Discuss in a group whether or not each of the following statements is true. If you come to the conclusion that a particular statement is true, list examples to support your conclusion.

1. There are many examples to show that two quantities can be connected by an exact relationship.
2. Sometimes two quantities are related, but not in an exact way. (If you can, list some quantities that are very closely related and others that are related but not strongly.)
3. In some cases, one quantity can be related to several different quantities at the same time.
4. There are cases where two quantities are not related in any way.

## Constructing formulas

Electricity bills are presented every quarter (every three months). They are made up of a fixed standing charge plus the cost of the number of units used in the quarter.

By using letters for the unknown quantities, we can construct a formula for a quarterly electricity bill.

If £$C$ is the total bill, £$R$ is the standing charge, the cost of one unit is $x$ pence and $N$ units are used, then

the cost of the units is $Nx$ pence or £$\frac{Nx}{100}$

therefore $C = R + \frac{Nx}{100}$

Notice that pounds are used throughout – the cost of the units is converted from pence to pounds so that we add pounds to pounds, not pounds to pence.

## Exercise 9a

### Worked example

→ A number $p$ is equal to the sum of a number $q$ and twice a number $r$. Write down a formula for $p$ in terms of $q$ and $r$.

$p = q + 2r$

In questions **1** to **7**, write a formula connecting the given letters.

1. A number $a$ is equal to the sum of two numbers $b$ and $c$.
2. A number $m$ is equal to twice the sum of two numbers $n$ and $p$.

181

3 A number z is equal to the product of two numbers x and y.

4 A number a is equal to twice the product of two numbers b and c.

5 A number d is equal to the difference of two numbers e and f, where e is greater than f.

6 A number n is equal to the sum of a number p and its square.

7 A number v is equal to the sum of a number u and the product of the numbers a and t.

8 Fabric is sold at £p per metre. The cost of N metres is £R. Find a formula for R in terms of N and p.

9 A shop sells two brands of tinned beans. It has N tins of beans altogether; y of them are one brand and z of them are the other brand. Find a formula for N in terms of y and z.

10 A ship moving with constant speed takes x minutes to cover one nautical mile. It takes X minutes to cover y nautical miles. Find a formula for X in terms of x and y.

11 A card has P metres of lace edging wound on it. Nerys buys n lengths of edging, each x centimetres long. If Q metres of edging are left on the card, find a formula for Q in terms of P, n and x.

12 Fertiliser is applied at the rate of a grams per square metre. It takes b kilograms to cover a field of area c square metres. Find a formula for b in terms of a and c.

13 A bag of coins contains x ten-pence coins and y twenty-pence coins. The total value of the coins is £R. Find a formula for R in terms of x and y.

## Substituting numbers into formulas

**Exercise 9b**

**Worked example**

→ Given that $s = ut - \frac{1}{2}gt^2$, find s when $u = 8$, $t = 6$ and $g = -10$

$s = ut - \frac{1}{2}gt^2$

When $u = 8$, $t = 6$ and $g = -10$

$s = (8)(6) - (\frac{1}{2})(-10)(6)^2$

$= 48 - (-5)(36)$

$= 48 - (-180)$

$= 48 + 180$

$= 228$

 Notice that we have put each number in brackets. This is particularly important when we are dealing with negative numbers.

1. Given that $v = \dfrac{u-t}{3}$, find $v$ when $u = 4$ and $t = -2$

2. Given that $z = \dfrac{1}{x} + \dfrac{1}{y}$, find $z$ when $x = 2$ and $y = 4$

3. If $C = rt$, find $C$ when $r = -3$ and $t = -10$

4. If $x = rt - v$, find $x$ when $r = 2$, $t = 10$ and $v = -4$

5. If $p = x + x^2$, find $p$ when
   a $\ x = 2$ \qquad b $\ x = 3.4$ \qquad c $\ x = 0.79$

6. Given that $s = \tfrac{1}{2}(a + b + c)$, find $s$ when
   a $\ a = 6, b = 9$ and $c = 5$
   b $\ a = 5.04, b = 7.35$ and $c = 4.83$

7. If $p = r(2t - s)$, find $p$ when $r = \tfrac{1}{2}, t = 3$ and $s = -2$

8. If $a = (b + c)^2$, find $a$ when
   a $\ b = 8$ and $c = -5$
   b $\ b = 4.1$ and $c = 7.8$

9. If $r = \dfrac{2}{s+t}$, find $r$ when $s = \tfrac{1}{2}$ and $t = \tfrac{1}{4}$

10. Given that $a = bc - \tfrac{1}{2}dc$, find $a$ when $b = 3, c = -4$ and $d = 7$

11. Given that $V = \tfrac{1}{2}(X - Y)^2$, find $V$ when $X = 3$ and $Y = -5$

12. If $P = 2Q + 5RT$, find $P$ when $Q = 8, R = -2$ and $T = -\tfrac{1}{2}$

13. Given that $a = (b - c)(c - d)$, find $a$ when $b = 2, c = 4$ and $d = 7$

14. The displacement ($D\,\text{cm}^3$) of an engine is given by the formula
$$D = \pi n \times \left(\dfrac{b}{2}\right)^2 \times s$$
where $n$ is the number of cylinders, $b$ is the bore (diameter) of each cylinder in centimetres and $s$ is the length of the stroke in centimetres.

Find the displacement of a 4-cylinder petrol engine that has a bore of 78 mm and a stroke of 84 mm.

15. A person's mean blood pressure ($P$), in millimetres of mercury, is calculated using the formula
$$P = D + \dfrac{(S - D)}{3}$$

where $D$ is the diastolic pressure and $S$ is the systolic pressure, both measured in millimetres of mercury (mm Hg). Use this formula to find Gwen's mean blood pressure if her diastolic pressure is 82 mm Hg and her systolic pressure is 146 mm Hg.

# STP Maths 9

**16** The gross yield of a government stock, as a percentage, is given by the formula

$$\text{Gross yield} = \frac{I}{P} \times 100$$

where $I$ is the annual rate of interest and $P$ is the price of the stock. Find the gross yield on

**a** Treasury $11\frac{3}{4}\%$ stock at 121

**b** Exchequer 4% stock at 94.

---

Sometimes the letter whose value we need to find is not isolated on one side of the formula. For example, to use the formula $v = u + at$ to find $a$ when $v = 20$, $u = 8$ and $t = 2$, we substitute the values for $v$, $u$ and $t$. This gives the equation $20 = 8 + 2a$, which we can then solve to find $a$.

## Exercise 9c

### Worked examples

Given that $p = q - 5r$, find the value of $q$ when $p = 5$ and $r = 2$

$p = q - 5r$

Replacing $p$ by 5 and $r$ by 2 gives
$5 = q - 10$
$15 = q$

*Add 10 to both sides.*

$\therefore$ if $p = 5$ and $r = 2$, $q = 15$

Given that $p = q - 5r$, find the value of $r$ when $p = 4$ and $q = 24$

$p = q - 5r$
If $p = 4$ and $q = 24$, $4 = 24 - 5r$

*Add 5r to both sides.*

$5r + 4 = 24$

*Subtract 4 from each side.*

$5r = 20$

*Divide both sides by 5.*

$r = 4$

$\therefore$ if $p = 4$ and $q = 24$, $r = 4$

**1** If $a = b + 2c$, find

  **a** $b$ when $a = 10$ and $c = 2$

  **b** $c$ when $a = 13$ and $b = 5$

**2** Given that $x = y - z$, find

  **a** $y$ when $x = 9$ and $z = 4$

  **b** $z$ when $x = 5$ and $y = 8$

3  If $p = qr$, find
   a  $q$ when $p = 36$ and $r = 9$
   b  $r$ when $p = 24$ and $q = 8$

4  Given that $v = u + 8t$, find
   a  $u$ when $v = 45$ and $t = 3.5$
   b  $t$ when $v = 64$ and $u = 0$

5  If $s = \frac{1}{2}(a + b + c)$, find
   a  $a$ when $s = 5.8$, $b = 4.5$ and $c = 3.4$
   b  $c$ when $s = 4.7$, $a = 2.7$ and $b = 3.8$

6  At an athletics meeting a points system operates for the long jump event. The number of points scored ($P$) is calculated using the formula
   $$P = a \times (L - b)^2$$
   where $L$ is the distance jumped in metres, measured correct to the nearest centimetre, and $a$ and $b$ are non-zero constants.
   a  The points score for a jump of 6 metres is 0. Which constant does this information enable you to find? What is the value of this constant?
   b  A jump of 8 metres gives a score of 400 points. Find the value of the other constant.
   c  What is the shortest jump that will score points?
   d  How many points are scored for a jump of
      i   6.5 m
      ii  7.34 m?

## Changing the subject of a formula

In question **16** in **Exercise 9b** the gross yield, $G$, of a government stock is given by the formula

$$G = \frac{I}{P} \times 100$$

where $I$ is the annual interest rate and $P$ the price of the stock. This formula can be used to find $G$ directly for different values of $I$ and $P$.

A financial adviser may want to use this formula to find the interest rate for a gross yield of 10% when the price of the stock is 120.

He can do this by substituting the numbers directly into the formula to give

$$10 = \frac{I}{120} \times 100$$

and then solving this equation for $I$. This gives $I = 12$.

The adviser may wish to find values of $I$ for several different values of $G$ and $P$. He can repeat the method used above or he can start by rearranging the formula so that $I$ is alone on the left-hand side.

We can do this by thinking of $G = \frac{I}{P} \times 100$ as an equation and solving it for $I$.

As with any equation, we start by getting rid of any fractions. Multiplying both sides by $P$ achieves this;

i.e. $\qquad P \times G = \cancel{P}^1 \times \frac{I}{\cancel{P}_1} \times 100 \Rightarrow PG = 100I$

> The symbol $\Rightarrow$ means 'gives' or implies.

Then dividing both sides by 100 gives $\frac{GP}{100} = I$

i.e. $\qquad I = \frac{GP}{100}$

Now we can find $I$ directly for different values of $G$ and $P$.

When the formula is in the form $G = \frac{I}{P} \times 100$, $G$ is called the **subject of the formula**.

When the formula is written as $I = \frac{GP}{100}$, $I$ is the subject of the formula.

Rearranging $G = \frac{I}{P} \times 100$ as $I = \frac{GP}{100}$ is called *changing the subject of the formula*.

Note that when we enter a formula into a spreadsheet, it has to be in the form where the required quantity is the subject of the formula. It is therefore important to be able to change the subject of a formula. Since changing the subject of a formula is like solving an equation, we start by solving some equations where the letter is the denominator of a fraction.

## Exercise 9d

### Worked example

→ Solve the equation $4 + \frac{2}{x} = 7$

$4 + \frac{2}{x} = 7$

$4 \times x + \frac{2}{\cancel{x}_1} \times \cancel{x}^1 = 7 \times x$

> Multiplying each term by $x$ gets rid of the fraction.

$4x + 2 = 7x$

$2 = 3x$, i.e. $x = \frac{2}{3}$

> Take $4x$ from both sides.

Check: LHS $= 4 + 2 \div \frac{2}{3} = 4 + 2 \times \frac{3}{2} = 7 =$ RHS

Solve

1. $\frac{4}{x} = 5$

2. $1 + \frac{2}{a} = 3$

3. $3 - \frac{4}{t} = 6$

4. $\frac{2}{3x} = 1$

5. $\frac{5}{y} = 2$

6. $\frac{3}{p} + 4 = 5$

7. $1 - \frac{7}{x} = 9$

8. $9 = \frac{4}{2t}$

9. $2 + \frac{5}{3x} = 7$

## Worked example

→ Make $t$ the subject of the formula $v = u + t$

$$v = u + t$$
Subtract $u$ from both sides: $v - u = t$
i.e. $t = v - u$

> We need to 'solve' the formula for $t$, so we use the same methods as for solving an equation for $x$.

Make the letter in brackets the subject of the formula.

10. $p = s + r$     $(s)$

11. $x = 3 + y$     $(y)$

12. $a = b - c$     $(b)$

13. $u = v - 5$     $(v)$

14. $z = x + y$     $(y)$

15. $X = Y - Z$     $(Y)$

16. $r = s + 2t$     $(s)$

17. $k = l + m$     $(m)$

18. $N = P - Q$     $(P)$

19. $v = u + 10t$     $(u)$

20. $x = 2y$     $(y)$

21. $v = \frac{1}{2}t$     $(t)$

22. $a = bc$     $(b)$

23. $t = \frac{u}{3}$     $(u)$

24. $l = \frac{m}{k}$     $(m)$

25. $a = 3b$     $(b)$

26. $X = \frac{1}{10}N$     $(N)$

27. $v = ut$     $(u)$

28. $z = \frac{w}{100}$     $(w)$

29. $n = \frac{p}{q}$     $(p)$

One operation only was needed to change the subject of a formula in the last exercise. In the following exercise, more than one operation is required.

## Exercise 9e

**Worked example**

→ Make $t$ the subject of the formula $v = u + 2t$

$$v = u + 2t$$

Take $u$ from each side $\quad v - u = 2t$

∴ $\quad 2t = v - u$

Divide each side by 2 $\quad t = \dfrac{(v - u)}{2}$

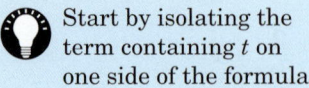 Start by isolating the term containing $t$ on one side of the formula.

Make the letter in brackets the subject of the formula.

1. $p = 2s + r$    $(s)$
2. $v = u - 3t$    $(t)$
3. $a = b - 4c$    $(c)$
4. $V = 2v + 3u$    $(v)$
5. $x = 2w - y$    $(w)$
6. $l = k + 4t$    $(t)$
7. $w = x - 6y$    $(y)$
8. $N = It - 2s$    $(s)$
9. $x = \dfrac{3y}{4}$    $(y)$
10. $u = v + 5t$    $(t)$
11. $A = P + \dfrac{1}{10}I$    $(I)$
12. $z = x - \dfrac{y}{3}$    $(y)$
13. $V = \dfrac{2R}{I}$    $(R)$
14. $p = 2r - w$    $(r)$
15. $a = b + \dfrac{1}{2}c$    $(c)$
16. $p = q - \dfrac{r}{5}$    $(r)$

17. Make $u$ the subject of the formula $v = u + at$.
    Find $u$ when $v = 80$, $a = -10$ and $t = 6$

18. Make $B$ the subject of the formula $A = \dfrac{C}{100} + B$
    Find $B$ when $A = 20$ and $C = 250$

19. Make $C$ the subject of the formula $P = \dfrac{C}{N}$
    Find $C$ when $N = 20$ and $P = \dfrac{1}{2}$

20. Make $x$ the subject of the formula $z = \dfrac{1}{2}x - 3t$
    Find $x$ when $z = 4$ and $t = -3$

21. A number $a$ is equal to the sum of a number $b$ and twice a number $c$.
    
    a. Find a formula for $a$ in terms of $b$ and $c$.
    
    b. Find $a$ when $b = 8$ and $c = -2$.
    
    c. Make $b$ the subject of the formula.

**22** A number $x$ is equal to the product of a number $z$ and twice a number $y$.

  **a** Find a formula for $x$ in terms of $z$ and $y$.

  **b** Find $x$ when $z = 3$ and $y = 2$.

  **c** Make $y$ the subject of the formula.

**23** A number $d$ is equal to the square of a number $e$ plus twice a number $f$.

  **a** Find a formula for $d$ in terms of $e$ and $f$.

  **b** Make $f$ the subject of the formula.

  **c** Find $f$ when $d = 10$ and $e = 3$.

**24** The heat setting on a gas oven is called its gas mark. The formula $F = 25G + 250$ will convert a gas mark $G$ into a temperature $F$ measured in degrees Fahrenheit.

  **a** Sally puts a joint into her gas oven, which she has previously set at gas mark 6. What is the temperature inside the oven?

  **b** Make $G$ the subject of this formula.

  **c** Gary wants to bake some bread. The recommended baking temperature is 450°F.
  What gas mark should Gary set on the oven?

**25** The length of a woman's femur ($f$ cm) and her height ($h$ cm) are approximately related by the formula $h = 3.5f + 40$

  **a** Part of the skeleton of a woman is unearthed at an archaeological dig. The length of the femur is 34 cm. Use the formula to estimate her height.

  **b** Rearrange the formula to make $f$ the subject.

  **c** A woman is 1.58 m tall. Calculate the length of her femur if her measurements fit the formula exactly.

  **d** Sophie was 50 cm long (tall) at birth. Use the formula to find the value for the length of her femur and state why this is impossible.

**26** The body mass index ($I$) for an adult is given by the formula $I = \dfrac{W}{H^2}$, where $W$ is the weight of the person in kilograms and $H$ the height in metres.

  **a** Use this formula to express $H$ in terms of $I$ and $W$.

  **b** Hence find the height of a person whose body mass index is 23.4 who weighs 94.2 kg. Give your answer correct to 2 decimal places.

## Substituting one formula into another

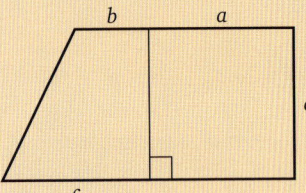

The area of this shape is given by the formula $A = \dfrac{a(b+c)}{2} + a^2$

It is also known that $a = 2b$. We can use this fact to simplify the formula for the area so that $A$ is given in terms of only two letters instead of the three, $a$, $b$ and $c$. We do this by replacing every $a$ in the formula by $2b$.

i.e.
$$A = \dfrac{(2b)(b+c)}{2} + (2b)^2$$

Notice that we place $2b$ in brackets. This reduces the likelihood of mistakes.

$$= \dfrac{2 \times b \times (b+c)}{2} + 4b^2$$

This can now be simplified by cancelling the common factor, 2, from the fraction.

$$= b^2 + bc + 4b^2$$

$\therefore \quad A = 5b^2 + bc$

### Exercise 9f

**Worked example**

→ Given $V = u + at$ and $u = 3t$, find $V$ in terms of $a$ and $t$.

$V = u + at$ and $u = 3t$

Substituting $3t$ for $u$ gives $V = (3t) + at$

i.e. $\qquad\qquad V = 3t + at$

To find $V$ in terms of $a$ and $t$ means that we have to eliminate $u$ from the formula for $V$. We can do this by replacing $u$ by $3t$, i.e. by substituting $3t$ for $u$.

1. Substitute $2u$ for $a$ in the formula $v^2 = u^2 + 2as$

2. Given that $p = \dfrac{r}{4}$ and $r = 2v$, find $p$ in terms of $v$.

3. If $p = r - nt$ and $n = 4t$, find a formula for $p$ in terms of $r$ and $t$.

4. Substitute $2a$ for $b$ in the formula $A = bc + \dfrac{a^2 + b^2}{2}$.

5. If $A = (b-c)(b-a)$ and $b = 4c$, find $A$ in terms of $a$ and $b$.

6. Use the formula $s = ut + 5t^2$ together with $t = 2s$ to give a relationship between $s$ and $u$.

7 Given that $P = (V - 2U)^2$ and that $U = \frac{3}{2}V$, find $P$ in terms of $V$.

8 Find a formula for $n$ in terms of $p$ and $q$, given that $n = \dfrac{r}{p - q}$ and $r = p + q$.

9 If $P = D + \dfrac{(S - D)}{3}$ and $D = 1 - S$, find $P$ in terms of $S$.

10 Given that $a = (b + c)^2 + (b - c)^2$ and that $b = \frac{1}{2}c$, find $a$ in terms of $c$.

11 The capacity of this tank is given by $V = \frac{1}{2}ab(c + d)$.

The surface area of the tank is given by $A = 2b(c + e) + a(c + d)$.

If $V = 300, A = 150, c = 5$ and $e = 1$,

show that $\quad 2b^2 - 25b + 100 = 0$.

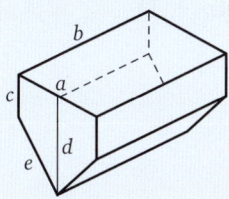

## The $n$th term of a sequence

In a sequence the terms occur in a particular order, that is, there is a first term, a second term, and so on. The value of each term depends on its position in the order. (This is the difference between a sequence of numbers and a set of numbers, which can be in any order.)

The letter $n$ is used for a **natural number**, so we can refer to the $n$th term of a sequence in the same way as we refer to the 4th term or the 8th term.

If we are given a formula such as

$\qquad n$th term $= n(n + 1)$

then we can find any term of the sequence by giving $n$ a numerical value. In this case, we find the first term by substituting 1 for $n$ in $n(n + 1)$,

i.e. $\qquad\qquad$ 1st term $= 1 \times 2 = 2$

Similarly, $\qquad$ 2nd term $= 2 \times 3 = 6 \quad$ (substituting 2)

$\qquad\qquad\qquad$ 3rd term $= 3 \times 4 = 12$

$\qquad\qquad\qquad$ ............

$\qquad\qquad\qquad$ 10th term $= 10 \times 11 = 110$

and so on.

# Exercise 9g

### Worked example

The $n$th term of a sequence is given by the formula
$$n\text{th term} = (n-1)^2$$
Give the first two terms and the eighth term of the sequence.

$n = 1$  1st term $= (1-1)^2 = 0$
$n = 2$  2nd term $= (2-1)^2 = 1^2 = 1$
$n = 8$  8th term $= (8-1)^2 = 7^2 = 49$

Write down the first four terms and the seventh term of the sequence for which the $n$th term is given.

1. $2n + 1$
2. $2n - 1$
3. $2^n$
4. $n^2$
5. $(n-1)(n+1)$
6. $n + 4$
7. $3 + 2n$
8. $\dfrac{1}{n}$

## Finding an expression for the $n$th term

When we know the pattern in a sequence, we can often find an expression for the $n$th term.

Consider the sequence    2, 4, 6, 8, ...

We need to find the relationship between each term and the number, $n$, that gives its position. It is helpful to start by making a table of values of $n$ and the corresponding terms in the sequence.

| $n$ | 1 | 2 | 3 | 4 | ... |
|---|---|---|---|---|---|
| $n$th term | 2 | 4 | 6 | 8 | ... |

The pattern here is that each term is twice its position number, so the 10th term will be $2 \times 10$ and the $n$th term is $2n$.

Now we can check that this does give the correct sequence, i.e.

if $n = 1$, 1st term $= 2 \times 1 = 2$,
if $n = 2$, 2nd term $= 2 \times 2 = 4$, and so on.

Each term in the sequence 2, 4, 6, 8, ... is 2 more than the term before it and it is an example of an **arithmetic progression**.

An arithmetic progression is one in which successive terms always differ by the same amount.

## Formula for the nth term of an arithmetic progression

If the first term is $a$ and the difference between successive terms is $d$, then the second term is $a + d$. Adding another $d$ gives the third term: $a + d + d = a + 2d$. Similarly, the fourth term is $a + 3d$, and so on. Note that $d$ is called the common difference.

We can now make a table like the one above.

| $n$ | 1 | 2 | 3 | 4 | ... |
|---|---|---|---|---|---|
| $n$th term | $a$ | $a + d$ | $a + 2d$ | $a + 3d$ | ... |

The pattern is that we add on one less $d$ than the position number.

So the 20th term is $a + 19d$, and so on, so the $n$th term is $a + (n - 1)d$.

**The $n$th term of an arithmetic progression whose first term is $a$ and whose common difference is $d$ is given by**
$$n\text{th term} = a + (n - 1)d$$

### Exercise 9h

**Worked example**

→ Find the $n$th term of the sequence 6, 4, 2, 0, −2, ....

The first term is 6, so $a = 6$.

Each term is 2 less than the term before it, so $d = -2$

Therefore, using $n$th term $= a + (n - 1)d$, gives

$$n\text{th term} = 6 + (n - 1)(-2)$$
$$= 6 - 2n + 2 = 8 - 2n$$

Check: when $n = 1$, 1st term $= 8 - 2(1) = 6$
when $n = 2$, 2nd term $= 8 - 2(2) = 4$ and so on.

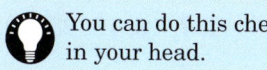 You can do this check in your head.

Find, in terms of $n$, an expression for the $n$th term of each of the following sequences.

1. 3, 6, 9, 12, ...
2. −1, −2, −3, −4, ...
3. 2, 3, 4, 5, ...
4. 0, 1, 2, 3, ...
5. 4, 8, 12, 16, ...
6. 5, 10, 15, 20, ...
7. 7, 9, 11, 13, ...
8. 0, 3, 6, 9, 12, ...
9. 3, 2, 1, 0, −1, ...
10. 3, 2.5, 2, 1.5, 1, ...

# STP Maths 9

## Geometric progressions and other sequences

A **geometric progression** is one where the ratio between successive terms is constant.

For example, 2, 6, 18, 54, 162, ... is a geometric sequence because each term is 3 times the term before it.

The table shows values of $n$ and the corresponding values of the terms.

| $n$ | 1 | 2 | 3 | 4 | ... |
|---|---|---|---|---|---|
| $n$th term | 2 | $6 (= 2 \times 3)$ | $18 (= 2 \times 3^2)$ | $54 (= 2 \times 3^3)$ | ... |

Now we can see that every term is 2 multiplied by a power of 3 that is one less than the value of $n$. Therefore the $n$ term $= 2 \times 3^{n-1}$.

## Exercise 9i

Some of the following sequences are arithmetic, some are geometric and some are neither. When a sequence is arithmetic, use the formula to find the $n$th term. For other sequences make a table like the one above and hence find an expression for the $n$th term.

1. 3, 6, 12, 24, 48, ...
2. 5, 15, 45, 135, 405, ...
3. 3, 3.2, 3.4, 3.6, 3.8, 4, ...
4. 10, 5, 2.5, 1.25, 0.625, ...
5. 0, 0.5, 1, 1.5, 2, ...
6. 2, −4, 8, −16, 32, ...
7. 5, 6, 7, 8, 9, ...
8. 2, −1, 0.5, −0.25. 0.125, ...
9. $\frac{1}{3}, \frac{1}{4}, \frac{1}{5}, \frac{1}{6}, ...$
10. $1 \times 3, 2 \times 4, 3 \times 5, ...$
11. 1, 8, 27, 64, ...
12. $1 \times 2, 2 \times 4, 3 \times 8, 4 \times 16, ...$

## Worked example

→ Find an expression for the $n$th term of the sequence
4, 7, 12, 19, 28, 39, ...

| $n$ | 1 | 2 | 3 | 4 | 5 | 6 |
|---|---|---|---|---|---|---|
| $n$th term | 4 | 7 | 12 | 19 | 28 | 39 |
| $n^2$ | 1 | 4 | 9 | 16 | 25 | 36 |

i.e. $n$th term $= n^2 + 3$

 Neither the differences nor the ratios between consecutive terms are constant, so we must try something different. Consider the square of each position number.

 Comparing the values of $n^2$ with the terms in the given sequence, we see that every term in the sequence is 3 more than the corresponding value of $n^2$.

In questions **13** to **16**, find an expression for the $n$th term of the sequence.

13. 6, 9, 14, 21, 30, ...
14. 3, 9, 19, 33, ...
15. 11, 8, 3, −4, ...

**16** 3, 10, 29, 66, ...

**17 a** Write down the $n$th term of the sequence 2, 4, 8, 16, 32,...
  **b** Show algebraically that the product of any two terms is also a term of the sequence.
  **c** What term will the product of the $n$th term and the next term give?

**18**

One of the races at a school sports day is set out with bean bags placed at 1 metre intervals along the track.

A competitor starts at S, runs to the first bag, picks it up and returns it to S. Then she runs to the second bag, picks it up and returns it to S, and so on.

How far has a competitor run when she has returned
  **a** 1 bean bag    **b** 4 bean bags    **c** $n$ bean bags?

## Mixed exercises

### Exercise 9j

**1** A number $z$ is equal to three times a number $x$ minus a number $y$. Write down a formula connecting $x$, $y$ and $z$.

**2** Use the formula $s = \frac{1}{2}(a + b + c)$ to find
  **a** $s$ when $a = 3.45$, $b = 2.76$ and $c = 4.27$
  **b** $a$ when $s = 15.1$, $b = 8.2$ and $c = 12.3$

**3** Make $d$ the subject of the formula
  **a** $C = \pi d$    **b** $a = c + d$

**4** Make $b$ the subject of the formula
  **a** $a = 7b + c$    **b** $a = c - \frac{b}{2}$

**5 a** Write down the first four terms and the twelfth term of the sequence for which the $n$th term is $n(n + 3)$.
  **b** Find, in terms of $n$, an expression for the $n$th term of the sequence 2, 6, 12, 20, ...

**6** Given that $z = \frac{x}{y}$, find
  **a** $x$ when $y = 4$ and $z = 12$    **b** $y$ when $x = 20$ and $z = 5$

**7** Use the formula $v = u + 10t$ together with $t = 2u$ to find $v$ in terms of $u$.

# STP Maths 9

### Exercise 9k

1. A rectangle is 5 cm longer than it is wide. If the perimeter of the rectangle is $P$ cm and the length is $x$ cm, find a formula connecting $P$ and $x$.

2. Given that $P = \dfrac{100I}{RT}$, find $P$ when
   a. $I = 3, R = 4$ and $T = 2$
   b. $I = 63.14, R = 5.74$ and $T = 2$

3. Given that $u = v - gt$, find $u$ when
   a. $v = 16, g = -10$ and $t = 4$
   b. $v = 13.2, g = 9.8$ and $t = 3.5$

4. Make $Q$ the subject of the formula
   a. $P = \dfrac{Q}{10}$
   b. $P = R + 3Q$

5. a. Write down the first four terms and the 20th term of the sequence for which the $n$th term is $5n - 3$.
   b. Find, in terms of $n$, an expression for the $n$th term of the sequence $3, -6, 12, -24, \ldots$

6. Solve the equation $4 - \dfrac{3}{P} = 6$

7. Substitute $r = h + 2$ into the formula $A = r^2 - 2rh$ to give $A$ in terms of $h$.

### Consider again

A newspaper headline reports that an ancient skull has been found in a peat bog. The scientists say that it belonged to a man who was between 1.78 m and 1.80 m tall. They worked this out by using data that shows that an adult's height is just over three times the distance measured round their head immediately above the eyebrows.

Now express this formula algebraically, using $H$ m for the height and $C$ m for the circumference of the head.

There is also a relationship between the length of the femur bone and height: the height in centimetres is approximately 54 plus 2.5 times the length of the femur (in centimetres).

Express this formula algebraically, using $H$ cm for the height and $L$ cm for the length of the femur.

Use these two formulas to find the length of the femur in terms of the circumference of the head.

Now can you solve these problems?

# 10 Simultaneous equations

A school governor has most, but not all, of the information she needs to carry out an audit of the equipment in Highfield School.

There are two types of desk in the school: those that seat two students and those that seat three students. The governor knows that there are 200 desks in total and that 550 students can be seated at these desks but she does not know how many there are of each type.

- The school governor can go round the school and count the numbers of each type of desk but this is likely to take an hour or longer.
- She can use algebraic methods to find the information she needs and this will take about one minute. Can you do this for her?

*You should be able to answer this question after you have worked through this chapter.*

## Class discussion

There are several ways in which we can obtain answers to the following problems. Do not attempt to solve these problems, but discuss at least two methods for solving them and the efficiency of each method, that is, how long it would take and how accurate an answer it would be likely to give.

1  Glasses suitable for hot drinks are sold either in packs of four with one 'free' glass holder at £9.50 a pack, or in packs of six with one 'free' glass holder at £13.20 a pack. Assuming that the glasses are the same price in both packs, is the 'free' holder really free?

2  A 20 m long rod is cut into two pieces so that one piece is 2 m longer than half the other piece. How long is the shorter piece?

## Solving simultaneous equations

Your discussion of the questions in the class discussion will show you that algebraic methods often give the quickest and most accurate answers to problems. In Book 8 we found that it may be possible to solve a problem that involves two unknown quantities by forming two equations and then solving them simultaneously. The method used was to eliminate one of the letters by addition or by subtraction of the given equations. This works when the **coefficient** (number) of one of the letters is the same in both equations. You can use the Summary of Years 7 and 8 and Revision exercise 6 at the front of this book to remind yourself of the method.

In this chapter we look at equations where we need to adapt one or both of the equations before we can add or subtract to eliminate one of the letters.

Consider this problem:

The seats in an aeroplane can be arranged in either groups of two or in groups of four. A total of 160 groups are to be used for 450 passengers. How many two-seat groups are needed?

There are two unknown numbers here; the number of two-seat groups and the number of four-seat groups.

If there are $x$ two-seat groups and $y$ four-seat groups, then, as there are 160 groups in total,

$$x + y = 160 \qquad [1]$$

The number of passengers in two-seat groups is $2x$ and the number of passengers in four-seat groups is $4y$. As there are 450 passengers in total, we have

$$2x + 4y = 450 \qquad [2]$$

We now have two **simultaneous equations**, but they do not have the same number of $x$s or of $y$s. But if we multiply equation [1] by 2, we get $2x + 2y = 320$. We can now solve this equation and equation [2] using the same method as in Book 8.

The solution can be set out as follows.

$$x + y = 160 \qquad [1]$$
$$2x + 4y = 450 \qquad [2]$$
$[1] \times 2$ gives $\quad 2x + 2y = 320 \qquad [3]$
$[2] - [3]$ gives $\quad\quad\quad 2y = 130$
so $\quad\quad\quad\quad\quad\quad y = 65$

Substituting 65 for $y$ in [1] gives

$$x = 95$$

Therefore, there are 95 two-seat groups.

Notice that we have numbered the equations and explained what we have done with them – this is essential so that other people can follow the argument.

## Exercise 10a

### Worked example

 Solve the equations $\quad 3x - 2y = 1$
$\quad\quad\quad\quad\quad\quad\quad\quad\quad\quad 4x + y = 5$

$3x - 2y = 1 \quad [1]$
$4x + y = 5 \quad [2]$

💡 Multiplying [2] by 2 gives us the same number of $y$s as in [1].

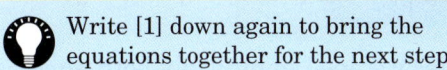

| | | |
|---|---|---|
| [2] × 2 gives | $8x + 2y = 10$ | [3] |
| | $3x - 2y = 1$ | [1] |
| [1] + [3] gives | $11x = 11$ | |
| ∴ | $x = 1$ | |

Write [1] down again to bring the equations together for the next step.

Substitute 1 for $x$ in [2]    $4 + y = 5$
Take 4 from both sides    $y = 1$

*Check* in [1]:    LHS $= 3 - 2 = 1 =$ RHS

Therefore, the solution is $x = 1, y = 1$

Solve the following pairs of equations.

1. $3x + 2y = 11$
   $2x + y = 7$

2. $5x - 4y = -3$
   $3x + y = 5$

3. $9x + 7y = 10$
   $3x + y = 2$

4. $5x + 3y = 21$
   $2x + y = 3$

5. $6x - 4y = -4$
   $5x + 2y = 2$

6. $4x + 3y = 25$
   $x + 5y = 19$

## Worked example

 Solve the equations    $5x + 3y = 7$
                                              $10x + 4y = 16$

$5x + 3y = 7$    [1]
$10x + 4y = 16$    [2]

[1] × 2 gives    $10x + 6y = 14$    [3]
                 $10x + 4y = 16$    [2]
[3] − [2] gives    $2y = -2$
                   $y = -1$

We will subtract [2] from [3] so we write [2] again underneath.

Substitute −1 for $y$ in [1]    $5x - 3 = 7$
Add 3 to both sides    $5x = 10$
                       $x = 2$

*Check* in [2]:    LHS $= 20 + (-4) = 16 =$ RHS

Therefore, the solution is $x = 2, y = -1$

Solve the pairs of equations.

7. $5x + 3y = 11$
   $4x + 6y = 16$

8. $2x - 3y = 1$
   $5x + 9y = 19$

9. $2x + 5y = 1$
   $4x + 3y = 9$

10. $9x + 5y = 15$
    $3x - 2y = -6$

11. $4x + 3y = 1$
    $16x - 5y = 21$

12. $7p + 2q = 22$
    $3p + 4q = 11$

199

## Exercise 10b

Sometimes we need to multiply both equations before we can add or subtract to eliminate a letter.

### Worked example

Solve the equations $\quad 3x + 5y = 6$
$\quad\quad\quad\quad\quad\quad\quad\quad\quad\quad 2x + 3y = 5$

$3x + 5y = 6 \quad$ [1]
$2x + 3y = 5 \quad$ [2]

 We can get $15y$ in both equations from [1] $\times$ 3 and [2] $\times$ 5. (Alternatively, we could get $6x$ in both equations from [1] $\times$ 2 and [2] $\times$ 3.)

[1] $\times$ 3 gives $\quad 9x + 15y = 18 \quad$ [3]
[2] $\times$ 5 gives $\quad 10x + 15y = 25 \quad$ [4]
$\quad\quad\quad\quad\quad\quad\quad 9x + 15y = 18 \quad$ [3]

 We will subtract [3] from [4] so we write [3] again underneath [4].

[4] − [3] gives $\quad\quad x = 7$

Substitute 7 for $x$ in [2] $\quad 14 + 3y = 5$
Take 14 from both sides $\quad 3y = -9$
Divide both sides by 3 $\quad\quad y = -3$

*Check* in [1]: LHS $= 21 - 15 = 6 =$ RHS

Therefore, the solution is $x = 7, y = -3$

Solve the following pairs of equations.

**1** $\quad 2x + 3y = 12$
$\quad\quad 5x + 4y = 23$

**2** $\quad 3x - 2y = -7$
$\quad\quad 4x + 3y = 19$

**3** $\quad 2x - 5y = 1$
$\quad\quad 5x + 3y = 18$

**4** $\quad 6x + 5y = 9$
$\quad\quad 4x + 3y = 6$

**5** $\quad 14x - 3y = -18$
$\quad\quad 6x + 2y = 12$

**6** $\quad 6x - 7y = 25$
$\quad\quad 7x + 6y = 15$

**7** $\quad 5x + 4y = 21$
$\quad\quad 3x + 6y = 27$

**8** $\quad 9x + 8y = 17$
$\quad\quad 2x - 6y = -4$

**9** $\quad 9x - 2y = 14$
$\quad\quad 7x + 3y = 20$

**10** $\quad 5x + 4y = 11$
$\quad\quad\; 2x + 3y = 3$

**11** $\quad 4x + 5y = 26$
$\quad\quad\; 5x + 4y = 28$

**12** $\quad 2x - 6y = -6$
$\quad\quad\; 5x + 4y = -15$

**13** $\quad 5x - 6y = 6$
$\quad\quad\; 2x + 9y = 10$

**14** $\quad 3p + 4q = 5$
$\quad\quad\; 2p + 10q = 18$

**15** $\quad 6x + 5y = 8$
$\quad\quad\; 3x + 4y = 1$

**16** $\quad 7x - 3y = 20$
$\quad\quad\; 2x + 4y = -4$

**17** $\quad 10x + 3y = 12$
$\quad\quad\; 3x + 5y = 20$

**18** $\quad 6x - 5y = 4$
$\quad\quad\; 4x + 2y = -8$

**19** $\quad 5x + 3y = 8$
$\quad\quad\; 3x + 5y = 8$

**20** $\quad 7x + 2y = 23$
$\quad\quad\; 3x - 5y = 4$

**21** $\quad 6x - 5y = 17$
$\quad\quad\; 5x + 4y = 6$

**22** $\quad 3x + 8y = 56$
$\quad\quad\; 5x - 6y = 16$

**23** $\quad 7x + 3y = -9$
$\quad\quad\; 2x + 5y = 14$

**24** $\quad 7x + 6y = 0$
$\quad\quad\; 5x - 8y = 43$

**25** $\quad 2x + 6y = 30$
$\quad\quad\; 3x + 10y = 49$

**26** $\quad 4x - 3y = -7$
$\quad\quad\; 3x + 2y = 16$

**27** $\quad 17x - 2y = 47$
$\quad\quad\; 5x - 3y = 9$

**28** $\quad 8x + 3y = -17$
$\quad\quad\; 7x - 4y = 5$

## Exercise 10c

In this exercise, you can solve some pairs of equations without having to multiply either equation, for some you will need to multiply one equation by a number and for others you will need to multiply both equations by numbers.

Solve the following pairs of equations.

1. $x + 2y = 9$
   $2x - y = -2$

2. $x + y = 4$
   $x + 2y = 9$

3. $2x + 3y = 0$
   $3x + 2y = 5$

4. $3x + 2y = -5$
   $3x - 4y = 1$

5. $x + y = 6$
   $x - y = 1$

6. $3x - 5y = 13$
   $2x + 5y = -8$

7. $3x - y = -10$
   $4x - y = -4$

8. $5x + 2y = 16$
   $2x - 3y = -5$

9. $2x - 5y = 1$
   $3x + 4y = 13$

10. $7x + 3y = 35$
    $2x - 5y = 10$

11. $9x + 2y = 8$
    $7x + 3y = 12$

12. $3x - 2y = -2$
    $5x - y = -15$

Sometimes the equations are arranged in an awkward way and need to be rearranged before we can solve them.

### Worked example

→ Solve the equations $\quad x = 4 - 3y$
$\qquad\qquad\qquad\qquad\quad 2y - x = 6$

We must first arrange the letters in the same order in both equations. By adding $3y$ to both sides, we can write equation [1] $3y + x = 4$.

$\quad x = 4 - 3y \qquad$ [1]
$2y - x = 6 \qquad$ [2]

[1] becomes $\quad 3y + x = 4 \qquad$ [3]
$\qquad\qquad\qquad 2y - x = 6 \qquad$ [2]

[3] + [2] gives $\quad 5y = 10$
$\qquad\qquad\qquad\quad y = 2$

Substitute 2 for $y$ in [1] $\quad x = 4 - 6$
$\qquad\qquad\qquad\qquad\qquad\quad x = -2$

Check in [2]: $\quad$ LHS $= 4 - (-2) = 6 = $ RHS

Therefore, the solution is $x = -2, y = 2$

Solve the following pairs of equations.

13. $y = 6 - x$
    $2x + y = 8$

14. $x - y = 2$
    $2y = x + 1$

15. $3 = 2x + y$
    $4x + 6 = 10y$

16. $9 + x = y$
    $x + 2y = 12$

17. $2y = 16 - x$
    $x - 2y = -8$

18. $3x + 4y = 7$
    $2x = 5 - 3y$

STP Maths 9

As long as the $x$ and $y$ and number terms are in corresponding positions in the two equations, they do not need to be in the order we have had so far.

### Worked example

→ Solve the equations $\quad y = x + 5$
$\qquad\qquad\qquad\qquad\quad y = 7 - x$

$$y = x + 5 \quad [1]$$
$$y = 7 - x \quad [2]$$
Rewrite [1] as $\quad y = 5 + x \quad [3]$
[2] + [3] gives $\quad 2y = 12$
$\qquad\qquad\qquad y = 6$
Substitute 6 for $y$ in [1] $\quad 6 = x + 5$
$\qquad\qquad\qquad\qquad\quad x = 1$

*Check* in [2]: $\quad$ LHS $= 6$
$\qquad\qquad\qquad$ RHS $= 1 + 5 = 6 =$ LHS

Therefore, the solution is $x = 1, y = 6$

Solve the following pairs of equations.

**19** $\;y = 9 + x$
$\quad\; y = 11 - x$

**20** $\;x = 3 + y$
$\quad\; 2x = 4 - y$

**21** $\;y = 4 - x$
$\quad\; y = x + 6$

**22** $\;2y = 4 + x$
$\quad\; y = x + 8$

**23** $\;x + 4 = y$
$\quad\; y = 10 - 2x$

**24** $\;x + y = 12$
$\quad\; y = 3 + x$

## Special cases

Some pairs of equations have no solution and some have an infinite number of solutions.

### Exercise 10d

Try solving the following pairs of equations. Comment on why the method breaks down.

**1** $\;x + 2y = 6$
$\quad\; x + 2y = 7$

**2** $\;3x + 4y = 1$
$\quad\; 6x + 8y = 2$

**3** $\;y = 4 + 2x$
$\quad\; y - 2x = 6$

**4** $\;9x = 3 - 6y$
$\quad\; 3x + 2y = 1$

**5** Make up other pairs of equations which either have no solution or have an infinite set of solutions.

# 10 Simultaneous equations

## Solving problems

We can often find answers to problems involving two unknown quantities by forming a pair of simultaneous equations and solving them.

In any problem, it is important to
- read the problem carefully and make sure you understand it
- identify the unknown quantities
- allocate letters to the unknown numbers when this is not done for you
- look for information that you can use to give relationships between the unknown numbers
- draw a diagram when the problem is about a geometric figure; this will often help you to identify information that is not spelled out in the problem
- check that the answer you obtain fits the information given.

## Exercise 10e

### Worked example

→ In a right-angled triangle, the two smaller angles are such that one is twice the size of the other. Find the smallest angle.

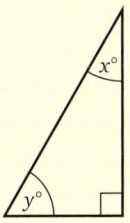

First we draw a right-angled triangle. The size of the two smaller angles is unknown, so we let one be $x°$ and the other be $y°$. There is no need to write this down; it is sufficient to show the letters clearly in the diagram.

This information is given in the question.

$y = 2x$     [1]

$y + x = 90$     [2]

We know that the sum of the angles in a triangle is 180°.

[1] becomes    $y - 2x = 0$     [3]
[2] − [3] gives    $3x = 90$
                   $x = 30$

The smallest angle is 30°.

*Check*: using the information in the question, if the smallest angle is 30°, this gives the other angles as 60° and 90° and they all add up to 180°.

1. The perimeter of a rectangle is 30 cm. The shorter side is $x$ cm long and is 3 cm less than the longer side, which is $y$ cm long. Use this information to form two equations in $x$ and $y$. Find the length of the shorter side.

2. In a pentagon, four of the angles are each $x°$. The fifth angle is $y°$ and is half the size of each of the others. Form two equations in $x$ and $y$. Find the size of the largest angles.

3. The perimeter of an isosceles triangle is 280 mm. The base of the triangle is 70 mm longer than the other two sides. Find the length of one of the equal sides.

203

STP Maths 9

**Worked example**

A shop makes a profit of 6 pence on each brown roll sold and a profit of 8 pence on each white roll sold. One Saturday the shop sells 180 rolls and makes a profit of £12.40 on these rolls.
How many brown rolls were sold?

If $x$ brown rolls and $y$ white rolls were sold,

then $\quad\quad 6x + 8y = 1240 \quad\quad$ [1]
and $\quad\quad\quad x + y = 180 \quad\quad\quad$ [2]

1 brown roll gives a profit of 6 p, so $x$ brown rolls gives a profit of $6x$ p. Likewise, $y$ white rolls gives a profit of $8y$ p. The profit on the rolls sold is £12.40, i.e. 1240 p.

[2] × 6 gives $6x + 6y = 1080 \quad\quad$ [3]
$\quad\quad\quad\quad\quad 6x + 8y = 1240 \quad\quad$ [1]
$\quad\quad\quad\quad\quad 6x + 6y = 1080 \quad\quad$ [3]
[1] − [3] gives $\quad 2y = 160$, so $y = 80$

180 rolls are sold.

Substituting 80 for $y$ in [2] gives $x + 80 = 180$
$\quad\quad\quad\quad\quad\quad\quad\quad\quad\quad\quad\quad\quad\quad x = 100$

100 brown rolls are sold.

*Check*: profit on 100 brown rolls is £6, profit on 80 white rolls is £6.40, so total profit is £12.40.

**4** When three times a number, $x$, is added to a second number, $y$, the total is 33. The first number added to three times the second number is 19.
   **a** Form two equations connecting $x$ and $y$.
   **b** Find the two numbers.

**5** Find two numbers such that twice the first added to the second is 26 and the first added to three times the second is 28.

**6** Find two numbers such that twice the first added to the second gives 27 and twice the second added to the first gives 21.

**7** A cutlery manufacturer packs 200 teaspoons into boxes holding either four spoons or six spoons. There are 48 full boxes.
   **a** If there are $a$ four-spoon boxes and $b$ six-spoon boxes, use the information to find two equations relating $a$ and $b$.
   **b** How many boxes hold four spoons?

**8** The manager of a bookshop ordered 70 copies of *Teen* magazine; $s$ of these were charged at £1.90 each and the remainder, $t$, were charged at a discounted price of £1.70 p each.
The bill for all 70 copies was £124.
How many magazines were charged at the discounted price?

9   Airdale Study Centre can accommodate 128 students in 20 rooms, $x$ of which have eight beds and $y$ of which have four beds. How many four-bed rooms are there?

10  The charge at a leisure centre for renting a tennis court for 30 minutes is £4 for non-members and £1 for members, who also pay an annual fee of £50.
    a  Write down an equation for $C$ in terms of $n$, where £$C$ is the cost for a non-member to hire the court for $n$ hours.
    b  Write down another equation for $C$ in terms of $n$, where £$C$ is the total cost for a member to hire the court for $n$ hours in one year.
    c  Hence find the number of hours a year a member would have to hire the court for it to be worth paying the membership fee.

11  The equation of a line is $y = mx + c$. The line goes through the points $(1, 3)$ and $(2, 5)$. Find the equation of this line.

Questions **12** and **13** were in the class discussion. Now solve them.

12  A 20 m long rod is cut into two pieces so that one piece is 2 m longer than half the other piece. How long is the shorter piece?

13  Glasses suitable for hot drinks are sold either in packs of four with one 'free' glass holder at £9.50 a pack, or in packs of six with one 'free' glass holder at £13.20 a pack. Assuming that the glasses are the same price in both packs, is the 'free' holder really free?

## Graphical solutions

In Book 8 we saw that simultaneous equations can be solved graphically. For example, to solve the equations

$$x + 3y = 6 \quad [1]$$
and $\quad 3x - y = 6 \quad [2]$

we start by rearranging each equation so that they are in the form $y = ...$, that is, we make $y$ the subject of the equation.

[1]   becomes   $3y = -x + 6$
      giving    $y = -\tfrac{1}{3}x + 2$ $\quad$ [3]

[2]   becomes   $3x = y + 6$
      giving $3x - 6 = y$, i.e. $y = 3x - 6$ $\quad$ [4]

Now we can plot the lines whose equations are

$y = -\tfrac{1}{3}x + 2$
and $\quad y = 3x - 6$

We can do this by drawing up a table of values and plotting the points on graph paper, or by using a graphics calculator or a graph drawing program on a computer.

Using the values $0 \leq x \leq 5$ and $0 \leq y \leq 5$ gives this display using a graph drawing program.

The solution is the point where the two lines intersect. The solution given by the program is $x = 2.4$ and $y = 1.2$.

We do not know whether this solution is exact or correct to 1 decimal place.

We can find out by solving the equations algebraically.

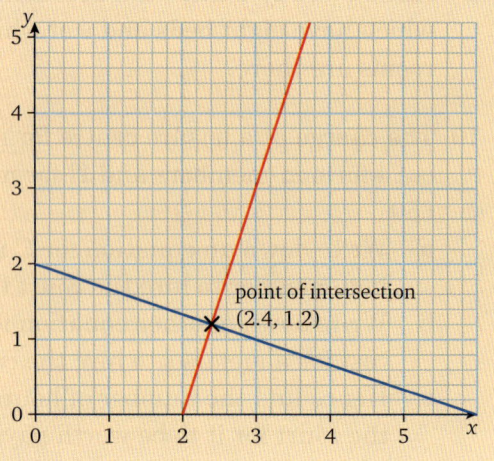

point of intersection (2.4, 1.2)

### Exercise 10f

1  In the above explanation we used a graphical method to solve the equations
$$x + 3y = 6 \quad \text{and} \quad 3x - y = 6$$
Solve these equations algebraically.

Comment on the advantages and disadvantages of each method of solution.

Solve the following equations graphically, either by plotting the graphs on graph paper or by using appropriate technology. In either case, use values for $x$ and $y$ in the ranges given. Check your answers by solving the equations algebraically.

2  $3x + 2y = 9$   $\quad 0 \leq x \leq 4, -2 \leq y \leq 5$
   $2x - 2y = 3$

3  $2x + 3y = 4$   $\quad -2 \leq x \leq 2, 0 \leq y \leq 4$
   $y = x + 2$

**The examples in this exercise show that, when using graphical methods, we cannot always tell whether solutions obtained are exact or not.**

### Consider again

There are two types of desk in Highfield School: those that seat two students and those that seat three students. The school governor knows that there are 200 desks in total and that 550 students can be seated at these desks but she does not know how many there are of each type.
- The governor can go round the school and count the numbers of each type of desk but this is likely to take an hour or longer.
- She can use algebraic methods to find the information she needs and this will take her about one minute.

Now can you use algebraic methods to solve this problem for the school governor?

# Summary 2

## Algebraic expressions

The number, including the sign, that is multiplied by a letter is called the *coefficient* of that letter. In the expression $3x - 4y$, for example, the coefficient of $x$ is 3 and the coefficient of $y$ is $-4$.

### Product of two brackets

$(a + b)(c + d)$ means $a \times (c + d) + b \times (c + d)$
$= ac + ad + bc + bd$

that is, each term in the second bracket is multiplied by each term in the first bracket. The order in which the terms are multiplied does not matter, but it is sensible to stick to the same order each time,

e.g. $(2x - 3)(4x + 5) = (2x)(4x) + (2x)(5) + (-3)(4x) + (-3)(5)$
$= 8x^2 + 10x - 12x - 15 = 8x^2 - 2x - 15$

In particular, when squaring a bracket we can use

$(x + a)^2 = (x + a)(x + a) = x^2 + 2ax + a^2$
or $(x - a)^2 = (x - a)(x - a) = x^2 - 2ax + a^2$

The product of two brackets that are the same except for the sign between the two terms is called the **difference between two squares.**

i.e. $(x + a)(x - a) = x^2 - a^2$

## Factorising

Factorising is the reverse of the process of expanding (multiplying out) an algebraic expression.

A *common factor* of two or more terms can be seen by inspection and can be 'taken outside a bracket',

e.g. the terms $2ab + 4bc$ both have $2b$ as factors,

so $2ab + 4bc = 2b(a + 2c)$. (We can check this by expanding the result.)

To factorise an expression such as $x^2 + 3x - 10$, we look for two brackets whose product is equal to the original expression.

We start by writing $x^2 + 3x - 10 = (x + \phantom{x})(x - \phantom{x})$

The sign in each bracket is determined by the signs in the original expression, i.e.

a +ve number term *and* a +ve $x$-term gives a '+' sign in both brackets

a +ve number term *and* a $-$ve $x$-term gives a '$-$' sign in both brackets

a $-$ve number term gives '+' in one bracket and '$-$' in the other bracket.

The numbers at the ends of the brackets have to satisfy two conditions:
- their product has to be equal to the number at the end of the original expression
- collecting the product of the outside terms in the brackets and the inside terms in the brackets must give the $x$ term in the original expression.

In the case of $x^2 + 3x - 10$, the product is $-10$, so the numbers could be 10 and $-1$, $-10$ and 1, 5 and $-2$ or $-5$ and 2.

We try each pair in turn until we find a pair (if there is one) which give the correct $x$ term when the brackets are expanded:
$$x^2 + 3x - 10 = (x + 5)(x - 2)$$

If we cannot find two numbers that satisfy the conditions, the expression does not factorise.

### Formulas

The formula $v = u + st$ gives $v$ in terms of $u$, $s$ and $t$; $v$ is called the *subject of the formula*.

When the formula is rearranged to give $s = \dfrac{v - u}{t}$, $s$ is the subject.

The process of rearranging $v = u + st$ as $s = \dfrac{v - u}{t}$ is called *changing the subject of the formula*. It is achieved by thinking of $v = u + st$ as an equation which has to be 'solved' to find $s$.

Start by isolating the term containing $s$ on one side of the formula:

take $u$ from both sides:     $v - u = st$

divide both sides by $t$:     $\dfrac{v - u}{t} = s$, i.e. $s = \dfrac{v - u}{t}$

### Sequences

The $n$th term of a sequence is sometimes expressed in terms of $n$, which is the position number of the term. We can then find any term of the sequence by giving $n$ a numerical value.

For example, when the $n$th term $= 3n - 2$,

the 10th term is given by substituting 10 for $n$, i.e. by $3(10) - 2 = 28$

An *arithmetic progression* is a sequence where successive terms always differ by the same amount, for example, 4, 4.5, 5, 5.5, 6, ...

If the first term is $a$ and the difference between successive terms is $d$, then the second term is $a + d$. Adding another $d$ gives the third term, $a + d + d = a + 2d$, and so on. The formula for the $n$th term is $a + (n - 1)d$

A *geometric progression* is one where the ratio between successive terms is constant, i.e. each term is multiplied by the same quantity to get the next term, for example, 3, 6, 12, 24, ...

### Simultaneous equations

We can solve a pair of simultaneous equations in two unknowns algebraically by eliminating one of the letters. It may be necessary to multiply one or both equations by numbers to make the coefficients of one of the letters the same in each equation.

For example, to solve $2y - x = 7$   [1]
and $3y + 4x = 5$   [2]
[1] × 4 gives $8y - 4x = 28$   [3]
then [2] + [3] eliminates $x$ to give $11y = 33$
so $y = 3$ and, from [1], $x = -1$

## Statistics

### Mean value

The *mean* value of a frequency distribution is given by $\dfrac{\Sigma fx}{\Sigma f}$

where $x$ is the value of an item and $f$ is its frequency, and $\Sigma$ means 'the sum of all such items'.

In the case of grouped data, $x$ is the *midclass value* and the mean obtained is an estimate.

### Cumulative frequency

*Cumulative frequency* is the sum of the frequencies of all values up to and including a particular value.

A *cumulative frequency polygon* is drawn by plotting the cumulative frequencies against the upper ends of the groups and joining the points with straight lines. A *cumulative frequency curve* is obtained by drawing a smooth curve through the points.

For example, the grouped frequency distribution given in this table of heights of tomato plants gives this cumulative frequency curve.

| Height, $h$ (cm) | $20 \leqslant h < 30$ | $30 \leqslant h < 40$ | $40 \leqslant h < 50$ | $50 \leqslant h < 60$ |
|---|---|---|---|---|
| Cumulative frequency | 5 | 20 | 45 | 55 |

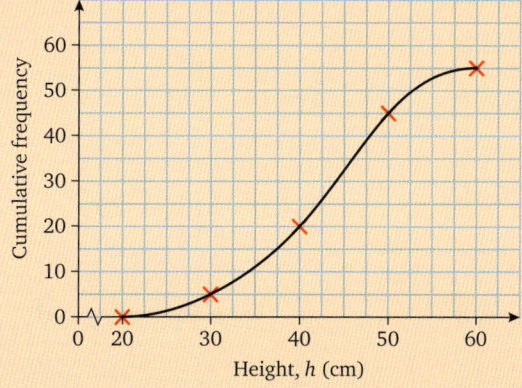

The *median* of a grouped distribution of $n$ values is the $\dfrac{n}{2}$th value in order of size and is denoted by $Q_2$.

The median can be estimated from a cumulative frequency curve.

The *lower quartile* is the value that is $\frac{1}{4}$ of the way through a set of values arranged in order of size and is denoted by $Q_1$.

The *upper quartile* is the value that is $\frac{3}{4}$ of the way through a set of values arranged in order of size and is denoted by $Q_3$.

For a grouped distribution, $Q_1$ is the $\frac{n}{4}$th value, and $Q_3$ is the $\frac{3n}{4}$th value.

$Q_1$ and $Q_3$ can be estimated from a cumulative frequency curve.

**Interquartile range**

The interquartile range is the difference between the upper and lower quartiles,

i.e. $Q_3 - Q_1$

The graph shows the median, the upper and lower quartiles and the interquartile range of the distribution described above.

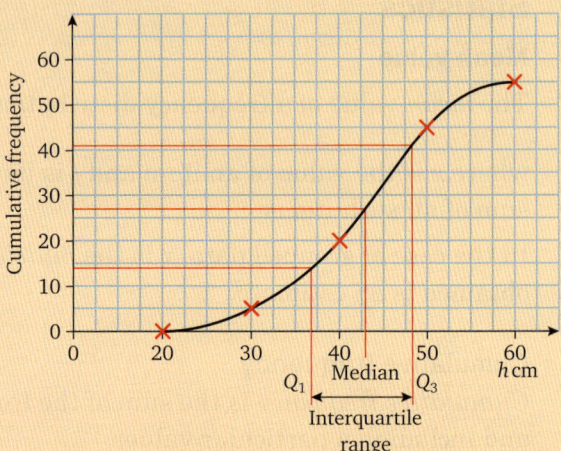

## REVISION EXERCISE 2.1 (Chapters 6 and 7)

**1** Expand
  **a** $3(1 - x)$
  **b** $(a + b)(a + 2b)$
  **c** $(x - 3)(x - 6)$

**2** Expand
  **a** $(x + 7)(x - 3)$
  **b** $(x - 4)(3x + 5)$
  **c** $(3x - 2)(5x + 1)$

**3** Expand
  **a** $(x + 5)^2$
  **b** $(2x + 3y)(2x - 3y)$
  **c** $(x + 1)(x + 2) + x(x + 3)$

**4** Factorise
  **a** $4a + 12$
  **b** $x^2 - 5x$
  **c** $b^3 - 3b^2$

**5** Factorise
  **a** $x^2 + 11x + 18$
  **b** $x^2 - 6x + 8$
  **c** $x^2 - 10x + 25$

**6** Factorise
  **a** $x^2 - 2x - 15$
  **b** $x^2 + 2x - 8$
  **c** $18 + 9x + x^2$
  **d** $14 + 5x - x^2$

**7** Factorise
  **a** $a^2 - b^2$
  **b** $p^2 - 4q^2$
  **c** $16x^2 - y^2$

**8** Find, without using a calculator
  **a** $67^2 - 33^2$
  **b** $7.31^2 - 2.69^2$

## REVISION EXERCISE 2.2 (Chapters 8 to 10)

1. Forty boxes of oranges were examined and the number of bad oranges in each box was noted.

   | Number of bad oranges | 0–5 | 6–11 | 12–17 | 18–23 |
   |---|---|---|---|---|
   | Frequency | 25 | 8 | 5 | 2 |

   a What is the modal group for this distribution?
   b Estimate the mean number of bad oranges per box.

2. Use the frequency table from question **1** to make a cumulative frequency table for this data.
   Draw the cumulative frequency curve and use it to find the median number of bad oranges per box, the upper and lower quartiles and the interquartile range.

3. The masses of a group of boys were measured and the results recorded in the following table.

   | Mass, $m$ (kg) | $60 \leqslant 64$ | $64 \leqslant 69$ | $69 \leqslant 74$ | $74 \leqslant 79$ | $79 \leqslant 84$ | $84 \leqslant 89$ |
   |---|---|---|---|---|---|---|
   | Frequency | 3 | 7 | 13 | 15 | 11 | 6 |

   a How many boys were there in the group?
   b How many boys had a mass of more than 79 kg?
   c Use the data to draw a cumulative frequency curve and use your curve to find
      i the median mass     ii the upper and lower quartiles.

4. Write down a formula connecting the given letters.
   a A number $x$ is equal to three times the sum of two numbers $p$ and $q$.
   b A number $z$ is twice the product of two numbers $x$ and $y$.
   c A number $z$ is the sum of $x$ and $y$ minus their product.

5. Given that $a = b^2 + c$, find $a$ when
   a $b = 2$ and $c = 3$
   b $b = -1$ and $c = 4$
   c $b = -3$ and $c = -2$

6. a Write down the first four terms and the tenth term of the sequence for which the $n$th term is $2n - 5$.
   b Find, in terms of $n$, an expression for the $n$th term of the sequence 4, 8, 12, 16, …

7. a Solve the equation $4 = \dfrac{12}{x}$
   b Given that $R = 3st$ and that $s = 5t$, find $R$ in terms of $t$.

8. Solve the simultaneous equations.
   a $7x + y = 25$
      $2x + y = 5$
   b $7x + 5y = 30$
      $4x - y = 21$

**9** Solve the simultaneous equations.

   **a**    $4x - 3y = 11$
           $6x + 4y = 8$

   **b**    $3x + 7y = 27$
           $5x + 2y = 16$

**10 a** Two numbers, $x$ and $y$, are such that three times the first number added to twice the second number is 24 and the first number added to twice the second number is 23.
     **i** Form two equations relating $x$ and $y$.
     **ii** Solve the equations to find the two numbers.

   **b** Solve the following simultaneous equations graphically, either by plotting the graphs on graph paper or by using a graphics calculator. In either case, use values for $x$ and $y$ in the ranges given.
       $3x + 2y = 11$     $2 \leqslant x \leqslant 4,\ -1 \leqslant y \leqslant 5$
       $5x - 2y = 10$

## REVISION EXERCISE 2.3 (Chapters 6 to 10)

**1** Expand
   **a**   $5(2 - 3a)$            **c**   $(p - 3)(p - 8)$
   **b**   $(3x + 2y)(1 - 2z)$    **d**   $(3x + 4)(3 - x)$

**2** Expand and simplify
   **a**   $(3x - 1)^2$           **b**   $(x + 2)(x + 3) + 4(x + 1)$

**3** Factorise
   **a**   $15 - 5x$           **c**   $x^2 - 10x + 21$
   **b**   $x^2 - 2x - 15$      **d**   $2x^2 + 4x + 8$

**4** Factorise
   **a**   $12 - 7x + x^2$      **c**   $9 - a^2$
   **b**   $12 + x - x^2$       **d**   $25 - x^2$

**5** The lengths of 35 leaves taken from the same tree are given in the table.

| Length, $h$ (cm) | $8 \leqslant h < 8.5$ | $8.5 \leqslant h < 9$ | $9 \leqslant h < 9.5$ | $9.5 \leqslant h < 10$ |
|---|---|---|---|---|
| Frequency | 3 | 10 | 18 | 4 |

   **a** Estimate the mean length of the leaves.
   **b** Construct a cumulative frequency table and draw a cumulative frequency curve.
   **c** Estimate
     **i** the median length of these leaves
     **ii** the upper and lower quartiles, and the interquartile range.

**6** Grapefruit are sold at $x$ pence each. The cost of $n$ grapefruit is $C$ pence. Find a formula for $C$ in terms of $x$ and $n$.

**7** If $R = (p + 2q)^2$, find $R$ when
   **a**   $p = 10$ and $q = -3$      **b**   $p = 2.1$ and $q = 3.7$

**8** Given that $V = 2xy^2 + \dfrac{y^4}{x}$ and that $y = 2x$, find $V$ in terms of $x$.

# Summary 2

**9** Solve the simultaneous equations.
  **a**  $3x + 4y = 17$
      $2x + y = 8$
  **b**  $x = 5 + 3y$
      $2y = 3x - 8$

## REVISION EXERCISE 2.4 (Chapters 1 to 10)

**1 a** Esther drives for $1\frac{3}{4}$ hours at an average speed of 72 km/h. How far does she travel?
  **b** Rohan cycles at 20 km/h. How long will it take him to cycle 36 km?
  **c** Penny runs 3 miles at 12 mph then walks 2 miles at 4 mph. Find her average speed for the whole journey.

**2** A coach leaves Wexley at noon to travel to Swanson, a town 95 miles away. It travels the first 40 miles at an average speed of 40 mph and, without stopping, completes the journey to Swanson arriving there at 2.30 p.m. A second coach leaves Swanson and travels at 60 mph, arriving at Wexley at 2.44 pm.
Taking 2 cm to represent 10 miles on one axis and 20 minutes on the other, draw travel graphs for the two journeys and use them to find
  **a** the time at which the second coach begins its journey
  **b** when and where the two coaches pass each other
  **c** the average speed of the first coach for the whole journey.

**3 a** Illustrate on a number line the range of possible values of a number if it is given as 0.7 correct to 1 decimal place.
  **b** One evening 100 people were asked if they had drunk coffee that morning. To the nearest ten, 60 people said that they had. What is the largest possible number who had not drunk coffee that morning?
  **c** Find $\left(2 - \dfrac{1}{1 + \frac{1}{3}}\right) \div \left(4 + \dfrac{1}{2 + \frac{3}{4}}\right)$

**4** Two dice are rolled one after the other. Find the probability of getting
  **a** a 3 on the first dice
  **b** a double 3
  **c** a double 3 or a double 4.

**5** After a pay rise of 3%, Sheila's weekly wage is £226.60. How much did she earn before the rise?

**6 a** A page of text in a book is 42 lines, with an average of 16 words per line. The text is rewritten with an average of 14 words per line. How many lines will be needed?
  **b** In a mixed choir of 156 people there are 72 men. What is the ratio of the number of men to the number of women?
  **c** The cost of a 25 kg bag of potatoes is £15.75. At the same price per kilogram, how much will a bag containing 12 kg cost?

**7** Expand
  **a** $5a(2b + 3c)$
  **b** $(1 - a)(a + 3b)$
  **c** $(7x + 3)(x + 1)$
  **d** $(x - 4y)^2$

8 Factorise
   a $x^2 + 6x$
   b $x^2 - 10x + 24$
   c $8 + x^2 - 6x$
   d $p^2 - 36$

9 The bar chart shows the number of flowers on a new variety of plant grown from the seeds in one packet.
   a Make a frequency table and estimate the mean number of flowers per plant.
   b Make a cumulative frequency table for this data and draw the cumulative frequency curve. Hence find estimates for the median, the upper and lower quartiles and the interquartile range.

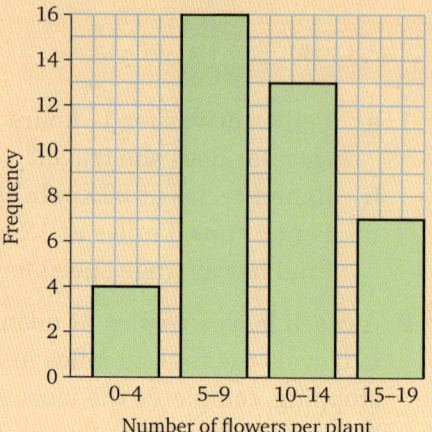

10 a Make $T$ the subject of the formula $I = \dfrac{PRT}{100}$.
     Find $T$ when $I = 28$, $P = 350$ and $R = 4$.
   b Make $s$ the subject of the formula $v^2 = u^2 + 2as$.

11 Solve the simultaneous equations.
   a  $6x - 7y = 31$
      $7x + 6y = 22$
   b  $y = x + 6$
      $2y = 15 + x$

## REVISION EXERCISE 2.5 (Chapters 1 to 10)

1  a  Write down the reciprocal of
       i 20    ii $\frac{1}{8}$    iii 1.8
   b  Use dot notation to write the following fractions as decimals.
       i $\frac{9}{11}$    ii $\frac{11}{12}$    iii $\frac{3}{13}$
   c  Express as a fraction
       i $0.0\dot{5}$    ii $0.00\dot{5}$    iii $0.0\dot{8}$    iv $0.0\dot{6}$
   d  Simplify $\left(\frac{3}{7} \times \frac{5}{9}\right) \div \left(5\frac{1}{4} - 4\frac{4}{15}\right)$

2 The probability that the postman calls before I leave for school in the morning is $\frac{5}{6}$. He never calls at the moment I am leaving. Find the probability that
   a the postman calls after I have left for school
   b next Monday and Tuesday the postman will call before I leave.

3 The selling price of a patio set is £177. This includes VAT at 18%. Find the price excluding VAT.

4  a  Express $3:7$ in the form $1:n$.
   b  A scale model of a car is such that the model car is 6.5 cm long and the actual car is 4.68 m long. The height of the car is 1.44 m. How high is the model?

**5** Expand
  **a** $3x(5y - 2z)$
  **b** $(a + 4)(a + 6)$
  **c** $(5x + 3y)(z + 4)$
  **d** $(a + 3b)^2$

**6** The graph shows Alex's journey from Antford to Combly via Beasley and Bianca's journey from Combly to Antford along the same road.
  **a** How far is it from Antford to Combly?
  **b** How far did Alex travel before he rested for the first time?
  **c** What was his average speed for this part of the journey?
  **d** How far is it from Beasley to Combly?
  **e** How long was Alex's second stop?
  **f** What was Alex's average speed between his two stops?
  **g** What time did Alex arrive at Combly?
  **h** How long did Bianca take to travel from Combly to Antford?
  **i** What was Bianca's average speed for her journey?
  **j** When and where did the two pass each other?
  **k** How far apart were they at 1.34 p.m.?

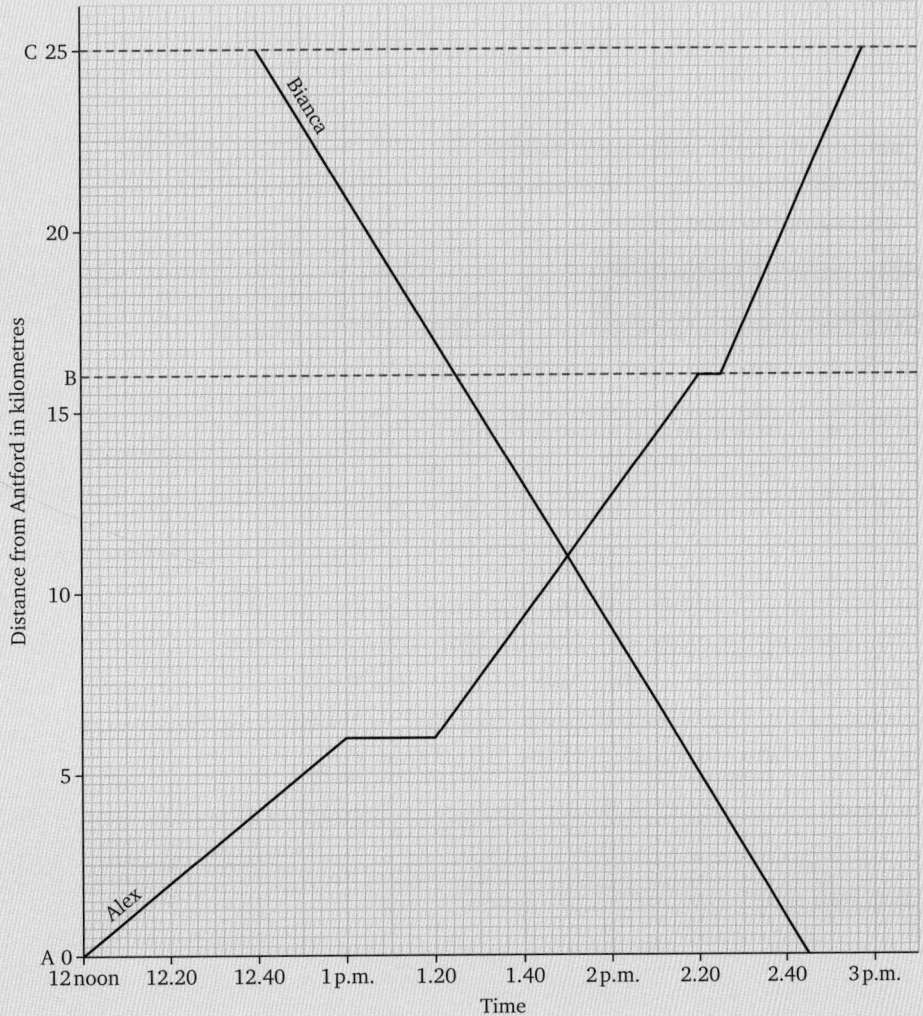

**7** Factorise

   **a**   $3t - 6t^2$

   **b**   $3x^2 + 9x - 12$

   **c**   $x^2 - 4x - 12$

   **d**   $x^2 + 8x + 16$

**8** The duration of the telephone calls made from an office on one particular day were recorded and the frequency polygon summarises the results.

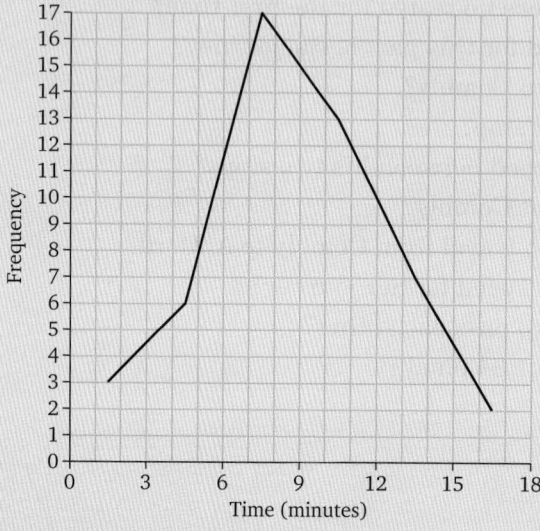

   **a**   How many telephone calls were made?

   **b**   Estimate the range of the times taken.

   **c**   Copy and complete the table.

| Middle value, $t$ (minutes) | 1.5 | | | | |
|---|---|---|---|---|---|
| Frequency | | | | | |

   **d**   Estimate the mean duration of a telephone call that day.

**9** Make the letter in brackets the subject of the formula.

   **a**   $z = x - 2y$     $(x)$

   **b**   $p = q + 2r$     $(r)$

   **c**   $a = 3b - c$     $(b)$

**10** The perimeter of a triangle is 172 mm. The shortest side is 50 mm shorter than each of the other two sides, which are equal. If the length of the shortest side is $x$ mm and the length of each of the longer sides is $y$ mm, form two equations in $x$ and $y$ and solve them simultaneously to find the lengths of the three sides.

## Mental arithmetic practice 2

1. Find $\frac{2}{5} + \frac{3}{8}$
2. What is the value of $x$, given $2x + y = 6$ and $x + y = 4$?
3. Three nails weigh 25 grams. What do nine similar nails weigh?
4. The radius of a circle is 6 cm. What, roughly, is its circumference?
5. What is the sum when 11 is added to the next prime number?
6. Divide 36 kg into two parts in the ratio $2:7$
7. Express the ratio $3:5$ in the form $n:1$
8. Express the ratio $5:9$ in the form $1:n$
9. Increase 60 cm in the ratio $2:3$
10. Express the ratio $\frac{2}{3}:\frac{3}{5}$ in the form $1:n$
11. What is the square root of 1.44?
12. What is the reciprocal of 0.25?
13. What is the sum of the exterior angles of a nine-sided polygon?
14. Express the ratio $\frac{1}{4}:3$ in the form $1:n$
15. Find the coordinates of the point where the line $y = 2 - x$ cuts the $y$-axis.
16. A rope is 5.4 m long. Two-thirds is cut off. What length remains?
17. A pen costs £1.40 and a pencil costs 80 p. What is the ratio of the cost of a pencil to the cost of a pen?
18. Express $\frac{7}{9}$ as a decimal correct to 3 decimal places.
19. What is $\frac{7}{11}$ of 88 m?
20. What is the value of $3x + 2$ when $x = -\frac{1}{2}$?
21. Find $1846 + 967$
22. It takes three men four days to build a wall. How long would it take two men to build the same wall?
23. A table costing £120 is sold at a loss of 20%. Find the sale price.
24. What is the formula for the area of a circle?
25. A copy of *Dead Easy* is 6 mm thick. What is the height, in centimetres, of a pile of 30 of these books?
26. Find $\frac{5}{8}$ as a decimal.
27. Expand $(x - 4)(x - 5)$
28. What is the largest prime number before 30?
29. Express 12 cm$^2$ in mm$^2$.
30. What is the solution of the inequality $x + 3 > 2$?

In questions **31** to **35**, expand

**31** $4x(3y + 2z)$

**32** $(x + a)(x + b)$

**33** $(x + a)^2$

**34** $(x - a)^2$

**35** $(x + a)(x - a)$

**36** Decrease 500 cm by 45%

**37** Expand $(a + 3)(a - 2)$

**38** Expand $(a - 3)(a + 2)$

**39** Give 15 as the sum of three prime numbers.

**40** What is the formula for the volume of a cylinder?

**41** Find £3 as a percentage of £60.

**42** A rectangle is $4x$ cm long and $x$ cm wide. Find an expression for its perimeter.

**43** Find $0.81 \div 0.09$

**44** Expand $(b - 4)(b - 7)$

**45** Expand $(b + 4)(b + 7)$

**46** What is the value of $3^2 - 4^0$?

**47** Find the density of a block weighing 200 g whose volume is 8 cm$^3$.

**48** Expand $(x + 4)^2$

**49** Express $3 : 7.5$ in the form $1 : n$

**50** What name is given to a quadrilateral with just one pair of opposite sides parallel?

**51** What is the next prime number after 60?

**52** Does $x = 4$ satisfy $x < 4$?

**53** Solve the inequality $3 - x > 5$

**54** How far will a car going at 60 mph travel in $2\frac{1}{4}$ hours?

**55** The $n$th term of a sequence is $5n - 3$. What is the 8th term?

**56** Solve $x + 2 < 8$

**57** Find $\frac{4}{5}$ of 75 m.

**58** Expand $(a + 3b)(a - 3b)$

**59** Find the length of the diagonal of a rectangle whose sides are 5 cm and 12 cm long.

**60** In a triangle, one angle is 74° and another is 68°. What is the third angle?

# 11 Quadratic equations

## Consider

Paul is designing an electronic chip and wants to make a prototype.

Because of the number of pins needed along each side, the chip must be at least 5 mm longer than it is wide.

For Paul's purposes, the larger the chip the better. The material he needs to use, however, is expensive so the surface area must be no more than 150 mm².

Paul needs to work out the length and width of the chip (excluding the pins).

- He can try various values for the width until he finds one that fits the constraints. As we saw in Year 8, this is a 'hit and miss' affair; it can take ages to find a suitable value.
- He can carry out a more organised search by forming an equation and using trial and improvement methods.
- Trial and improvement methods give solutions that are as accurate as are needed for the purpose. The disadvantage is that they involve a lot of arithmetic, which is tedious without the help of suitable technology.

Use an algebraic method to find the dimensions of the chip.

*You should be able to answer this question after you have worked through this chapter.*

### Class discussion

1. Amy needs to find a value for $x$ that satisfies the equation
   $$x^2 - 6x = 17.01$$
   Discuss the advantages and disadvantages of these methods:
   a. trial and improvement
   b. drawing a graph
   c. using a suitable graph drawing program on a calculator or computer.

2. There are two values of $x$ that satisfy the equation
   $$x^2 - 6x = 17.01$$
   a. Discuss which of the methods in question **1** will find both values of $x$.
   b. Discuss whether any of the methods in question **1** can tell you whether the solutions are exact values or whether they can only be given correct to a number of significant figures.

## Quadratic equations

The class discussion shows that there are disadvantages in using numerical and graphical methods for solving equations. One disadvantage is the time it can take to find a solution so, when it is possible, using algebraic methods is more efficient.

We can always solve equations such as $2x - 3.5 = 5x + 8.5$. Equations like this, which do not involve powers of the unknown letter, are called **linear equations**.

Equations such as $x^2 - 6x = 17.01$, which have a term involving the square of the unknown letter, are called **quadratic equations**.

In this chapter we shall solve some straightforward quadratic equations. The method we shall use depends on the product of two numbers being zero.

### Exercise 11a

**Worked example**

→ Find the value of $(x + 3)(x - 7)$ if
   **a** $x = 8$    **b** $x = 7$    **c** $x = -3$

**a**   If $x = 8$, $(x + 3)(x - 7) = (8 + 3)(8 - 7)$
$$= (11)(1)$$
$$= 11$$

**b**   If $x = 7$, $(x + 3)(x - 7) = (10)(0)$
$$= 0$$

**c**   If $x = -3$, $(x + 3)(x - 7) = (-3 + 3)(-10)$
$$= (0)(-10)$$
$$= 0$$

**1**   Find the value of $(x - 4)(x - 2)$ if
   **a** $x = 6$        **b** $x = 4$        **c** $x = 2$

**2**   Find the value of $(x - 5)(x - 9)$ if
   **a** $x = 5$        **b** $x = 10$       **c** $x = 9$

**3**   Find the value of $(x - 7)(x - 1)$ if
   **a** $x = 1$        **b** $x = 8$        **c** $x = 7$

**4**   Find the value of $(x - 3)(x + 5)$ if
   **a** $x = 6$        **b** $x = 3$        **c** $x = -5$

**5**   Find the value of $(x - 4)(x + 6)$ if
   **a** $x = 0$        **b** $x = -6$      **c** $x = 4$

**6**   Find the value of $(x + 4)(x + 5)$ if
   **a** $x = 4$        **b** $x = -5$      **c** $x = 0$

# 11 Quadratic equations

When we consider the product of two factors, questions **1** to **6** show that, if one or both of them are zero, so is the product.

Further, if neither factor is zero, the product is *not* zero.

In general, we can say

    **if**               $A \times B = 0$

    **then either**    $A = 0$    **or**    $B = 0$

Remember that '$A = 0$ or $B = 0$' is not mutually exclusive, so it is possible that $A = 0$ and $B = 0$.

In questions **7** to **12**, find, if possible, the value or values of $A$. Note that if $A \times 0 = 0$, then $A$ can have any value.

**7**  $A \times 6 = 0$          **9**  $A \times 4 = 0$          **11**  $3 \times A = 12$

**8**  $A \times 7 = 0$          **10**  $0 \times A = 0$          **12**  $8 \times A = 8$

**13**  If $AB = 0$, find
   **a**  $A$ if $B = 2$
   **b**  $B$ if $A = 10$

**14**  If $AB = 0$, find
   **a**  $A$ if $B = 5$
   **b**  $B$ if $A = 5$

## Worked example

→ Find $a$ and $b$ if $a(b - 3) = 0$

Either          $a = 0$    or    $b - 3 = 0$

i.e.     either     $a = 0$    or    $b = 3$

Find $a$ and $b$ if

**15**  $a(b - 1) = 0$       **17**  $a(b - 2) = 0$       **19**  $(a - 3)b = 0$

**16**  $a(b - 5) = 0$       **18**  $a(b - 4) = 0$       **20**  $(a - 9)b = 0$

## Solving quadratic equations

Consider the equation $(x - 1)(x - 2) = 0$

This is true either when $x - 1 = 0$ or when $x - 2 = 0$,

i.e.     either when $x = 1$ or when $x = 2$

There are, therefore, two values of $x$ that satisfy the equation

$$(x - 1)(x - 2) = 0$$

Expanding the left-hand side gives $x^2 - 3x + 2 = 0$

This is a quadratic equation and we have solved it.

# STP Maths 9

We can solve any quadratic equation written in the form
$$(x - a)(x - b) = 0$$
because we know that
either $\quad x - a = 0 \quad$ and/or $\quad x - b = 0$
i.e. $\quad\quad x = a \quad\quad$ and/or $\quad x = b$

## Exercise 11b

### Worked example

→ What values of $x$ satisfy the equation $x(x - 9) = 0$?

If $\quad\quad x(x - 9) = 0$
either $\quad\quad x = 0 \quad$ or $\quad x - 9 = 0$
i.e. either $\quad x = 0 \quad$ or $\quad x = 9$

What values of $x$ satisfy the following equations?

1. $x(x - 3) = 0$
2. $x(x - 5) = 0$
3. $(x - 3)x = 0$
4. $x(x + 4) = 0$
5. $(x + 5)x = 0$
6. $x(x - 6) = 0$
7. $x(x - 10) = 0$
8. $(x - 7)x = 0$
9. $x(x + 7) = 0$
10. $(x + 9)x = 0$

### Worked example

→ What values of $x$ satisfy the equation $(x - 3)(x + 5) = 0$?

If $\quad\quad (x - 3)(x + 5) = 0$
either $\quad\quad x - 3 = 0 \quad$ or $\quad x + 5 = 0$
i.e. either $\quad x = 3 \quad$ or $\quad x = -5$

What values of $x$ satisfy the following equations?

11. $(x - 1)(x - 2) = 0$
12. $(x - 5)(x - 9) = 0$
13. $(x - 10)(x - 7) = 0$
14. $(x - 4)(x - 7) = 0$
15. $(x - 6)(x - 1) = 0$
16. $(x - 8)(x + 11) = 0$
17. $(x - 3)(x + 5) = 0$
18. $(x + 7)(x - 2) = 0$
19. $(x + 2)(x + 3) = 0$
20. $(x + 4)(x + 9) = 0$
21. $(x + 1)(x + 8) = 0$
22. $(x - p)(x - q) = 0$

**23** $(x + a)(x + b) = 0$
**24** $(x - 4)(x + 1) = 0$
**25** $(x + 9)(x - 8) = 0$
**26** $(x + 6)(x + 7) = 0$

**27** $(x + 10)(x + 11) = 0$
**28** $(x - a)(x - b) = 0$
**29** $(x + a)(x - b) = 0$
**30** $(x - c)(x + d) = 0$

**31** $(2x - 3)(2x + 1) = 0$
**32** $(4x + 7)(3x - 1) = 0$

## Solution by factorisation

Quadratic equations do not often come in the form
$$(x - 4)(x - 1) = 0$$
They are more likely to arise in the form $x^2 - 5x + 4 = 0$

**Exercise 11b** shows that if the left-hand side of such a quadratic equation can be expressed as two linear factors, we can use these factors to solve the equation.

e.g. $x^2 - 5x + 4 = 0$
gives $(x - 4)(x - 1) = 0$
from which we see that either $x - 4 = 0$ or $x - 1 = 0$
i.e. $x = 4$ or $x = 1$

## Exercise 11c

### Worked example

→ Solve the equation $x^2 - 10x + 9 = 0$

If $x^2 - 10x + 9 = 0$
then $(x - 1)(x - 9) = 0$
∴ either $x - 1 = 0$ or $x - 9 = 0$
i.e. $x = 1$ or $9$

*Check:* when $x = 1$, LHS $= 1 - 10 + 9 = 0 =$ RHS
when $x = 9$, LHS $= 81 - 90 + 9 = 0 =$ RHS

Solve the equations.

**1** $x^2 - 3x + 2 = 0$
**2** $x^2 - 8x + 7 = 0$
**3** $x^2 - 5x + 6 = 0$
**4** $x^2 - 7x + 10 = 0$
**5** $x^2 - 7x + 12 = 0$

**6** $x^2 - 6x + 5 = 0$
**7** $x^2 - 12x + 11 = 0$
**8** $x^2 - 6x + 8 = 0$
**9** $x^2 - 8x + 12 = 0$
**10** $x^2 - 13x + 12 = 0$

# STP Maths 9

## Worked example

→ Solve the equation $x^2 + 2x - 8 = 0$

If $\qquad x^2 + 2x - 8 = 0$
then $\qquad (x + 4)(x - 2) = 0$
∴ either $\quad x + 4 = 0 \qquad$ or $\qquad x - 2 = 0$
i.e. $\qquad x = -4 \qquad$ or $\qquad 2$

*Check*: when $x = -4$, LHS $= 16 - 8 - 8 = 0 =$ RHS
$\qquad$ when $x = 2$, LHS $= 4 + 4 - 8 = 0 =$ RHS

Solve the equations.

**11** $x^2 + 6x - 7 = 0$

**12** $x^2 - 2x - 8 = 0$

**13** $x^2 - x - 12 = 0$

**14** $x^2 - 2x - 15 = 0$

**15** $x^2 - 7x - 18 = 0$

**16** $x^2 - 12x - 13 = 0$

**17** $x^2 + x - 6 = 0$

**18** $x^2 - 4x - 12 = 0$

**19** $x^2 + x - 20 = 0$

**20** $x^2 - 5x - 24 = 0$

## Worked example

→ Solve the equation $x^2 + 9x + 8 = 0$

If $\qquad x^2 + 9x + 8 = 0$
then $\qquad (x + 1)(x + 8) = 0$
∴ either $\quad x + 1 = 0 \qquad$ or $\qquad x + 8 = 0$
i.e. $\qquad x = -1 \qquad$ or $\qquad -8$

Solve the equations.

**21** $x^2 + 3x + 2 = 0$

**22** $x^2 + 8x + 7 = 0$

**23** $x^2 + 8x + 15 = 0$

**24** $x^2 + 8x + 12 = 0$

**25** $x^2 + 11x + 18 = 0$

**26** $x^2 + 7x + 6 = 0$

**27** $x^2 + 7x + 10 = 0$

**28** $x^2 + 14x + 13 = 0$

**29** $x^2 + 16x + 15 = 0$

**30** $x^2 + 9x + 18 = 0$

## Worked example

→ Solve the equation $x^2 - 49 = 0$

If $\qquad x^2 - 49 = 0$
then $\qquad (x + 7)(x - 7) = 0$
∴ either $\quad x + 7 = 0 \qquad$ or $\qquad x - 7 = 0$
i.e. $\qquad x = -7 \qquad$ or $\qquad 7$

Solve the equations.

**31** $x^2 - 1 = 0$
**32** $x^2 - 9 = 0$
**33** $x^2 - 16 = 0$
**34** $x^2 - 81 = 0$
**35** $x^2 - 169 = 0$

**36** $x^2 - 4 = 0$
**37** $x^2 - 25 = 0$
**38** $x^2 - 100 = 0$
**39** $x^2 - 144 = 0$
**40** $x^2 - 36 = 0$

The equations we have solved by factorising have all been examples of the equation $ax^2 + bx + c = 0$ when $a = 1$. We consider next the case when $c = 0$,

e.g. the equation  $\quad 3x^2 + 2x = 0$

Since $x$ is common to both terms on the left-hand side, we can rewrite this equation as
$\quad x(3x + 2) = 0$

Then, either  $\quad x = 0 \quad$ or $\quad 3x + 2 = 0$
i.e.  $\quad\quad\quad\quad x = 0 \quad$ or $\quad 3x = -2$
i.e.  $\quad\quad\quad\quad x = 0 \quad$ or $\quad -\frac{2}{3}$

## Exercise 11d

Solve the equations.

**1** $x^2 - 2x = 0$
**2** $x^2 - 10x = 0$
**3** $x^2 + 8x = 0$
**4** $2x^2 - x = 0$
**5** $4x^2 - 5x = 0$
**6** $x^2 - 5x = 0$
**7** $x^2 + 3x = 0$
**8** $x^2 + x = 0$
**9** $3x^2 - 5x = 0$
**10** $5x^2 - 7x = 0$

**11** $2x^2 + 3x = 0$
**12** $8x^2 + 5x = 0$
**13** $x^2 - 7x = 0$
**14** $3x^2 + 5x - 0$
**15** $7x^2 - 12x = 0$
**16** $6x^2 + 7x = 0$
**17** $12x^2 + 7x = 0$
**18** $x^2 + 4x = 0$
**19** $7x^2 - 2x = 0$
**20** $14x^2 + 3x = 0$

Sometimes a quadratic equation has two answers, or **roots**, that are exactly the same.
Consider  $\quad x^2 - 4x + 4 = 0$
then  $\quad\quad (x - 2)(x - 2) = 0$
i.e. either  $\quad x - 2 = 0 \quad$ or $\quad x - 2 = 0$
i.e.  $\quad\quad\quad\quad x = 2 \quad$ or $\quad x = 2$
i.e.  $\quad\quad\quad\quad x = 2 \quad$ (twice)

Such an equation involves a **perfect square**. As with any quadratic equation, it has two answers, or roots, but they are equal. We say that such an equation has a repeated root.

# STP Maths 9

### Exercise 11e

**Worked example**

→ Solve the equation $x^2 + 14x + 49 = 0$

If $\quad\quad x^2 + 14x + 49 = 0$
then $\quad\quad (x + 7)(x + 7) = 0$
∴ either $\quad x + 7 = 0 \quad$ or $\quad x + 7 = 0$
i.e. $\quad\quad x = -7 \quad\quad$ (twice)

Solve the equations.

1. $x^2 - 2x + 1 = 0$
2. $x^2 - 10x + 25 = 0$
3. $x^2 + 8x + 16 = 0$
4. $x^2 + 6x + 9 = 0$
5. $x^2 - 6x + 9 = 0$
6. $x^2 - 8x + 16 = 0$
7. $x^2 + 2x + 1 = 0$
8. $x^2 + 20x + 100 = 0$
9. $x^2 + 18x + 81 = 0$
10. $x^2 - 22x + 121 = 0$
11. $x^2 - x + \frac{1}{4} = 0$
12. $x^2 + 10x + 25 = 0$
13. $x^2 - 12x + 36 = 0$
14. $x^2 - 40x + 400 = 0$
15. $x^2 - 16x + 64 = 0$
16. $x^2 + \frac{4}{3}x + \frac{4}{9} = 0$

Quadratic equations do not always present themselves already arranged in the form $ax^2 + bx + c = 0$

Consider the equation $\quad x(x - 10) = 39$

This equation needs to be rearranged so that it is in the form
$\quad\quad (x^2 \text{ term}) \quad$ then $\quad (x \text{ term}) \quad$ then $\quad$ (number) $= 0$

We start by multiplying out the bracket: $\quad\quad x^2 - 10x = 39$
Then we take 39 from each side: $\quad\quad x^2 - 10x - 39 = 0$
Now we are ready to factorise, giving: $\quad\quad (x + 3)(x - 13) = 0$
Therefore, $\quad x = -3$ or $13$

### Exercise 11f

Solve the equations.

1. $x^2 - x = 30$
2. $x^2 - 6x = 16$
3. $x^2 + 9x = 36$
4. $x^2 = 2x + 8$

**5** $x^2 = 2x + 24$
**6** $x^2 = 12x - 35$
**7** $x^2 - x = 6$
**8** $x^2 + 6x = 7$
**9** $x^2 - x = 12$
**10** $x^2 = 3x + 10$
**11** $x^2 = 6x - 8$
**12** $x^2 = 5x + 50$

## Worked example

Solve the equation $21 = 10x - x^2$

If $\qquad 21 = 10x - x^2$
then $\quad 21 - 10x + x^2 = 0$
$\qquad\quad x^2 - 10x + 21 = 0$
$\qquad\quad (x - 7)(x - 3) = 0$

Either $\quad x - 7 = 0, \quad$ so $\quad x = 7$
or $\qquad x - 3 = 0, \quad$ so $\quad x = 3$

> Collect the terms on the LHS; this makes the $x^2$ term positive. Now rearrange so that the $x^2$ term is first, followed by the $x$ term, then the number term.

Solve the equations.

**13** $10 = 7x - x^2$
**14** $7 = 8x - x^2$
**15** $8 = 6x - x^2$
**16** $3x = 28 - x^2$
**17** $12 = 8x - x^2$
**18** $20 = 9x - x^2$
**19** $35 = 12x - x^2$
**20** $15 = 8x - x^2$

## Worked example

Solve the equation $2x^2 + 10x - 12 = 0$

If $\qquad 2x^2 + 10x - 12 = 0$
then $\quad 2(x^2 + 5x - 6) = 0$
$\qquad\quad x^2 + 5x - 6 = 0$
$\qquad\quad (x + 6)(x - 1) = 0$

Either $\quad x + 6 = 0, \quad$ giving $\quad x = -6$
or $\qquad x - 1 = 0, \quad$ giving $\quad x = 1$

> The LHS has a common factor, 2. We start by taking it out.

> $x^2 + 5x - 6$ must be zero since 2 is not.

Solve the equations.

**21** $2x^2 - 8x = 0$
**22** $5x^2 - 15x + 10 = 0$
**23** $3x^2 + 9x + 6 = 0$
**24** $3x^2 - 24x + 36 = 0$
**25** $2x^2 - 4x = 0$
**26** $3x^2 - 9x = 0$
**27** $6x^2 + 18x + 12 = 0$
**28** $8x^2 - 4x = 0$

**Worked example**

→ Solve the equation $x(x - 2) = 15$

If $\qquad x(x - 2) = 15$
then $\qquad x^2 - 2x = 15$
$\qquad\qquad x^2 - 2x - 15 = 0$
$\qquad\qquad (x - 5)(x + 3) = 0$
Either $\qquad x - 5 = 0 \quad \Rightarrow \quad x = 5$
or $\qquad\quad x + 3 = 0 \quad \Rightarrow \quad x = -3$
*Check*: when $x = 5$, LHS $= 5(3) = 15 =$ RHS
$\qquad\;\;$ when $x = -3$, LHS $= (-3)(-5) = 15 =$ RHS

Solve the equations.

**29** $x(x + 1) = 12$

**30** $x(x - 1) = x + 3$

**31** $x(x - 5) = 24$

**32** $x(x + 3) = 5(3x - 7)$

**Worked example**

→ Solve the equation $(x - 3)(x + 2) = 6$

If $\qquad (x - 3)(x + 2) = 6$
then $\qquad x^2 - x - 6 = 6$
$\qquad\qquad x^2 - x - 12 = 0$
$\qquad\qquad (x - 4)(x + 3) = 0$
Either $\qquad x - 4 = 0 \quad$ or $\quad x + 3 = 0$
i.e. $\qquad\;\; x = 4 \qquad$ or $\quad x = -3$

Solve the equations.

**33** $(x + 2)(x + 3) = 56$

**34** $(x + 9)(x - 6) = 34$

**35** $(x - 2)(x + 6) = 33$

**36** $(x + 3)(x - 8) + 10 = 0$

**37** $(x - 5)(x + 2) = 18$

**38** $(x + 8)(x - 2) = 39$

**39** $(x + 1)(x + 8) + 12 = 0$

**40** $(x - 1)(x + 10) + 30 = 0$

## Mixed exercise

In general, to solve a quadratic equation by factorising:
- multiply out brackets
- collect all the terms on one side of the equal sign (choose the side where $x^2$ is positive)
- arrange the terms in the order ($x^2$ term), ($x$ term), (number) and then look for factors.

### Exercise 11g

Solve the equations.

1. $x^2 - x - 20 = 0$
2. $x^2 = 4x - 4$
3. $x^2 - 36 = 0$
4. $x^2 = 7 - 6x$
5. $2x^2 + 12x + 18 = 0$
6. $2 - x = x^2$
7. $5x(x - 1) = 4x^2 - 4$
8. $15 - x^2 + 2x = 0$
9. $x^2 + 12x + 32 = 0$
10. $x(x + 8) = x + 30$
11. $12x^2 + 16x = 0$
12. $3x^2 + 5x = 0$
13. $x^2 - 2x - 15 = 0$
14. $(2x - 3)(x + 1) = 0$
15. $x^2 = 2x + 35$
16. $x^2 + 13x + 12 = 0$
17. $12x^2 - 48 = 0$
18. $x(x + 6) = 3x + 10$
19. $2x(x + 9) = 0$
20. $(x - 4)^2 = 25$

### Forming equations

When trying to form an equation from information given in a problem, start by identifying unknown numbers.

Use a letter to represent an unknown number. State what this letter stands for, either by writing a short sentence or by marking it clearly on a diagram. Use the information given in the problem to find a relationship between the letter and other numbers.

When you have found the solutions of an equation, remember to check that they fit the information given. Remember also that not all solutions of an equation give solutions to a problem.

## Exercise 11h

**Worked example**

→ I think of a positive number, $x$. I square it and then add three times the number I first thought of. The answer is 54.

Form an equation in $x$ and solve it to find the number I first thought of.

$$x^2 + 3x = 54$$
i.e. $\quad x^2 + 3x - 54 = 0$
$$(x - 6)(x + 9) = 0$$

💡 The square of $x$ is $x^2$ and three times $x$ is $3x$. Adding them together gives $x^2 + 3x$ and this must be 54.

Either $\quad x - 6 = 0 \quad$ or $\quad x + 9 = 0$
i.e. $\quad\quad x = 6 \quad\quad$ or $\quad\quad -9$

Reject $-9$.
The number I first thought of was therefore 6.

💡 The number must be positive.

*Check*: $\quad 6^2 + 3 \times 6 = 54$

💡 Use the number found with the original instructions.

**1** The square of a number $x$ is 16 more than six times the number.
   **a** Write 'the square of a number $x$' in symbols.
   **b** Write '16 more than six times $x$' in terms of $x$.
   **c** Hence form an equation in $x$ and solve it.

**2** When five times a number $x$ is subtracted from the square of the same number, the answer is 14.
   **a** Write 'five times a number $x$ is subtracted from the square of $x$' using mathematical notation.
   **b** Hence form an equation in $x$ and solve it.

**3** Dylan had $x$ marbles.
The number of marbles Shetha had was six less than the square of the number Dylan had.
   **a** Write down the number of marbles Shetha had in terms of $x$.
   **b** Write down, in terms of $x$, the total number of marbles they both had.
   **c** Together they had 66 marbles. Form an equation in $x$ and solve it.
   **d** How many marbles did Shetha have?

**4** I think of a positive number, $x$. If I square it and add it to the number I first thought of, the total is 42. Find the number I first thought of.

5   Ahmed is $x$ years old and his father is $x^2$ years old. The sum of their ages is 56 years. Form an equation in $x$ and solve it to find the age of each.

 6   Owen is $y$ years old and his sister is 5 years older. If the product of their ages is 84, form an equation in $y$ and solve it to find Owen's age.

 7   Kathryn is $x$ years old. If her mother's age is two years more than the square of Kathryn's age, and the sum of their ages is 44 years, form an equation in $x$ and solve it to find the ages of Kathryn and her mother.

### Worked example

→ A rectangle is 4 cm longer than it is wide. If it is $x$ cm wide and has an area of 77 cm², form an equation in $x$ and solve it to find the dimensions of the rectangle.

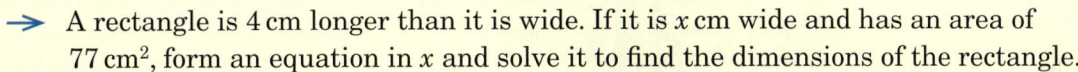

Area $= (x + 4) \times x$ cm²

i.e.  $(x + 4)x = 77$

$x^2 + 4x = 77$

$x^2 + 4x - 77 = 0$

$(x - 7)(x + 11) = 0$

Either   $x - 7 = 0$   or   $x + 11 = 0$

i.e.   $x = 7$   or   $-11$

The area of a rectangle is given by length × breadth.

We know that the area is 77 cm².

The breadth of a rectangle cannot be negative, so reject $-11$,

∴   $x = 7$

The rectangle measures $(7 + 4)$ cm by 7 cm,

i.e.   11 cm by 7 cm.

 8   A rectangle is $x$ cm wide and is 3 cm longer than it is wide.
   a   Draw a diagram showing this information, clearly marking the length of the rectangle in terms of $x$.
   b   Write down the area of the rectangle in terms of $x$.
   c   If the area is 28 cm², form an equation in $x$ and solve it to find the dimensions of the rectangle.

9  The base of a triangle is $x$ cm long and its perpendicular height is 5 cm less than the length of its base.
   a  Draw a diagram showing this information, clearly marking the perpendicular height of the triangle in terms of $x$.
   b  Write down an expression for the area of the triangle in terms of $x$.
   c  The area of the triangle is 25 cm². Form an equation in $x$ and solve it.
   d  What is the height of the triangle?

10  A rectangle is 5 cm longer than it is wide and its area is 66 cm². Find the dimensions of the rectangle.

11  A skier is sliding down a gentle slope. The distance she covers in $t$ seconds is given by $(t^2 + t)$ metres. Find how long it takes her to cover 110 metres.

12  A rectangular lawn is bordered on two adjacent sides by a path, as shown in the diagram.

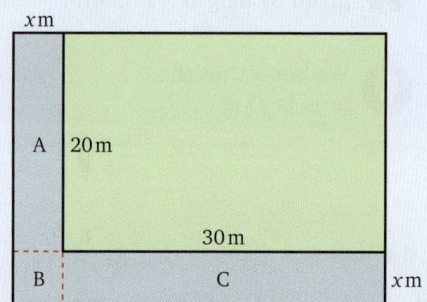

   a  Express in terms of $x$ each of the areas marked by the letters A, B and C.
   b  The area of the path is 104 m². Form an equation in $x$ and solve it to find the width of the path.

13  The formula $P = n(n + 2)$ is used to work out the cost, £$P$, for storing $n$ tonnes of butter for a year. Find the number of tonnes of butter that can be stored for a cost of £2600.

14  a  The square of an unknown positive integer, $x$, and the square of the consecutive (the next larger) integer are added. Write an expression for this sum in terms of $x$.
    b  The sum of these two squares is 181. Find the smaller integer.

# 11 Quadratic equations

## Solving equations by trial and improvement

In this chapter we have solved quadratic equations by factorisation. Not all quadratic equations can be solved in this way.

There are many other types of equation for which there is no method of finding an exact solution. In such cases we start by giving $x$ a value that looks reasonable and *try* it in the equation. Then we adjust that value to give an *improvement*.

Solution by **trial and improvement** was introduced in Year 8. The next worked example is a reminder of the process.

## Exercise 11i

### Worked example

→ Find a solution to the equation $x^2 - \dfrac{2}{x} = 6$ correct to 1 decimal place.

1 is clearly too small, so we try 2 as the first trial value for $x$.
If this gives the value of $x^2 - \dfrac{2}{x}$ as less than 6 (i.e. too small), we try next a value greater than 2, and vice versa.
We can keep track of results by placing them in a table.

| Try $x =$ | Value of $x^2 - \dfrac{2}{x}$ | Compared with 6 |
|---|---|---|
| 2 | $2^2 - \dfrac{2}{2} = 3$ | too small |
| 3 | $3^2 - \dfrac{2}{3} = 8.33...$ | too big |
| 2.5 | $2.5^2 - \dfrac{2}{2.5} = 5.45$ | too small |
| 2.6 | $2.6^2 - \dfrac{2}{2.6} = 5.99...$ | too small |
| 2.61 | $2.61^2 - \dfrac{2}{2.61} = 6.04...$ | too big |

This shows that $2 < x < 3$

But not by much, so try $x = 2.6$ next.

Only just too small.

Now we see that $2.6 < x < 2.61$

$x = 2.6$ correct to 1 decimal place.

For each equation, use trial and improvement to find a positive solution correct to 1 decimal place.

**1** $x^3 + 7x = 12$ (Start by trying $x = 1$)

**2** $x^3 - 2x - 5 = 0$ (Start by trying $x = 2$)

**3** $x^2 + x + \dfrac{1}{x} = 11$ (Start by trying $x = 2$)

**4** $x^3 + 3x = 7$

**5** $x + \dfrac{8}{x} = 12$

Sometimes when a trial number gives a result that is too small, trying a larger trial number gives a result that is even smaller. When this happens, try a number smaller than the first one.

### Worked example

The equation $2x - x^3 + 1 = 0$ has a solution between $x = 1$ and $x = 2$. Find the solution correct to 1 decimal place.

| Try $x =$ | $2x - x^3 + 1$ | Compared with 0 |
|---|---|---|
| 1.5 | 0.625 | $> 0$ |
| 1.3 | 1.403 | $> 0$ |
| 1.6 | 0.104 | $> 0$ |
| 1.7 | $-0.513$ | $< 0$ |
| 1.65 | $-0.192\ldots$ | $< 0$ |

This is even bigger than the value given by 1.5, so try a value larger than 1.5

This is nearer to 0 than the value given by $x = 1.5$, so try a number a little larger than 1.6

Now we see that $1.6 < x < 1.7$. We shall try 1.65 next.

Now we can see that $1.6 < x < 1.65$

i.e. $x = 1.6$ correct to 1 decimal place.

For each equation, use trial and improvement to find a positive solution correct to 1 decimal place.

**6** $\dfrac{10}{x} - x^2 = 6$  (Start with $x = 2$)

**7** $24 - x^3 + 2x = 0$

**8** $x(7 - x)(2x + 1) = 8$  (Start with $x = 6$)

### Consider again

Paul is designing an electronic chip and wants to make a prototype.

The chip must be at least 5 mm longer than it is wide. The surface area must be no more than 150 mm².

Paul needs to work out the length and width of the chip (excluding the pins).

Now can you use an algebraic method to find the dimensions of the chip?

# 12 Graphs

## Consider

Freda hopes to find a relationship between the distance a puck slides down a slope and the time for which it has been sliding.

She has collected the following data.

| Time, in seconds, after leaving the top of the slope | 1 | 2.7 | 3.5 | 4.5 | 4.9 |
|---|---|---|---|---|---|
| Distance, in metres, from the top of the slope | 0.05 | 0.4 | 0.6 | 1 | 1.2 |

To find a relationship between the time and distance, Freda could
- guess and see whether the figures in her table fit the guess. This is not likely to give an answer quickly, if ever.
- hope to spot a relationship from the figures in her table. This is not usually possible unless the relationship is a simple one.
- plot the points on graph paper. Freda may then recognise the points as lying on a curve whose equation is of a form she knows.

Using $t$ seconds for the time and $d$ metres for the distance, can you find a likely algebraic relationship between $t$ and $d$?

*You should be able to answer this question after you have worked through this chapter.*

The ability to recognise the kind of equation that gives a particular shape of curve requires a wide knowledge of different forms of graph.

This knowledge is also useful when we want to draw a graph from its equation because, if we know what shape to expect, we are less likely to make mistakes.

## Class discussion

1. Georgia needs to solve all these equations:  $x^2 - 2x - 4 = 0$
   $$x^2 - 2x - 4 = 6$$
   $$2x^2 - 4x + 1 = 0$$
   $$1 + 2x - x^2 = 0$$

   Discuss what these equations have in common.

   Discuss also the advantages and disadvantages of methods for solving them that you are familiar with.

2. Chloe knows that, for a given voltage, the current flowing in a circuit is inversely proportional to the resistance.

   She takes the following readings during an experiment.

   | Resistance (ohms) | 0.5 | 0.7 | 1 | 1.2 | 1.5 | 2 | 2.4 |
   |---|---|---|---|---|---|---|---|
   | Current (amps) | 5.6 | 4 | 2.8 | 2.3 | 1.9 | 1.4 | 1.5 |

   Chloe then has to plot the values on a graph to confirm the relationship between them. Discuss what knowledge would help her to decide whether the relationship is confirmed and if she has made a mistake in any of her readings.

3. Jordan is designing a swimming pool for a school project. He wants to include a sketch showing how the depth of water in the pool increases when it is filled at a constant rate.

   This is the cross-section of the pool. Discuss how he could do this and what he needs to know.

The class discussion shows that graphs have many uses. In this chapter we look at some of these and extend our knowledge of curves.

## Straight lines

From Book 8 we know that an equation of the form $y = mx + c$ gives a straight line, where $m$ is the **gradient** of the line and $c$ is the $y$-intercept. (The $y$-intercept is the point where the line crosses the $y$-axis.)

Sometimes a situation involves straight lines with a change of gradient.

For example, in places where water is in short supply, the cost per cubic metre used goes up after a given quantity is used.

This graph shows the cost of water used each month in one such place.

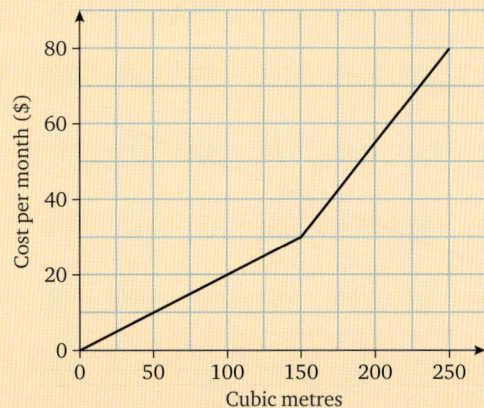

The gradient of the line from 0 to 150 m³ is $\frac{30}{150} = 0.2$ and this gives the cost, in dollars, for each cubic metre of water used up to 150 m³. The gradient of the line above 150 m³ is $\frac{50}{100} = 0.5$ so the cost of each cubic metre used over 150 m³ is $0.50.

If Meena uses 200 m³ of water in one month, it costs her $55.

Graphs like this one are called **piece-wise linear graphs**. (You have already used such graphs in Chapter 1. For example, questions **3** to **8** in **Exercise 1f**.)

### Exercise 12a

1   The graph illustrates the costs per month of using electricity on a two-part electricity tariff.

   **a**  In August, Mohamed used 200 units of electricity. What was the cost of using this number of units?

   **b**  Find the cost of each unit used up to 100 units.

   **c**  Find the cost of each unit used over 100 units.

   **d**  Abdul is on the same tariff as Mohamed. Work out the cost for a month in which he used 400 units of electricity.

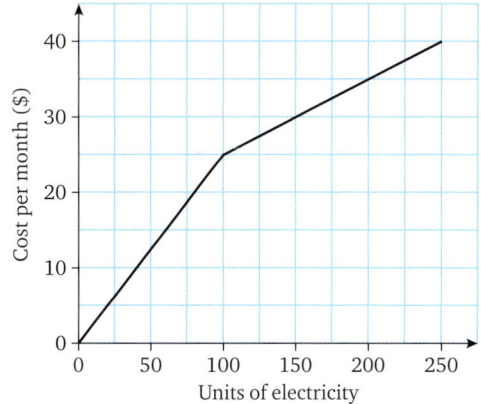

237

**2** The graph below shows how the depth of water in a swimming pool increases when water is pumped in at a constant rate.

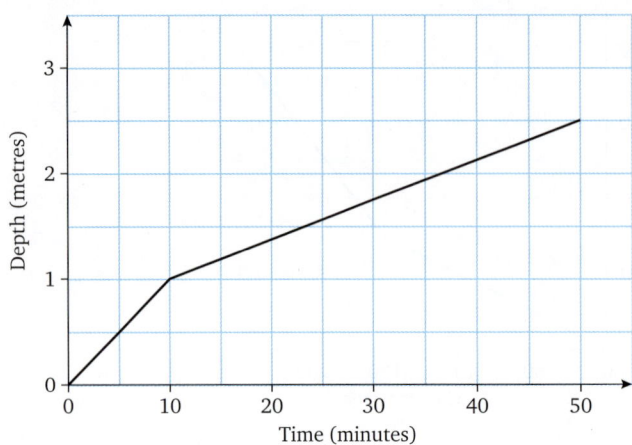

**a** How deep was the water in the pool after 15 minutes?

**b** Assuming the pool was full after 50 minutes, how deep was
  **i** the shallow end
  **ii** the deep end?

**c** Assuming that the pool was not full after 50 minutes but needed another 10 minutes to fill it, how deep was the deep end?

**3** This graph shows the journey of a bus from A to B.

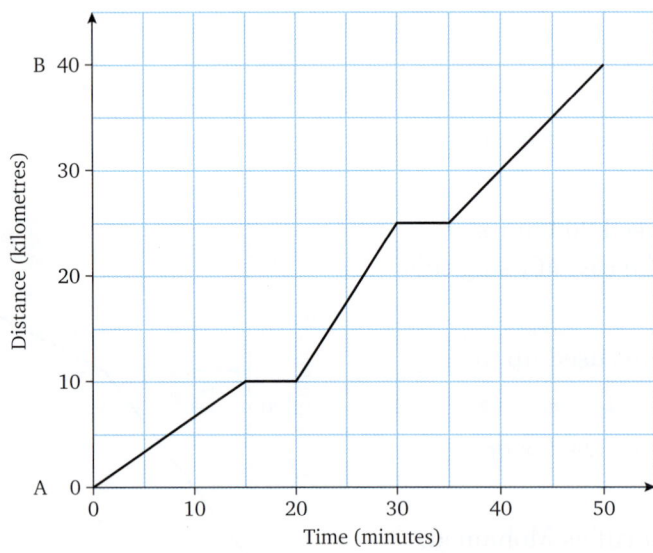

**a** What was the speed of the bus when travelling between A and its first stop?

**b** How far from A is B?

**c** What was the average speed of the bus between A and B?

4  The graph shows the amount a printer charges to print copies of a new book. There is a fixed charge for setting up the machinery, plus a charge for each book which depends on the total number printed.

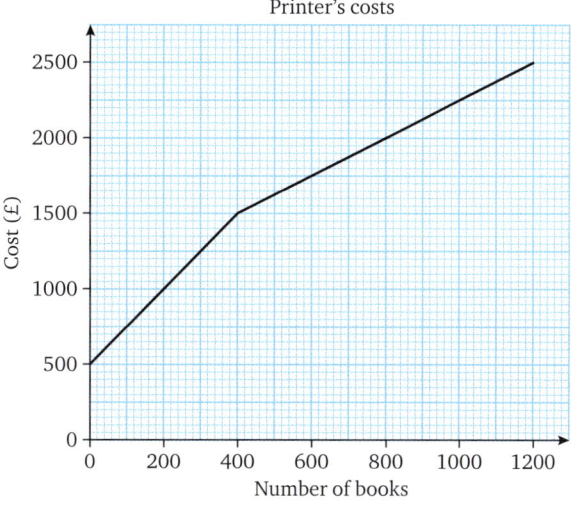

Use the graph to find

a  the fixed charge

b  the total cost if the numbers of book printed is   i 300   ii 400   iii 840

c  Find the cost per book if the number printed is   i 200   ii 660   iii 1000

d  How many books can be printed for   i £2000   ii £1200   iii £700?

## Parabolas

We know that an equation of the form $y = ax^2 + bx + c$ (a quadratic expression in $x$) gives a curve whose shape is called a **parabola** and looks like this.

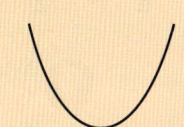

When the $x^2$ term is negative, the curve is 'upside down'.

## Using graphs to solve quadratic equations

One of the problems in the class discussion concerned solving the equations

$x^2 - 2x - 4 = 0$

$x^2 - 2x - 4 = 6$

$2x^2 - 4x + 1 = 0$

$1 + 2x - x^2 = 0$

You may have noticed that, in one form or another, the terms '$x^2 - 2x$' appear in all these equations. This suggests that we can use one graph to solve all these equations. The worked example in the next exercise shows how we can do this.

### Exercise 12b

**Worked example**

→ Use the graph of $y = x^2 - 2x - 4$ to solve the equations

   **a**   $x^2 - 2x - 4 = 0$          **c**   $2x^2 - 4x + 1 = 0$

   **b**   $x^2 - 2x - 10 = 0$     **d**   $1 + 2x - x^2 = 0$

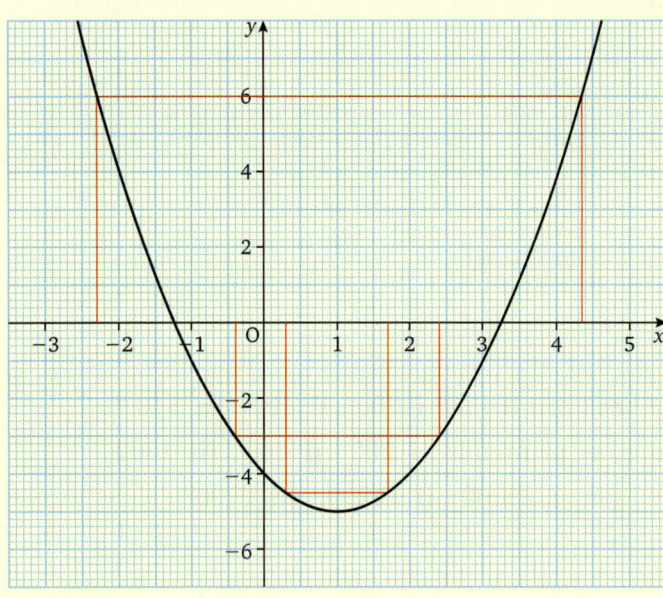

**a**   $x^2 - 2x - 4 = 0$

> 💡 When this graph crosses the $x$-axis, the value of $y$ is 0, i.e. $x^2 - 2x - 4 = 0$. The graph crosses the $x$-axis where $x = -1.25$ and $x = 3.25$.

$x = -1.25$ or $3.25$

> 💡 This agrees with what we discovered in Chapter 11: a quadratic equation has two solutions.

**b**   $x^2 - 2x - 10 = 0$

> 💡 To use the graph of $y = x^2 - 2x - 4$ to solve the equation $x^2 - 2x - 10 = 0$, we must convert the LHS to $x^2 - 2x - 4$; we can do this by adding 6 to both sides.

$x^2 - 2x - 10 + 6 = 6$, i.e. $x^2 - 2x - 4 = 6$

> 💡 From the equation of the graph of $y = x^2 - 2x - 4$, we see that if $x^2 - 2x - 4 = 6$, then $y = 6$. Therefore, the values of $x$ when $y = 6$ give the solutions to the equation.

$x = -2.3$ or $4.3$

**c** $2x^2 - 4x + 1 = 0$

>  To use the graph, we must convert the LHS to $x^2 - 2x - 4$. We can do this by dividing both sides by 2, then subtracting $4\frac{1}{2}$ from both sides.

$x^2 - 2x + \frac{1}{2} = 0$ or

$x^2 - 2x - 4 = -4\frac{1}{2}$

From the graph, when $y = -4\frac{1}{2}$, $x = 0.3$ or $1.7$

**d** $1 + 2x - x^2 = 0$

>  To convert $1 + 2x - x^2 = 0$ to the form $x^2 - 2x - 4 = ?$, first add $x^2 - 2x$ to both sides, then subtract 4 from both sides.

$1 = x^2 - 2x$ or

$-3 = x^2 - 2x - 4$

i.e. $x^2 - 2x - 4 = -3$ so $y = -3$

From the graph, when $y = -3$, $x = -0.4$ or $2.4$

**1** Use the graph of $y = x^2 - 3x - 3$, which is given below, to solve the equations.

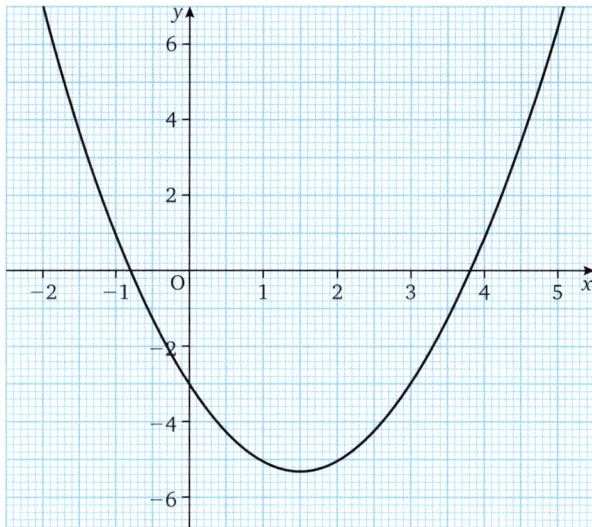

**a** $x^2 - 3x - 3 = 0$
**b** $x^2 - 3x - 3 = 5$
**c** $x^2 - 3x - 7 = 0$
**d** $x^2 - 3x + 1 = 0$

**2** The graph of $y = 4 + 4x - x^2$ is given here.
Use the graph to solve the equations

  **a** $4 + 4x - x^2 = 0$
  **b** $4 + 4x - x^2 = 5$
  **c** $1 + 4x - x^2 = 0$
  **d** $x^2 - 4x + 2 = 0$
  **e** $x^2 = 4x - 3$
  **f** $2x^2 - 8x - 1 = 0$
  **g** Use the graph to solve the equation $4 + 4x - x^2 = 8$. What do you notice?
  **h** Is it possible to use this graph to solve the equation $x^2 - 4x - 14 = 0$?

  If it is possible, give the solutions.

  If it is not possible, explain why.

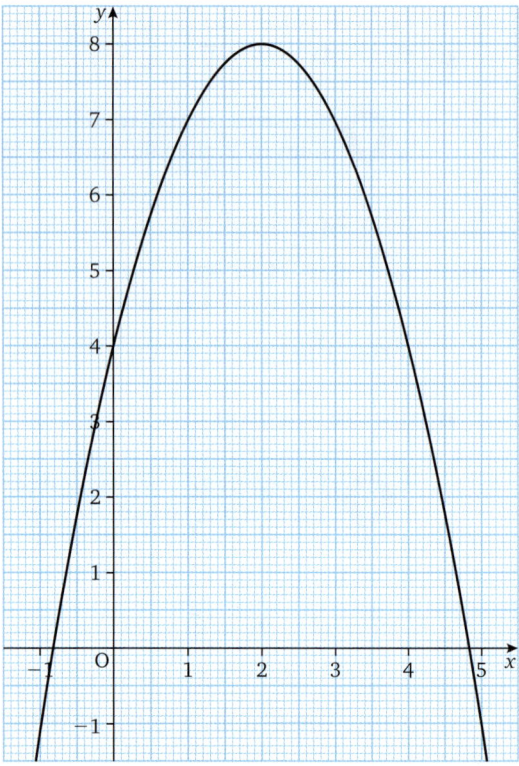

**3** Copy and complete the following table, which gives values of $(2 - x)(x + 1)$ for values of $x$ from $-3$ to $4$.

| $x$ | $-3$ | $-2\frac{1}{2}$ | $-2$ | $-1$ | $-\frac{1}{2}$ | $0$ | $\frac{1}{2}$ | $1$ | $1\frac{1}{2}$ | $2$ | $3$ | $3\frac{1}{2}$ | $4$ |
|---|---|---|---|---|---|---|---|---|---|---|---|---|---|
| $(2 - x)$ | $5$ | $4\frac{1}{2}$ | | $3$ | $2\frac{1}{2}$ | | $1\frac{1}{2}$ | $1$ | | $0$ | $-1$ | $-1\frac{1}{2}$ | $-2$ |
| $(x + 1)$ | $-2$ | $-1\frac{1}{2}$ | | $0$ | $\frac{1}{2}$ | | $1\frac{1}{2}$ | $2$ | | $3$ | $4$ | $4\frac{1}{2}$ | $5$ |
| $(2 - x)(x + 1)$ | $-10$ | $-6\frac{3}{4}$ | | $0$ | $1\frac{1}{4}$ | | $2\frac{1}{4}$ | $2$ | | $0$ | $-4$ | $-6\frac{3}{4}$ | $-10$ |

Hence draw the graph of $y = (2 - x)(x + 1)$ for values of $x$ from $-3$ to $4$. Use 2 cm as 1 unit for $x$ and 1 cm as 1 unit for $y$.

Use your graph to solve these equations.

  **a** $x^2 - x - 2 = 0$  **b** $2x^2 - 2x + 1 = 0$  **c** $x^2 = x + 1$

**4** The table gives values of $y$ for certain values of $x$ on the curve given by the equation $y = 2x^2 - 7x + 8$. Copy and complete this table.

| $x$ | $0$ | $\frac{1}{2}$ | $1$ | $1\frac{1}{2}$ | $2$ | $2\frac{1}{2}$ | $3$ | $3\frac{1}{2}$ | $4$ |
|---|---|---|---|---|---|---|---|---|---|
| $y$ | $8$ | $5$ | | $2$ | $2$ | $3$ | | $8$ | $12$ |

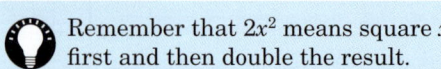 Remember that $2x^2$ means square $x$ first and then double the result.

Use the table to draw the graph of $y = 2x^2 - 7x + 8$ for values of $x$ from $0$ to $4$. Use 4 cm as 1 unit for $x$ and 2 cm as 1 unit for $y$.

Use the graph to solve the equation $2x^2 - 7x + 8 = 4$.

**5** What graph would you draw to solve the equation
  **a** $x^2 + 2x - 4 = 0$
  **b** $3 - 5x - x^2 = 0$?

6  Copy and complete the following table.

| $x$ | $-4$ | $-3\frac{1}{2}$ | $-3$ | $-2\frac{1}{2}$ | $-2$ | $-1\frac{1}{2}$ | $-1$ | $-\frac{1}{2}$ | $0$ | $\frac{1}{2}$ | $1$ |
|---|---|---|---|---|---|---|---|---|---|---|---|
| $7$ | $7$ | $7$ | | $7$ | $7$ | $7$ | | $7$ | $7$ | $7$ | |
| $-6x$ | $24$ | $21$ | | $15$ | $12$ | $9$ | | $3$ | $0$ | $-3$ | |
| $-2x^2$ | $-32$ | $-24\frac{1}{2}$ | | $-12\frac{1}{2}$ | $-8$ | $-4\frac{1}{2}$ | | $-\frac{1}{2}$ | $0$ | $-\frac{1}{2}$ | |
| $7 - 6x - 2x^2$ | $-1$ | $3\frac{1}{2}$ | | $9\frac{1}{2}$ | $11$ | $11\frac{1}{2}$ | | $9\frac{1}{2}$ | $7$ | $3\frac{1}{2}$ | |

Hence draw the graph of $y = 7 - 6x - 2x^2$ for values of $x$ from $-4$ to 1. Use 4 cm as 1 unit on the $x$-axis and 1 cm as 1 unit on the $y$-axis.

   a  Use your graph to solve the equation $7 - 6x - 2x^2 = 0$.
   b  Draw the line $y = 5$. Write down the $x$ values of the points where the line $y = 5$ meets the curve $y = 7 - 6x - 2x^2$. Find, in as simple a form as possible, the equation for which these $x$ values are the roots.

7  Using squared paper and values of $x$ from $-6$ to $0$, draw a graph to show that the equation $x^2 + 6x + 10 = 0$ cannot be solved.

# Cubic graphs

When the equation of a curve contains $x^3$ (and possibly terms involving $x^2$, $x$ or a number), the curve is called a **cubic curve**.

These equations all give cubic curves: $\quad y = x^3 + x \quad y = 2x^3 - 5 \quad y = x^3 - 2x^2 + 6$

We start by plotting the simplest cubic graph, whose equation is $y = x^3$.
The table gives values of $x^3$ for some values of $x$ from $-3$ to 3.

| $x$ | $-3$ | $-2$ | $-1.5$ | $-1$ | $-0.5$ | $0$ | $0.5$ | $1$ | $1.5$ | $2$ | $3$ |
|---|---|---|---|---|---|---|---|---|---|---|---|
| $x^3$ | $-27$ | $-8$ | $-3.4$ | $-1$ | $-0.1$ | $0$ | $0.1$ | $1$ | $3.4$ | $8$ | $27$ |

Plotting these points and joining them with a smooth curve gives this curve.

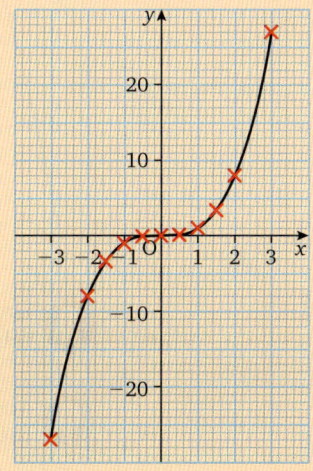

# Exercise 12c

1. Copy and complete the following table, which gives values of $\frac{1}{5}x^3$, correct to 1 decimal place, for values of $x$ from $-3$ to $+3$.

| $x$ | $-3$ | $-2.5$ | $-2$ | $-1.5$ | $-1$ | $-0.5$ | 0 | 0.5 | 1 | 1.5 | 2 | 2.5 | 3 |
|---|---|---|---|---|---|---|---|---|---|---|---|---|---|
| $x^3$ | $-27$ | $-15.6$ | $-8$ | $-3.4$ | | $-0.13$ | | $0.13$ | 1 | | | | 27 |
| $\frac{1}{5}x^3$ | $-5.4$ | $-3.1$ | $-1.6$ | $-0.7$ | | $-0.03$ | | $0.03$ | $0.2$ | | | | $5.4$ |

Hence draw the graph of $y = \frac{1}{5}x^3$ for values of $x$ from $-3$ to $+3$.

Use 2 cm as 1 unit on each axis.

Use your graph to solve the equations

   **a**  $\frac{1}{5}x^3 = 4$     **b**  $x^3 = -15$

2. Make your own copy of the graph of $y = x^3$

   **a**  On the same axes draw the line $y = x + 6$

   **b**  Give the values of $x$ where the curve and straight line intersect.

## Worked example

The graph of $y = (x - 1)(x + 1)(x - 3)$ is given here.

→ Use the graph to find the values of $x$ when $y = 1$ and when $y = -8$

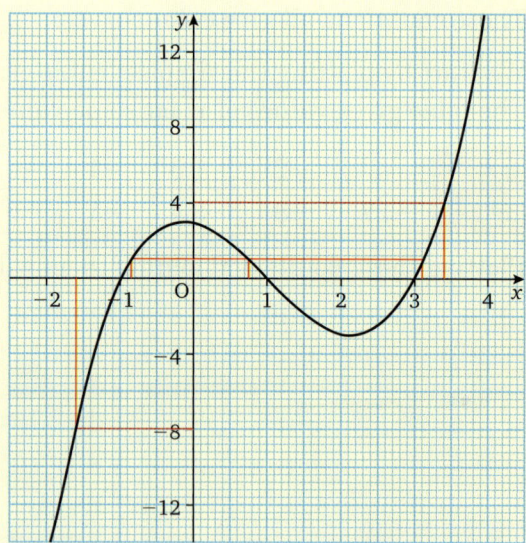

$y = 1$ where $x = -0.85, 0.75, 3.1$

$y = -8$ where $x = -1.6$

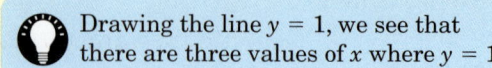

Drawing the line $y = 1$, we see that there are three values of $x$ where $y = 1$

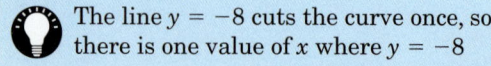

The line $y = -8$ cuts the curve once, so there is one value of $x$ where $y = -8$

> Use the graph to find the solution to this equation:
> $(x - 1)(x + 1)(x - 3) = 0$

$(x - 1)(x + 1)(x - 3) = 0$ where $y = 0$,
i.e. where the curve crosses the $x$-axis, which is when $x = -1, 1$ and $3$.

> Use the graph to find the solution to this equation:
> $(x - 1)(x + 1)(x - 3) = 4$

$(x - 1)(x + 1)(x - 3) = 4$ where the curve cuts the line $y = 4$, i.e. when $x = 3.4$

3  Copy and complete the table, which gives the values of $y$ when $y = x(x - 2)(x - 4)$.

| $x$ | 0 | 0.5 | 1 | 1.5 | 2 | 2.5 | 3 | 3.5 | 4 |
|---|---|---|---|---|---|---|---|---|---|
| $x - 2$ | | $-1.5$ | | | | | | $1.5$ | |
| $x - 4$ | | $-3.5$ | | | | | | $-0.5$ | |
| $y$ | | $2.625$ | | | | | | $-2.625$ | |

Hence draw the graph of $y = x(x - 2)(x - 4)$, using 4 cm for 1 unit on each axis.

a  Use your graph to find
   i  the lowest value    ii  the highest value
   of $x(x - 2)(x - 4)$ within the given range of values for $x$.
b  Find the solutions of the equation $x(x - 2)(x - 4) = 2$ within the given range.
c  If the values of $x$ were extended, do you think there may be another solution to the equation in part b?

4  Copy and complete the table, which gives the value, correct to 2 decimal places, of $\frac{1}{3}x^3 - 2x + 3$ for values of $x$ from $-2$ to $2$.

| $x$ | $-2$ | $-1.5$ | $-1$ | $-0.5$ | 0 | 0.5 | 1 | 1.5 | 2 |
|---|---|---|---|---|---|---|---|---|---|
| $\frac{1}{3}x^3$ | $-2.67$ | $-1.13$ | $-0.33$ | | 0 | | 0.33 | 1.13 | |
| $-2x$ | 4 | 3 | 2 | | 0 | | $-2$ | $-3$ | |
| $+3$ | 3 | 3 | 3 | | 3 | | 3 | 3 | |
| $\frac{1}{3}x^3 - 2x + 3$ | 4.33 | 4.87 | 4.67 | | 3 | | 1.33 | 1.13 | |

Hence draw the graph of $y = \frac{1}{3}x^3 - 2x + 3$ using 4 cm for 1 unit on each axis. Estimate the value(s) of $x$ where the graph crosses the $x$-axis.

5  The graph of $y = 1 - x + 2x^2 - x^3$ is given here.

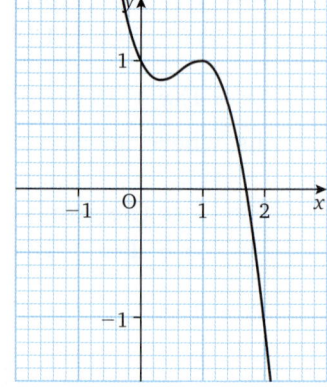

  a  Write down the value(s) of $x$ where the curve crosses the $x$-axis. What can you deduce about the number of solutions to the equation $1 - x + 2x^2 - x^3 = 0$?

  b  How many solutions are there to the equation $1 - x + 2x^2 - x^3 = -1$?
Explain your answer and give the solutions.

  c  Give a value of $c$ if the equation $1 - x + 2x^2 - x^3 = c$
    i  has only one solution    ii  has three solutions.

  6  Without drawing the curve or working out a table of values, explain how you can state the values of $x$ where the curve $y = (x - 2)(x - 3)(x - 4)$ crosses the $x$-axis. *Sketch* the curve.

### The shape of a cubic curve
From the last exercise we can state that a cubic curve looks like

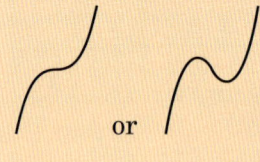

or            when the $x^3$ term is positive

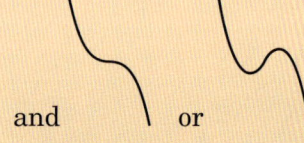

and    or    when the $x^3$ term is negative.

### Reciprocal graphs

The equation $y = \dfrac{a}{x}$, where $a$ is a number, is called a **reciprocal equation**.

The simplest reciprocal equation is $y = \dfrac{1}{x}$.

Making a table showing values of $y$ for some values of $x$ from $-4$ to $-\dfrac{1}{4}$ and from $\dfrac{1}{4}$ to 4 gives

| $x$ | $-4$ | $-3$ | $-2$ | $-1$ | $-\dfrac{1}{2}$ | $-\dfrac{1}{4}$ | $\dfrac{1}{4}$ | $\dfrac{1}{2}$ | $1$ | $2$ | $3$ | $4$ |
|---|---|---|---|---|---|---|---|---|---|---|---|---|
| $y$ | $-0.25$ | $-0.33$ | $-0.5$ | $-1$ | $-2$ | $-4$ | $4$ | $2$ | $1$ | $0.5$ | $0.33$ | $0.25$ |

Plotting these points on a graph and joining them with a smooth curve gives this distinctive two-part shape.

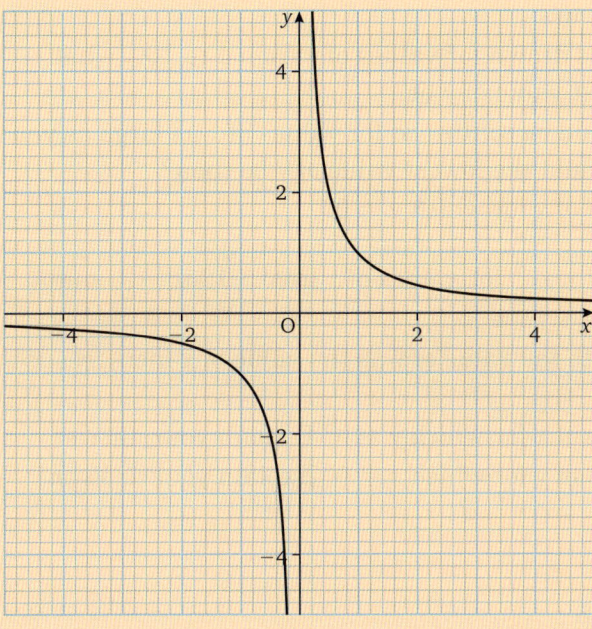

## Exercise 12d

1. Discuss why we did not use the value $x = 0$ in the table made to draw the graph of $y = \dfrac{1}{x}$.

2. Use the graph of $y = \dfrac{1}{x}$, drawn above, to answer these questions.
   a. Give the value of $y$ when $x = 2.5$
   b. What is the value of $x$ when $y = 2.5$?
   c. What is the value of $y$ when $x = 0.2$?
   d. What happens to the value of $y$ when $x$ gets smaller than 0.2?
   e. Why is there no point on the curve shown when $x = 0$?
   f. How many forms of symmetry does the graph have?

3. Draw the graph of $y = \dfrac{2}{x}$ for values of $x$ from $-4$ to $-\frac{1}{2}$ and from $\frac{1}{2}$ to 4. Use 2 cm for 1 unit on both axes.
   a. Why is there no point on the graph where $x = 0$?
   b. Give the value of $y$ when
      i. $x = 2.6$
      ii. $x = -1.8$

# STP Maths 9

4  Draw the graph of $y = \dfrac{12}{x}$ for values of $x$ from 1 to 12. Use 1 cm for 1 unit on both axes.
   a  Give the lowest value of $y$ in the given range, and the value of $x$ at which it occurs.
   b  If the graph was drawn for values of $x$ from 1 to 100, what would the lowest value of $y$ be?
   c  If the graph could be continued for values of $x$ as large as you choose, what would the lowest value of $y$ be then?
   d  Is there a value of $x$ for which $y = 0$?

5  *Sketch* the graph of $y = \dfrac{1}{x}$ for values of $x$ from $-10$ to $-\dfrac{1}{10}$ and from $\dfrac{1}{10}$ to 10.
   a  What happens to the value of $y$ as the value of $x$ increases beyond 10?
   b  Is there a value of $x$ for which $y = 0$? Explain your answer.
   c  Is there a value of $y$ for which $x = 0$? Explain your answer.

## Reciprocal curves

An equation of the form $y = \dfrac{a}{x}$, where $a$ is a constant (that is, a number), gives a distinctive two-part curve called a **reciprocal curve**. The shape is also called a **hyperbola**.

Notice that there is a break in the graph where $x = 0$.

This is because there is no value for $y$ when $x = 0$;

**we cannot divide by zero.**

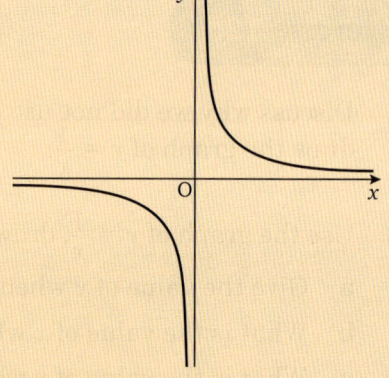

## Recognising curves

We are now in a position to look at a curve and recognise that its equation could be
- $y = mx + c$
- $y = ax^2 + bx + c$
- $y = ax^3 + bx^2 + cx + d$
- $y = \dfrac{a}{x}$
- none of these.

## Exercise 12e

For questions **1** to **4**, write down the letter that corresponds to the correct answer.

1. The equation of this curve could be
   - **A** $y = x^2$
   - **B** $y = \dfrac{1}{x}$
   - **C** $y = x^3$
   - **D** $y = 4x - x^2$

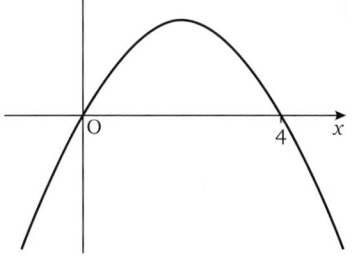

2. The equation of this curve could be
   - **A** $y = x^2 + x - 9$
   - **B** $y = (x - 3)(x + 3)(x + 1)$
   - **C** $y = \dfrac{9}{x}$
   - **D** $y = x^3$

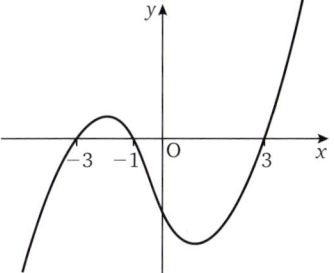

3. The equation of this curve could be
   - **A** $y = \dfrac{12}{x}$
   - **B** $y = x^2 - 9$
   - **C** $y = 9 - x^2$
   - **D** $y = x^3$

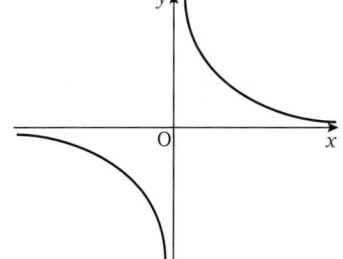

4. The equation of this curve could be
   - **A** $y = x^2$
   - **B** $y = 4 - x^2$
   - **C** $y = x^2 - 2x + 6$
   - **D** $y = x^3 - 4x^2 + 3$

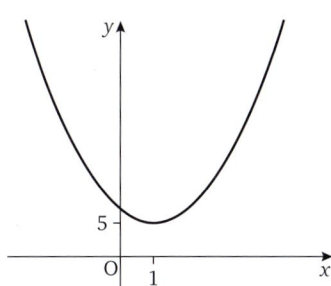

For question **5** to **13**, use squared paper. Draw $x$- and $y$-axes but do not scale them. *Sketch* the graph for each of the following equations.

5. $y = 3$
6. $y = x^2$
7. $y = \dfrac{1}{x}$ $\;(x \ne 0)$
8. $y = x$
9. $y = x^2 + 1$
10. $y = x^3$
11. $y = x + 1$
12. $y = 1 - x^3$
13. $x = 3$

For questions **14** to **20**, write down the letter that corresponds to the correct answer.

**14** The graph representing $y = x^2 - 1$ could be

A 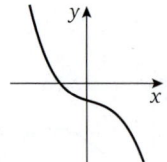 B 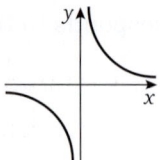 C 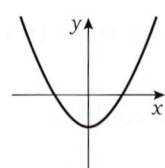 D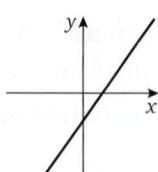

**15** The graph representing $y = \dfrac{3}{x}$ could be

A 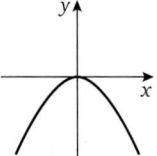 B 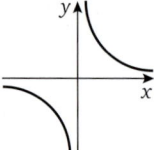 C 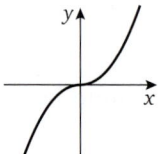 D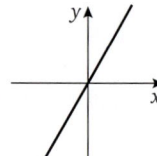

**16** The graph representing $y = 10x^2$ could be

A 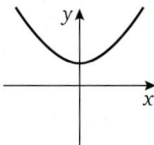 B 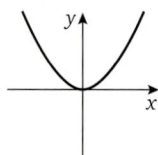 C 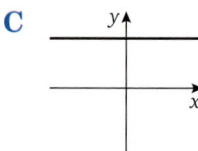 D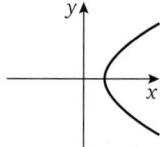

**17** The graph representing $y = x + 2$ could be

A 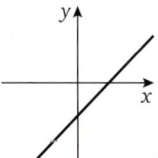 B 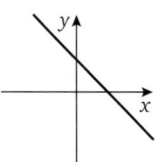 C 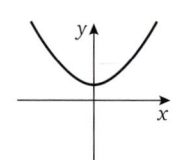 D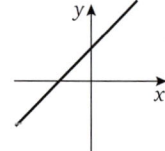

**18** The graph representing $y = -x^2$ could be

A 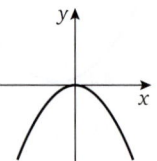 B 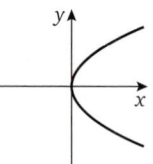 C 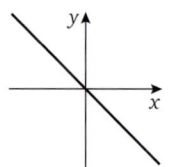 D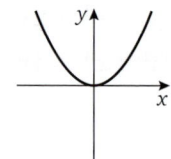

**19** The graph representing $y = x^3 + 1$ could be

A 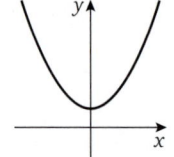 B 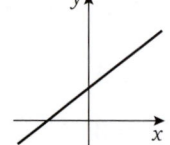 C 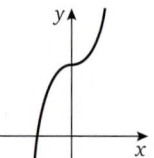 D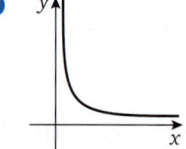

**20** The graph representing $y = x(x - 1)(x - 2)$ could be

A 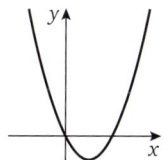  B 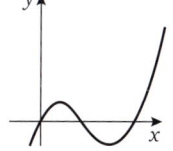  C 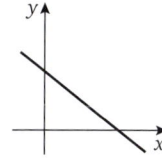  D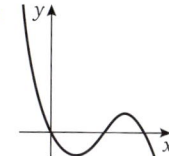

**21** For questions **1** to **4**, explain why you cannot be certain that the equation you chose is the equation of the graph.

## Using graphs

For question **2** in the class discussion, page 236, Chloe needed to use the following readings to confirm that the current flowing in a circuit is inversely proportional to the resistance.

| Resistance (ohms) | 0.5 | 0.7 | 1 | 1.2 | 1.5 | 2 | 2.4 |
|---|---|---|---|---|---|---|---|
| Current (amps) | 5.6 | 4 | 2.8 | 2.3 | 1.9 | 1.4 | 1.5 |

If the resistance is $R$ ohms and the current is $I$ amps, Chloe wants to confirm that $I$ is inversely proportional to $R$, that is that the product of corresponding values of $R$ and $I$ is constant.

This means that the relationship between $I$ and $R$ should be of the form $IR = k$.

Rearranging to make $I$ the subject gives $I = \frac{k}{R}$, where $k$ is a constant.

Chloe also knows that a relationship of the form $y = \frac{k}{x}$ gives a curve whose shape is

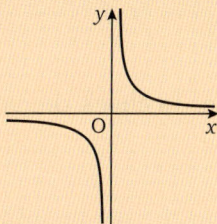

when $x$ and $y$ can have positive and negative values.

If $x$ and $y$ can only take positive values, we get just the right-hand side of this curve, i.e.

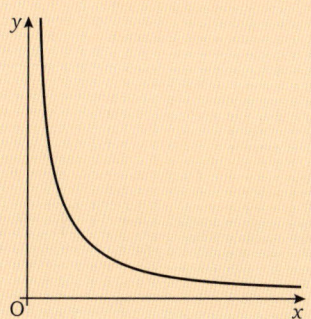

The resistance and the current have positive values only, so this is the shape that Chloe expects to find when she plots the values of $I$ against the values of $R$ in the table.

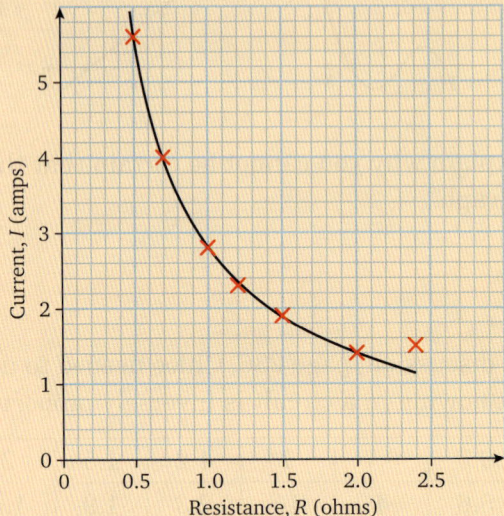

Allowing for slight errors in readings, these points do lie on a curve of the shape expected, except for one. Either a mistake has been made with this point (Chloe may have made an error with her readings), or it could be that the relationship between $R$ and $I$ no longer applies when $R = 2.4$.

Assuming that $I = \dfrac{k}{R}$, we can substitute a pair of values *from the graph* to find the value of $k$

i.e. when $R = 2$, $I = 1.4$ so $1.4 = \dfrac{k}{2}$

$\Rightarrow \quad k = 2 \times 1.4 = 2.8$

$\therefore \quad I = \dfrac{2.8}{R}$

(We can use another pair of coordinates to check the value found for $k$.)

## Inverse proportion

The quantities in the example above are inversely proportional.

**Any two quantities, $x$ and $y$, that are inversely proportional, are related by the equation $y = \dfrac{k}{x}$ and the graph representing them is a hyperbola.**

### Exercise 12f

1. This is a graph relating pressure and volume. The equation of the graph is $p = \dfrac{k}{v}$

   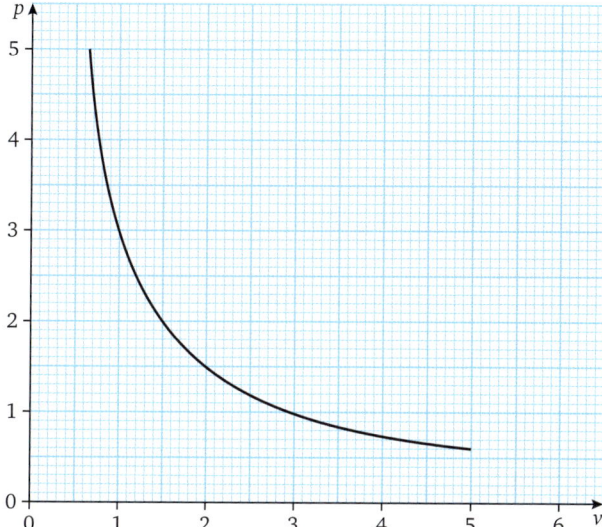

   a. Use the graph to find
   
   i. the value of $p$ when $v = 1.2$
   
   ii. the value of $v$ when $p = 3.6$

   b. Use the answer to part **a i** to find the value of $k$. Check your answer using the values found in part **a ii**.

2. Water in a plastic bottle cools down in the freezer. The graph shows how its temperature is changing.

   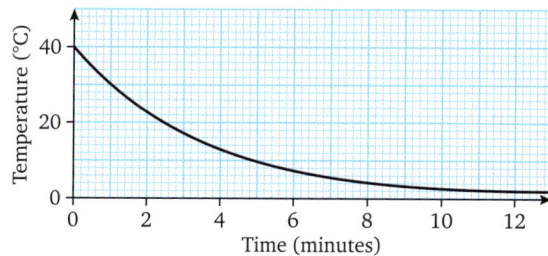

   a. What is the temperature of the water after
   
   i. 4 minutes
   
   ii. 8 minutes?

   b. How long does it take to cool down to 14 °C?

   c. Explain why this graph shows that the temperature is not inversely proportional to the time.

3  This graph shows the height of a cricket ball above the ground at different times during its flight.

   Use the graph to find
   a  the height of the ball above the ground 1.5 seconds after it has been hit
   b  the greatest height above the ground reached by the ball
   c  the times at which the ball is 10 m above the ground
   d  the total time of the ball's flight.

4  In question **3**, the relationship between the height, $h$ metres, and the time, $t$ seconds, is $h = at^2 + bt + c$
   a  Use the point on the curve where $t = 0$ to show that $c = 0$.
   b  Use the point on the curve where $t = 5$ to find a relationship between $a$ and $b$.
   c  Use the point on the curve where $t = 1$ to find another relationship between $a$ and $b$.
   d  Use your answers to parts **b** and **c** to find the values of $a$ and $b$. Hence write down the equation of the curve.

5  Under ideal conditions, the number of bacteria growing in a Petri dish doubles every hour. So if one bacterium is initially placed on the dish, there are 2 bacteria after 1 hour, 4 bacteria after 2 hours, 8 bacteria after 3 hours, and so on.

   The graph illustrates this relationship.

   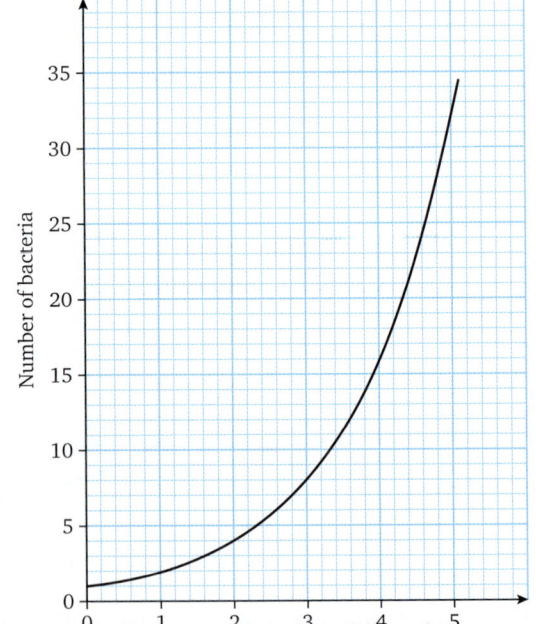

   a  Copy and complete the table, which shows the number, $n$, of bacteria $t$ hours after the first bacteria is placed on the dish.

| $t$ | 0 | 1 | 2 | 3 | 4 | 5 |
|---|---|---|---|---|---|---|
| $n$ | 1 | 2 | 4 | 8 | | |
| $n$ as a power of 2 | $2^0$ | | $2^2$ | | | |

   b  Use the table to deduce the relationship between $n$ and $t$.

6   A second-hand car is bought for £10 000. Its value then depreciates every year by half its value at the start of the year. The graph shows this relationship.

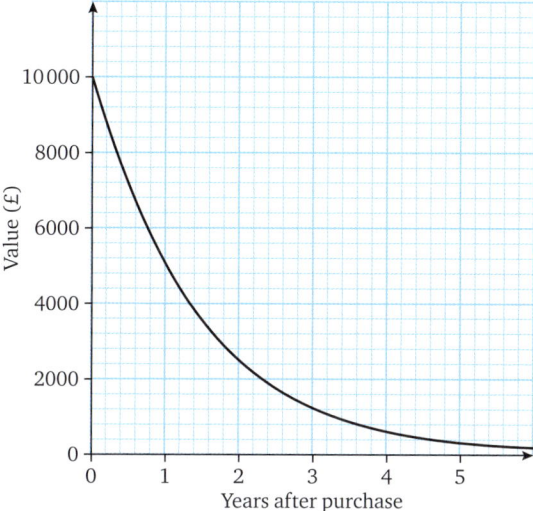

a   What is the car worth 3 years after it was bought?

b   Copy and complete the table.

| Years after purchase, $n$ | 0 | 1 | 2 | 3 | 4 | 5 |
|---|---|---|---|---|---|---|
| Value, £$y$ | 10 000 | 5000 $(= 10\,000 \times \frac{1}{2})$ | 2500 $(= 10\,000 \times (\frac{1}{2})^2)$ | | | |

c   Use the table to deduce the relationship between $n$ and $y$.

d   Express the initial value of the car as £10 000 times a power of $\frac{1}{2}$.

e   Explain why the value of the car will never be zero.

In question **5**, the relationship between $t$ and $n$ involves 2 to the power $n$. In question **6**, the relation between $y$ and $n$ involves $\frac{1}{2}$ to the power $n$. These are examples of **exponential graphs**.

Any relationship between variables $x$ and $y$ that involves $x$ in the power of a number gives an exponential curve.

7  The diagram shows a sketch of the curve $y = 3^x$.

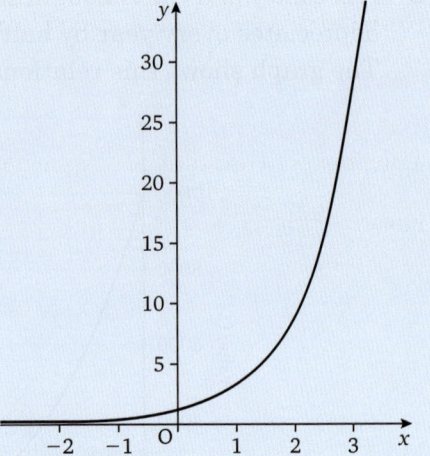

a  Give the coordinates of the point where the curve cuts the $y$-axis.

b  Explain why the curve never cuts the $x$-axis.

c  Copy the diagram and add a sketch of the curve $y = 2^x$.

d  Write down the coordinates of the point that is common to both curves.

e  Explain why any curve whose equation is $y = a^x$ will also go through this point.

8  Gravity pulls objects towards the surface of the Earth, making the speed of the objects increase. The effect of this pull decreases as the distance from the surface of the Earth increases.

When an object is $d$ metres from the centre of the Earth, the rate at which its speed increases is $g$ m/s per second, where $g$ is given by

$$g = \frac{4.1 \times 10^{14}}{d^2}$$

(Note that this relationship applies only when $d$ is greater than the radius of the Earth, i.e. outside the Earth's surface.)

This graph illustrates the relationship.

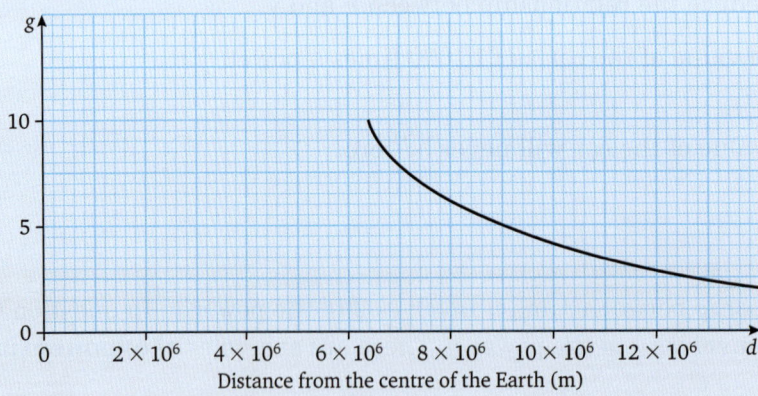

a  What is the radius of the Earth?

b  What is the value of $g$ when an object is $4.1 \times 10^6$ metres above the *surface* of the Earth?

c  When $g = 2.5$, what is the value of $d$? How far is the object above the Earth's surface when $g = 2.5$?

# 12 Graphs

Note that, in question **8**, the relationship between $g$ and $d$ is of the form

$$g = \frac{\text{number}}{d^2}$$

This form of the relationship is called the **inverse square law**.

For example, if $y = \frac{4}{x^2}$, $x$ and $y$ are related by the inverse square law.

9  The relationship between $x$ and $y$ for the graph shown is
$y = ax^3 + b$. Use the information on the graph to find the
values of $a$ and $b$.

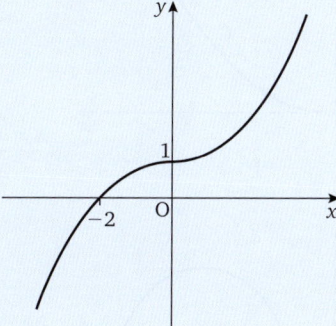

## Gradients of curves

The gradient of a straight line is the same at any point on the line, that is, the gradient is constant. The gradient, or slope, of a curve however, changes from point to point.

We can give an estimate of the gradient of a curve but it is difficult to calculate because it changes as we move along the curve.

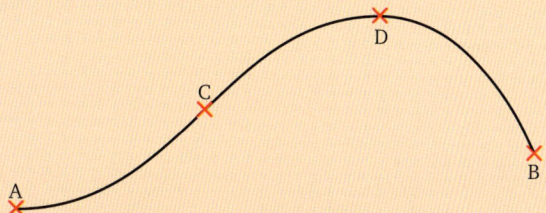

Moving along this curve from left to right, the gradient is zero at A, increases gradually to reach its maximum value at C and then decreases to zero again at D. The gradient then becomes negative and gets more negative (that is, the downhill slope increases) as we approach B.

257

# STP Maths 9

## Exercise 12g

Describe the way in which the gradient changes as we move along each curve from A to B.

1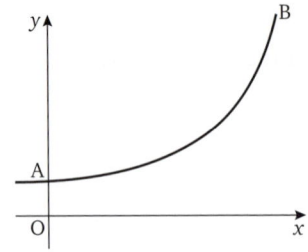

4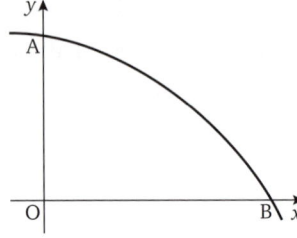

2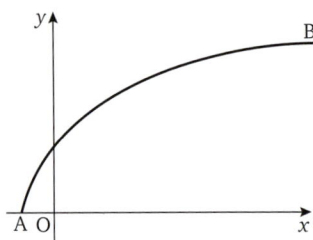

5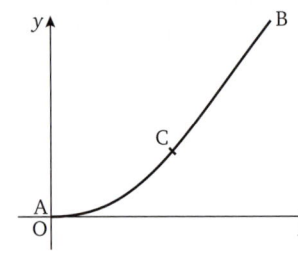

3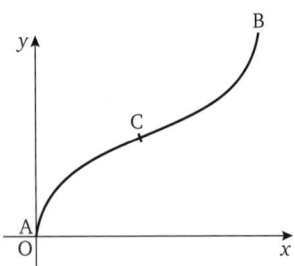

6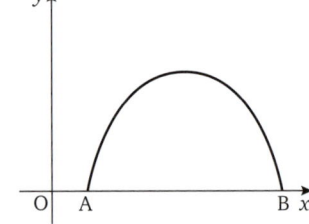

## Interpretation of gradient

*Gradient* gives the rate at which the quantity on the vertical axis is changing as the quantity on the horizontal axis increases.

For example, this graph shows the temperature of a saucepan of water.

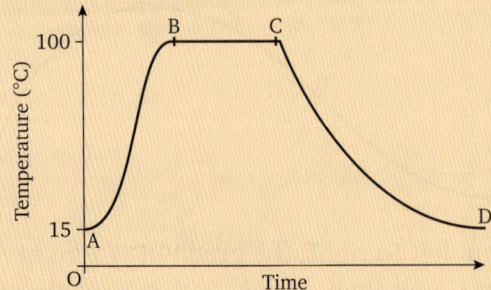

Although there are only two numbers on the axes, we can describe what is happening in general terms.

The water temperature starts at 15 °C at A, increases slowly at first but then more rapidly until it levels off at its maximum value at B. The gradient is positive for the whole of this section of the graph.

From B to C the temperature does not change (the water is boiling and its temperature cannot go above 100 °C). The gradient for this section is zero.

From C to D the temperature is falling, fairly fast to begin with and then more slowly. For this section of the curve the gradient is negative.

### Exercise 12h

1  The graph shows the temperature of a bowl of soup, from the moment it is served.

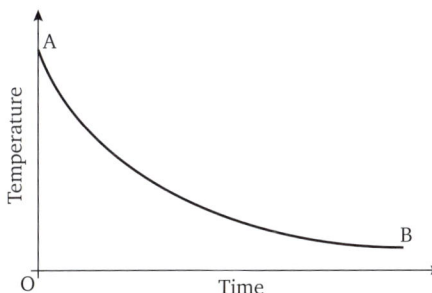

   a  Describe how the temperature of the soup changes with time.
   b  Roughly, how is the temperature changing near to B?
   c  What roughly, do you think is the temperature of the soup at B?

2  The graph shows, over several years, the number of people travelling by air.

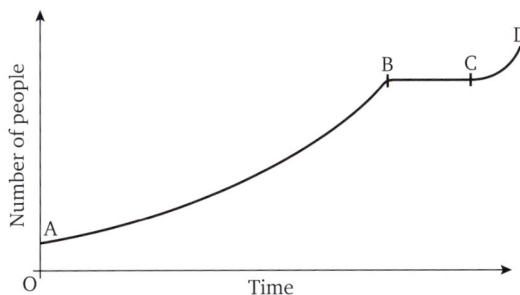

   a  Does the section of the curve from A to B represent an increase or a decrease in the number of people travelling by air? Is this rate of change constant?
   b  Describe how the number of people travelling by air changes for the sections from B to C and from C to D.

**3** The graph shows the speed of a car between two sets of traffic lights.

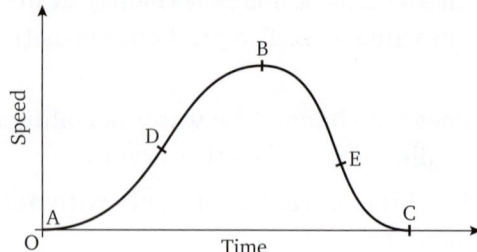

**a** Which section of the curve indicates
  **i** increasing speed
  **ii** decreasing speed?
**b** What can you say about the speed of the car at the point B?
**c** What colour is the second set of lights? Justify your answer.
**d** When is the increase in speed greatest?
**e** When is the car losing speed at the greatest rate?

**4** The graph shows how the surface area changes as a balloon is blown up.

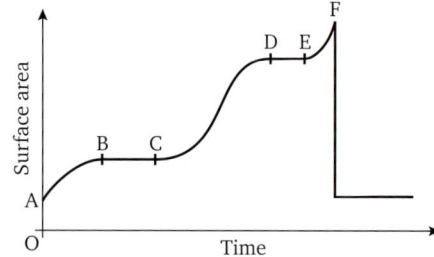

**a** What do you think happens at F?
**b** Describe how the surface area changes between
  **i** A and B
  **ii** B and C.
**c** What do you think accounts for the gradient of the curve for the sections B to C and D to E?

**5** A stone is thrown into the air. The graph shows the distance of the stone from the ground.

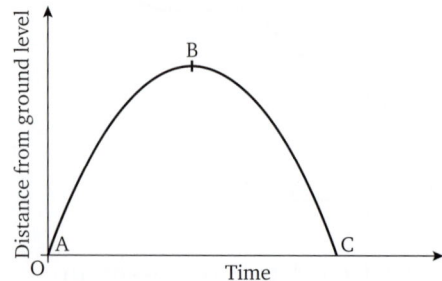

**a** Describe how the gradient of this curve varies throughout the stone's flight.
**b** Where on the curve is the speed of the stone
  **i** greatest
  **ii** least?

6 The graph shows the height of an aircraft above sea-level as the aircraft flies away from the airport. Describe how the height of the aircraft changes for the whole curve.

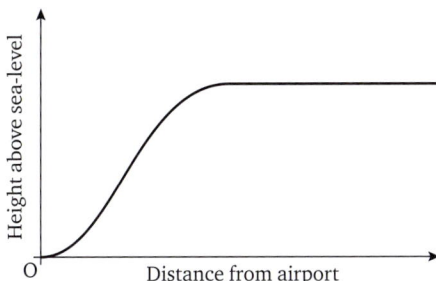

7

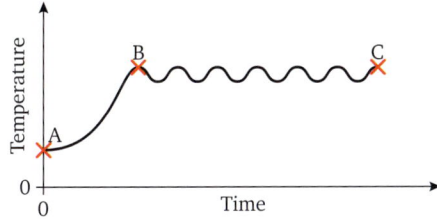

The graph shows the temperature of an oven controlled by a thermostat.

  a  Why does the curve not start at zero on the vertical axis?
  b  Describe what is happening for the part of the curve between A and B.
  c  What do you think can account for the shape of the curve between B and C?

8 The graph shows the speed of a motorcycle on a section of a cross-country race.

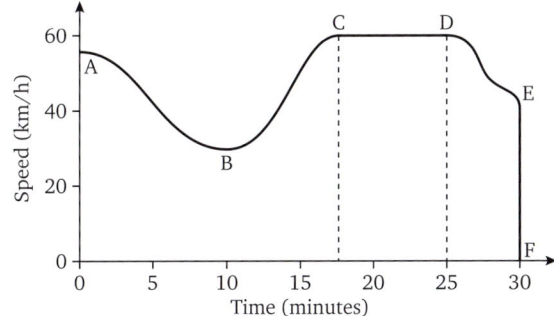

  a  Which sections of the curve represent the rider
      i  going uphill
      ii going downhill
      iii falling off?
  b  For roughly how long is the speed constant?
  c  What, roughly, is the rate at which the speed is changing for the section from B to C?

261

**9** When an aircraft takes off it accelerates rapidly from zero speed until it reaches roughly half its cruising speed. It then accelerates more and more slowly until it reaches its cruising speed of 400 km/h. Sketch a graph showing how the speed of the aircraft changes during this time.

**10** A car stops at one set of traffic lights. When these lights go green it moves off but has to stop again at the next set of traffic lights, which are 500 yards from the first set. Sketch a graph showing the distance of the car from the *first* set of lights during the time that the car takes to move between the two sets.

**11** The graph illustrates a 100 metre race between Phoebe and Gita.

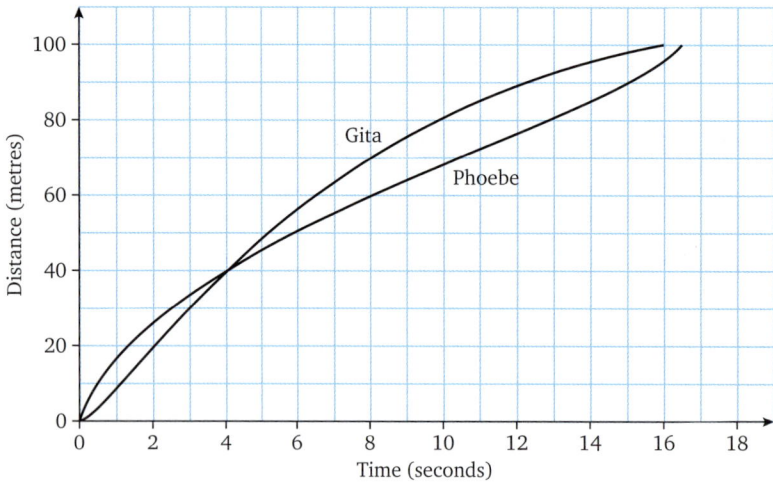

**a** Who won the race and in what time?
**b** What was the difference in the times taken to finish the race?
**c** Who ran faster at the start?
**d** How far from the start did one competitor overtake the other?

**12** Liquid is poured at a constant rate into each of the containers whose cross-sections are given below. Sketch a graph showing how the depth of liquid in each container increases as the liquid is poured in.

a  b 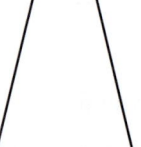 c  d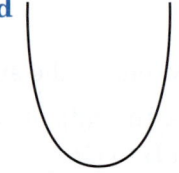

13  In the class discussion, you were asked to discuss how Jordan could sketch a graph showing how the depth of water in a swimming pool varied as the pool was filled at a constant rate. This is the cross-section of the pool.
Draw the graph for Jordan.

14  The diagram shows some grain storage tanks with different shapes. Grain is drawn off at the same constant rate from each tank.

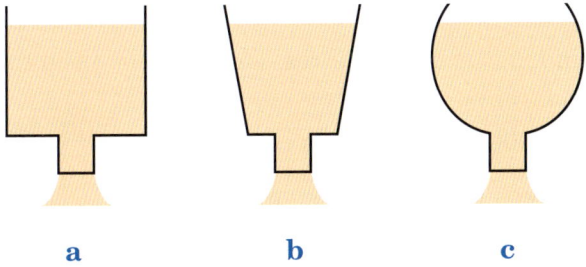

Make three copies of this diagram:

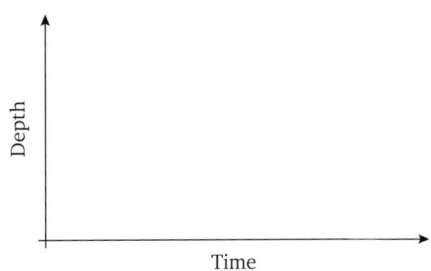

Sketch a graph, one on each diagram, showing how the depth of grain in the tank varies with time.

## Consider again

Freda hopes to find a relationship between the distance a puck slides down a slope and the time for which it has been sliding.

She has collected the following data.

| Time, in seconds, after leaving the top of the slope | 1 | 2.7 | 3.5 | 4.5 | 4.9 |
|---|---|---|---|---|---|
| Distance, in metres, from the top of the slope | 0.05 | 0.4 | 0.6 | 1 | 1.2 |

Using $t$ seconds for the time and $d$ metres for the distance, now can you find a likely algebraic relationship between $t$ and $d$?

# 13 Areas and volumes

## Consider

One of the components in the design for a new computer desk is a conduit (a channel) for cables.

These are the drawings for the conduit, which has a uniform cross-section.

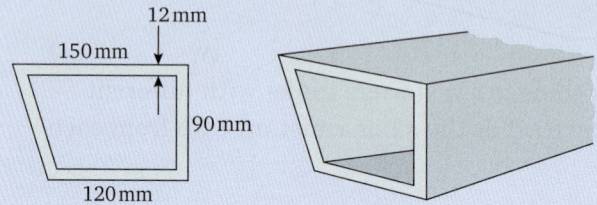

The conduit is to be manufactured in lengths of 2.5 metres. The material used to make this conduit has a **density** of 0.75 g/cm³.

Anna is ordering shelving for storing 5000 lengths of this conduit.

To do this, Anna needs to know how much space they take up, and their mass.

Can you find the volume of these 5000 lengths of conduit and their mass?

*You should be able to answer these questions after you have worked through this chapter.*

## Area of a trapezium

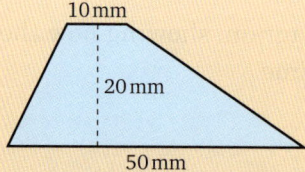

We can find the area of this **trapezium** by dividing it into two triangles, but trapeziums occur often enough in real objects to justify finding a formula for the area of any trapezium.

Any trapezium can be divided into two triangles.

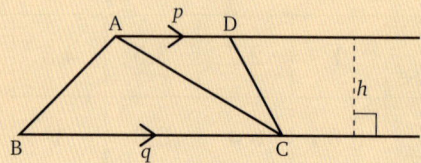

Triangles ABC and ADC both have the same height, as the height of each is the distance, $h$, between the parallel sides.

**13 Areas and volumes**

Now,       area △ADC = $\frac{1}{2}$ base × height = $\frac{1}{2} p \times h$

           area △ABC = $\frac{1}{2}$ base × height = $\frac{1}{2} q \times h$

Therefore,   area ABCD = $\frac{1}{2} ph + \frac{1}{2} qh$

                     = $\frac{1}{2}(p + q) \times h$

This formula is easier to remember in words: $(p + q)$ is the sum of the parallel lengths, so

**the area of a trapezium is equal to**

**$\frac{1}{2}$(sum of the parallel sides) × (distance between them).**

## Exercise 13a

### Worked example

→ Find the area of this trapezium.

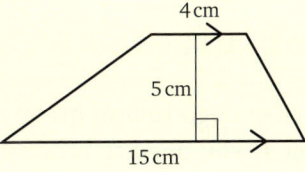

Area = $\frac{1}{2}$(sum of parallel sides) × (distance between them)

= $\frac{1}{2}(15 + 4) \times 5$ cm²

= $\frac{1}{2} \times 19 \times 5$ cm² = 47.5 cm²

In questions **1** to **4**, find the areas of the trapeziums.

1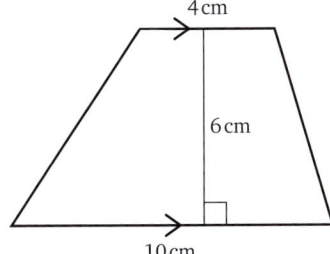

2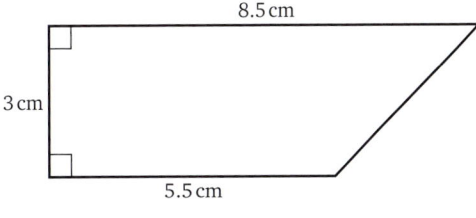

**3** 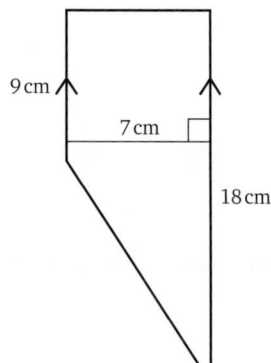   **4**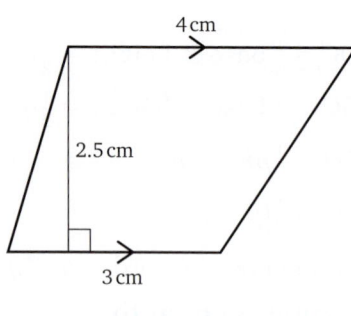

For questions **5** to **7**, use squared paper and draw axes for $x$ and $y$ using ranges $-6 \leq x \leq 6$ and $-6 \leq y \leq 6$, and a scale of one square to 1 unit. Plot the points and join them up in alphabetical order to form a closed shape. Find, in square units, the area of the resulting figure.

**5** A (6, 1), B (4, −3), C (−2, −3), D (−3, 1)

**6** A (4, 4), B (−2, 2), C (−2, −2), D (4, −3)

**7** A (3, 5), B (−4, 4), C (−4, −2), D (3, −5)

For questions **8** to **12**, you will also need to find areas of squares, rectangles, triangles and circles. If you need to remind yourself of the formulas for these, they are on page 6.

**8** The diagram shows the end wall of a lean-to conservatory. The shaded area is covered with cedar wood, and the hatched area is glazed.

Find

  **a** the total area of the end wall

  **b** the area of the door

  **c** the area of the glazed section

  **d** the area which is covered with cedar wood.

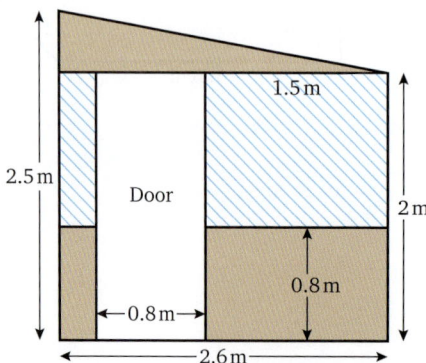

13 Areas and volumes

9  Find the areas of these cork gaskets.

a

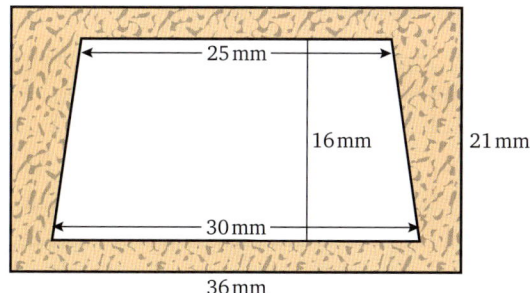

b

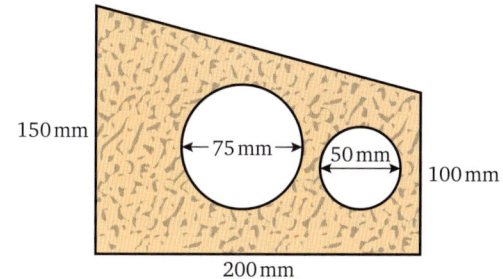

10 This is a classic patchwork quilt pattern called the 'captain's wheel'. It is based on a 4 unit by 4 unit square.

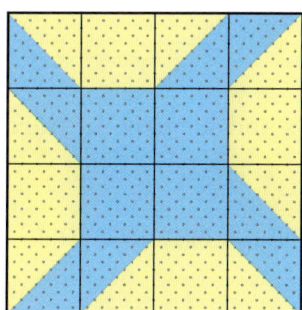

a  Find the area of the patterned yellow part of the design if the length of one side of the large square is 12 cm.
b  A quilt is made from 800 of these squares. What area of the quilt is blue?

11 This company logo is made from a semicircle on the shorter parallel side of a trapezium. Find its area.

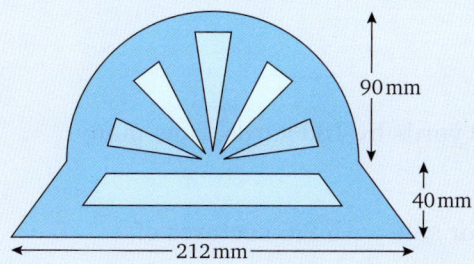

# STP Maths 9

 12 Wood is used to make the top of this child's posting box toy. The plan has been drawn on a 1 cm grid and then scaled down.

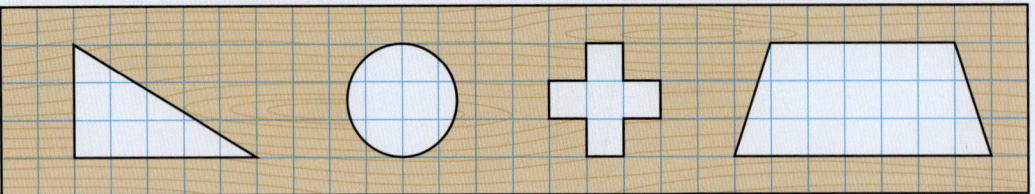

The wood weighs 0.35 grams per square centimetre. Find the mass of the top of this posting box.

## Units of area

Area is measured in standard sized squares. These squares are defined by the length of their sides. In metric units, for example, we may use the square centimetre ($cm^2$). In Imperial units, we use the square inch (sq in), the square foot (sq ft) and the square yard (sq yd).

For large areas of land, such as countries, the units used are usually the square kilometre or the square mile. For smaller areas, such as those of fields, the metric unit used is the **hectare**, where

   **1 hectare = 10 000 square metres**

The Imperial unit used for the areas of fields and so on, is the **acre**, where

   **1 acre = 4840 square yards**

Acres and hectares can be interchanged using the approximation

   **1 hectare ≈ 2.5 acres**

(This relationship is correct to 2 significant figures. More accurately, 1 hectare = 2.471 acres, which is correct to 3 decimal places.)

## Exercise 13b

1. Given that 1 yard = 3 feet and that 1 foot = 12 inches, express
   a  2.5 sq yd in square feet
   b  1.5 sq ft in square inches.

2. A rectangular field measures 250 yards by 130 yards. How many acres does the field cover?

3. A house in France is advertised for sale with 2.6 hectares of ground. How many acres is this?

# 13 Areas and volumes

**4** The diagram shows a plot of land in the shape of a trapezium.
  **a** Find the area of the plot in hectares, correct to 1 decimal place.

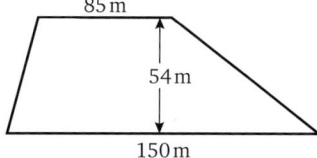

  **b** How many acres does this plot cover?

**5** The area of the United Kingdom is 94 214 square miles. Express this area in
  **a** acres   **b** hectares   **c** square kilometres.

**6** The area of Jamaica is 4240 square miles.
Use the relationship

  1 mile = 1.6093 km correct to 4 decimal places

to give the area of Jamaica in square kilometres to the nearest square kilometre. State with reasons whether your answer is accurate to the nearest whole number.

## Arcs and sectors of circles

Another shape that occurs frequently is a slice of a circle.

This wood beading, for example, has a cross-section of this shape.

Part of a circle is called an **arc**.

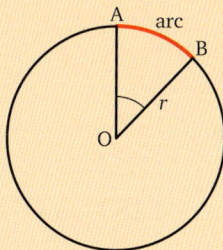

The length of an arc depends on the radius of the circle and the angle enclosed at the centre of the circle by the radii from the two ends of the arc. This angle is $A\hat{O}B$ in the diagram above; it is called the **angle subtended by the arc** AB at the centre, O, of the circle.

The length of the arc AB as a fraction of the circumference is $\frac{A\hat{O}B}{360°}$,

i.e.   length of arc AB = $\frac{A\hat{O}B}{360°}$ of the circumference.

STP Maths 9

The circumference of a circle is $2\pi r$, so

$$\text{length of arc AB} = \frac{A\hat{O}B}{360°} \times 2\pi r$$

The slice of the circle enclosed by the arc and the radii is called a **sector**.

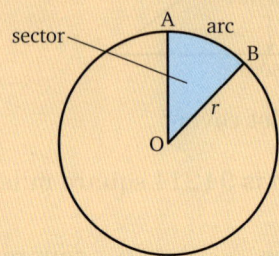

The area of the sector AOB as a fraction of the area of the circle is also equal to $\frac{A\hat{O}B}{360°}$

i.e. $\quad$ **area of sector AOB** $= \dfrac{A\hat{O}B}{360°} \times \pi r^2$

## Exercise 13c

### Worked example

→ Find the length of this red arc.

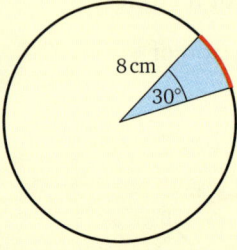

Length of arc $= \dfrac{30°}{360°} \times$ circumference

$= \dfrac{30}{360} \times 2 \times \pi \times 8 \text{ cm} = 4.188... \text{ cm}$

The length of the arc is 4.19 cm correct to 3 s.f.

→ Find the area of the blue sector.

Area of sector $= \dfrac{30°}{360°} \times$ area of circle

$= \dfrac{30}{360} \times \pi \times 8^2 = 16.75... \text{ cm}^2$

The area of the sector is 16.8 cm² correct to 3 s.f.

## 13 Areas and volumes

In each of questions **1** to **6**, find
- **a** the length of the arc
- **b** the area of the sector.

**1**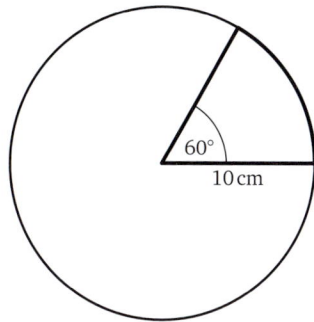

**4**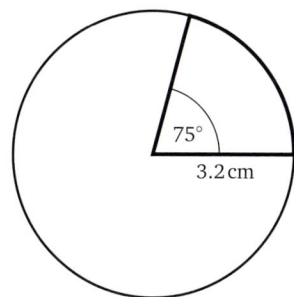

**2**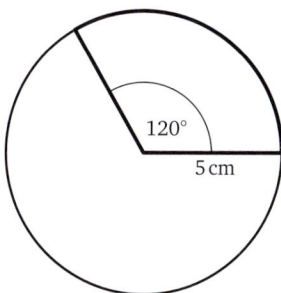

**5**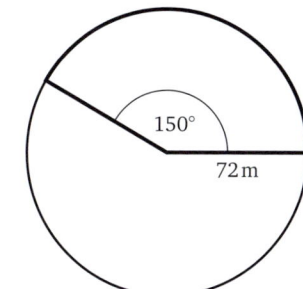

**3**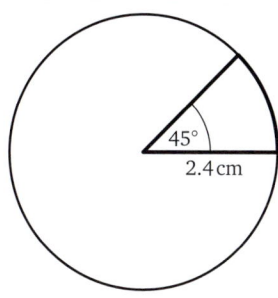

**6**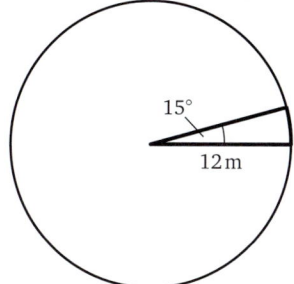

**7** A flower bed is a quadrant of a circle.
- **a** Find the length of edging needed for the curved edge of the bed.
- **b** Find the area of the flower bed.

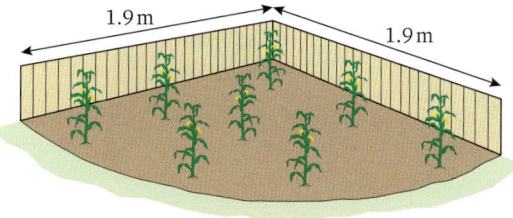

8 The diagram shows the plan of a herb garden. The beds are sectors of the circle, each one of which contains an angle of 40° at the centre. The radius of the circle is 0.9 m.

a  What length of edging is required to surround all these beds?

b  One handful of fertiliser covers one quarter of a square metre of soil.
How many handfuls are needed to cover all the beds?

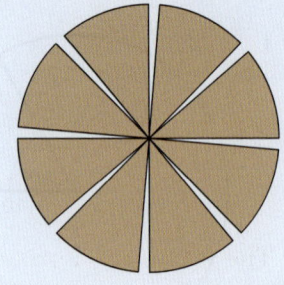

9 A silver earring pendant is part of a sector of a circle. Find the area of silver.

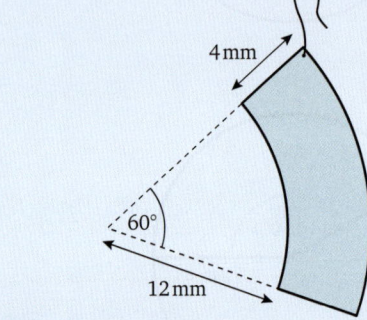

10 In this pattern, the blue sections are **congruent** sectors of a circle of radius 9 cm and the grey sections are congruent sectors of a circle of radius 7 cm.

Which of the grey or blue sections of the pattern covers the greater area? Justify your answer.

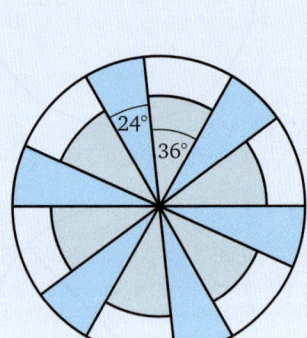

11 Harry is making a marquetry pattern.
The diagram shows one of the pieces he has cut.
The arcs are quadrants of circles.
Find  a  the perimeter of the piece
      b  its area.

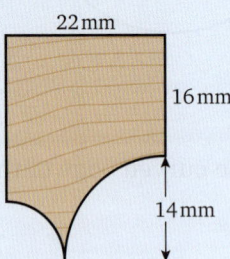

12 This is part of a map showing a wood.
The scale on the map is 2 cm to 1 km.
Estimate the area of the wood in
a  hectares      b  acres.

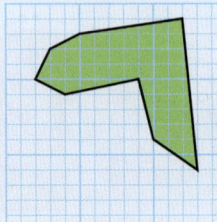

## 13 Areas and volumes

## Volume of a prism

A **prism** is a solid whose cross-section is the same all the way along its length. These are prisms.

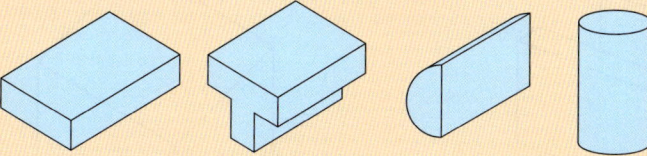

The volume of a prism is equal to

(**area of cross-section**) × (**length**)

## Imperial units of volume

Volume is measured in standard sized cubes. In Imperial units, these are the **cubic inch** (cu in), the **cubic foot** (cu ft) and the **cubic yard** (cu yd).

## Exercise 13d

### Worked example

→ Find the volume of the solid in the diagram.

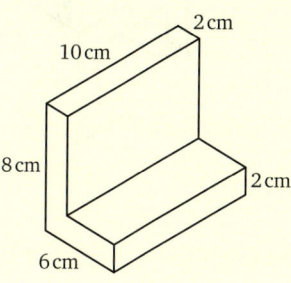

 We draw the cross-section, not the whole solid.
We can find its area by dividing the shape into two rectangles.

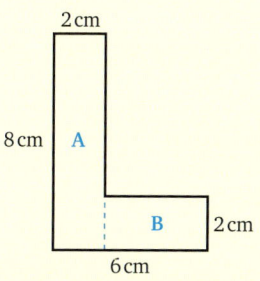

Area A = 8 × 2 cm² = 16 cm²

Area B = 4 × 2 cm² = 8 cm²

Area of the cross-section = 24 cm²

Volume = area of cross-section × length
       = 24 × 10 cm³ = 240 cm³

273

# STP Maths 9

Find the volumes of the solids illustrated in questions **1** to **6**. In each case, draw a diagram of the cross-section but do not draw a picture of the solid.

**1**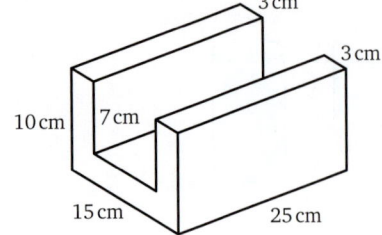

**4**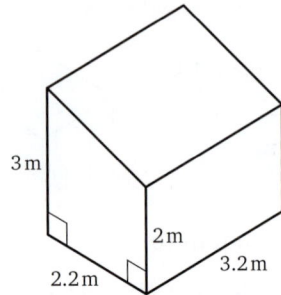

**2**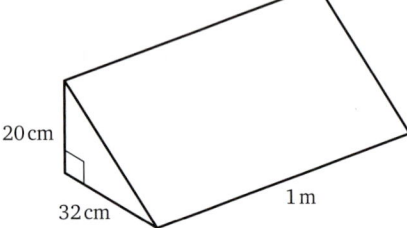

**5**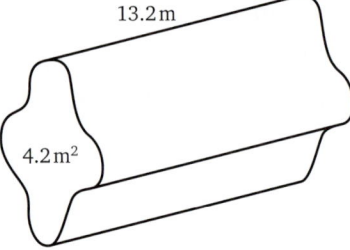

**3**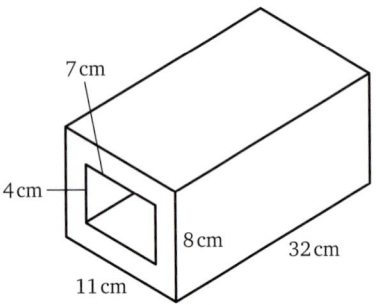

**6**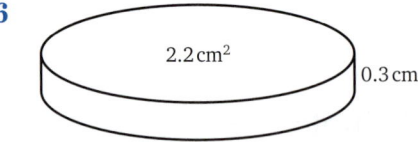

In each of questions **7** to **10**, find the volume of the solid whose cross-section and length are given. Give the answer in the unit indicated in brackets.

**7**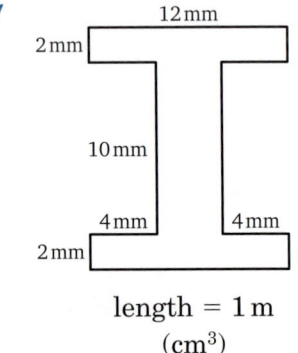

length = 1 m
(cm$^3$)

**8**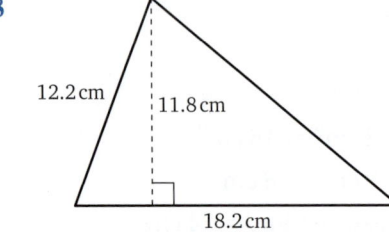

length = 24 cm
(cm$^3$)

9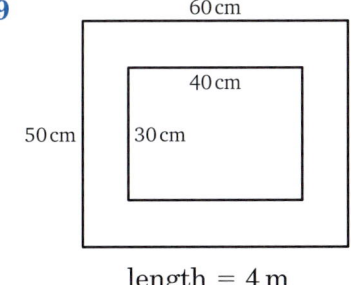

length = 4 m
(m³)

10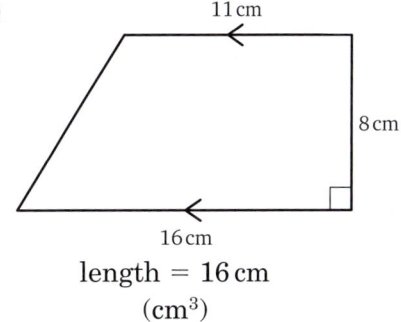

length = 16 cm
(cm³)

## Worked example

→ The volume of the solid shown in the diagram is 144 cm³ and the area of its cross-section is 14 cm². Find its length.

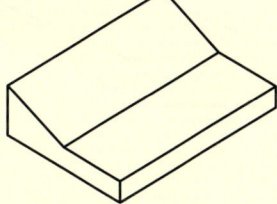

Let its length be $l$ cm.
Volume = area of cross-section × length
$144 = 14 \times l$
$l = \dfrac{144}{14} = 10.28...$

i.e. the length is 10.3 cm correct to 3 s.f.

11 The volume of a solid of uniform cross-section is 72 cm³. The area of its cross-section is 8 cm². Find the length of the solid.

12 The volume of a solid of uniform cross-section is 32 m³. Its length is 10 m. Find the area of its cross-section.

13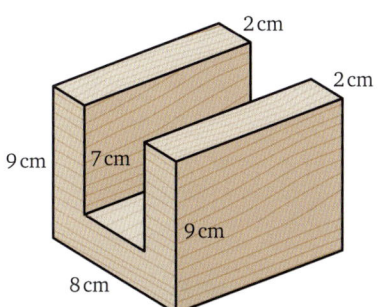

The volume of this block is 396 cm³.

Find  a  the area of the cross-section

  b  the length of the block.

**14**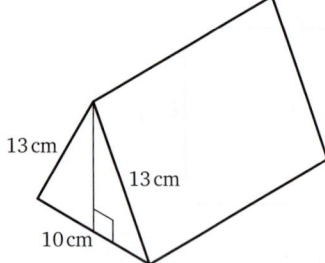

The cross-section of the solid is an isosceles triangle. The volume of the solid is 1200 cm³.

Find  **a** the area of the triangle

**b** the length of the solid.

**15**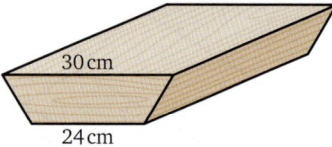

The cross-section of the block of wood is a trapezium. The height of the trapezium is 10 cm and the volume of the block is 7800 cm³. Find the length of the block.

## Mixed exercise

### Exercise 13e

1. A drop of oil of volume 2.5 cm³ is dropped on to a flat surface and spreads out to form a circular pool of even thickness with an area of 50 cm².

   How thick is the oil
   **a** in centimetres
   **b** in millimetres?

2. A cuboid of metal measuring 6 in by 8.2 in by 9.5 in is recast into the shape of a prism. The cross-section of the prism is shown in the diagram. How long is the prism?

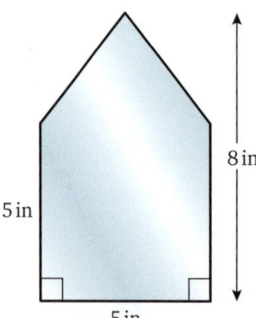

## Worked example

→ Water comes out of a pipe of cross-section 3.2 cm² at a speed of 0.5 m/s. What volume of water is delivered by the pipe in one second?

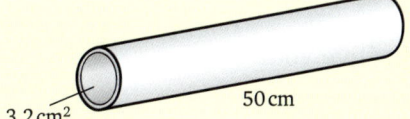

Imagine 0.5 m, i.e. 50 cm, of pipe being emptied in 1 second.

Volume = area of cross-section × length
= 3.2 × 50 cm³
= 160 cm³

160 cm³ of water is delivered in 1 second.

→ How many litres per minute is this?

Volume of water delivered in one minute is 160 × 60 cm³

$= \dfrac{160 \times 60}{1000}$ litres

= 9.6 litres

3. The cross-section of a pipe is 4.8 cm². Water comes out of the pipe at 30 cm/s. How much water is delivered in 1 second? Give your answer in litres.

4. Water comes out of a pipe at 60 cm/s. The cross-section of the pipe is a circle of radius 0.5 cm.
   How many litres of water are delivered
   a in 1 second
   b in 1 minute?

5. The diagram shows the net for a prism with an isosceles triangular cross-section. Find
   a the lengths of the sides of the triangular cross-section
   b the area of the triangular ends
   c the volume of the prism
   d the surface area of the prism.

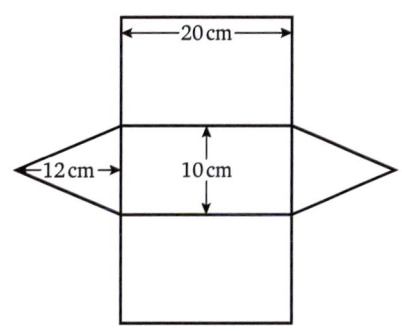

6. A paint manufacturer claims that 1 litre of paint will cover 10 m² of wall. What is the thickness of the paint applied to the wall?

**STP Maths 9**

7 The diagram shows the side view of a swimming pool of width 25 m.

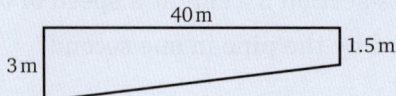

a  Find the volume of water in the pool when it is full. Give your answer in cubic metres.

b  The pool is emptied through a pipe whose cross-sectional area is 200 cm². The water runs out at 1.5 m/s. What volume of water is removed in 1 second?

c  Find how long it would take to empty the pool if four similar pipes were used, each removing water at the same steady rate as in part **b**.

8 Three identical cylindrical tumblers of height 9 cm and base radius 3 cm are completely filled with water. Their contents are poured, without any spillage, into a cylindrical jug with a diameter of 10 cm. How deep is the water in the jug?

9 A cylindrical tin has a height of 11 cm and the diameter of its base is 11.5 cm. The tin contains 1 litre of paint. If the can is opened, how far below the lid would you expect the level of the paint to be? Give your answer to the nearest millimetre.

10 The diagrams show two parcels, each of length 30 cm. The cross-sections of the two parcels are equal in area.

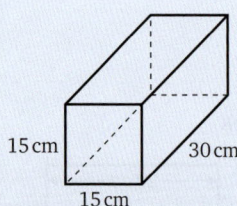 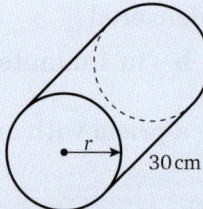

a  If the cross-section of the first parcel is a square of side 15 cm, find, correct to three significant figures, the radius of the cross-section of the second parcel.

b  Compare the volumes of the two parcels.

String is tied around the parcels. The string passes around each parcel in the middle of the length and once from end to end.

c  Which parcel requires the greater length of string, and by how much?

**11** The diagram shows the cross-section through a coping-stone which is used on the top of a wall. Each stone is 50 cm long. Find the mass of the stone if the density of the material it is made from is 3.5 g/cm³.

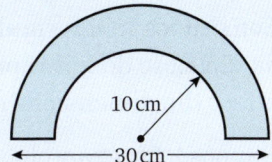

**12**

The diagram shows a prism whose cross-section is a trapezium. Find, in terms of $a$, $b$, $c$ and $h$, a formula for
  **a** the volume of the prism
  **b** the surface area of the prism.

**13** A cylindrical tin is 15 inches high and holds 5 gallons of paint when it is full. Find the diameter of the tin.
(1 gallon = 0.1605 cubic feet.)

**14**

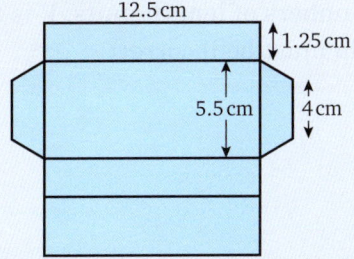

The diagram (not drawn to scale) shows the part of the wrapper round a bar of chocolate which is in contact with the bar. The density of the chocolate is 1.5 g/cm³. What is the mass of the bar?

## The dimensions of a formula

On page 265 we found a formula for the area of a trapezium. It is possible to check whether such a formula does give an area.

Formulas for finding lengths, areas and volumes all contain letters, each of which represents a number of units of length or area or volume.

An expression that has only one 'length unit' letter (or the sum of two or more such letters) is *one-dimensional* and gives a *length* as a number of miles, centimetres, kilometres, etc.

When an expression contains the product of two length symbols it is *two-dimensional* and represents an *area* as a number of (length unit)² such as cm², m², etc.

An expression with three length symbols multiplied together (or an area symbol multiplied by a length symbol) is *three-dimensional*. This gives a *volume* as a number of (length unit)³, e.g. m³, mm³, etc.

# STP Maths 9

Sometimes in a formula there is also a number or a symbol that stands for a number such as π. These do not represent a number of units of length, area or volume and so do not affect the dimensions of an expression.

Suppose, for example, that $d$ is a number of length units, then

$2d$   is one-dimensional

$4\pi d^2$   is two-dimensional

$\pi d^3$   is three-dimensional.

Checking units and **dimensions of a formula** can help us to decide whether a given quantity represents length, area or volume.

For example, if a sentence refers to '$z\,\text{cm}^3$', then $z$ must be a number of volume units.

Similarly, if a sentence contains '$a$ cm', '$b$ cm' and '$X = ab$' then $ab$ must be (a number of cm) × (a number of cm), that is, a number of cm². Therefore, $X$ represents a number of area units.

On the other hand, suppose we are told that the formula for the volume, $V$ cubic units, of a container is $V = 3\pi ab$ where $a$ and $b$ are numbers of length units. $V$ is three-dimensional but $3\pi ab$ is two-dimensional. So the formula must be incorrect.

## Exercise 13f

**1** State whether each of the following quantities is a length, an area or volume.

- **a** 10 cm
- **b** 21 cm³
- **c** 85 cm²
- **d** 4 m³
- **e** 630 mm
- **f** 93 km²

**2** State whether each of the following quantities should be measured in length or area or volume units.

- **a** diameter of a circle
- **b** amount of air in a room
- **c** space inside a sphere
- **d** perimeter
- **e** region inside a square
- **f** surface of a sphere

**3** $a$, $b$ and $c$ represent numbers of centimetres. Give a suitable unit (e.g. cm²) for $X$.

- **a** $X = a + b$
- **b** $X = abc$
- **c** $X = 4ab$
- **d** $X = \pi c$
- **e** $X = 4\pi a^2$
- **f** $X = \frac{4}{3}\pi a^3$

In questions **4** and **5**, $a$, $b$ and $c$ represent numbers of length units; $A$, $B$ and $C$ represent numbers of area units; $V$ represents a number of volume units.

**4** State whether each of the letters $P$ to $Y$ used in the following formulas represents numbers of length or area or volume units.

**a** $P = \pi bc$

**b** $S = \pi a^2 b$

**c** $W = 2a + 3b$

**d** $Q = A + B$

**e** $T = 4bA$

**f** $X = a^2 + b^2$

**g** $R = \dfrac{ab}{c}$

**h** $U = \dfrac{2A}{c}$

**i** $Y = \dfrac{V}{a}$

**5** Some of the following formulas have been constructed incorrectly. State which formulas are incorrect and justify your statement.

**a** $B = ac$

**b** $C = \pi a^2$

**c** $C = a^2 + b^3$

**d** $V = 6a^2 b$

**e** $V = ab + c$

**f** $A = a(b + c)$

**6** Cameron had to find the area of a circle whose diameter was 18 cm. He wrote down

$\quad$ Area $= 2\pi r \, \text{cm}^2$

$\qquad\quad = 2 \times \pi \times 9 \, \text{cm}^2$

$\qquad\quad = 56.5 \, \text{cm}^2$

Yasmina knew nothing about circle formulas but was able to tell Cameron that he was wrong. How did she know?

**7** Jessica copied a formula that she was supposed to use to find the volume of a solid, but she found that she could not read the index number. All she had was $V = \pi x^? y$. She knew that $x$ and $y$ were numbers of length units. What is the index number?

## Worked example

→ This bollard is a cylinder with a hemisphere (half a sphere) on top.

From the list given below, pick out the formula for
- **a** the overall height of the bollard
- **b** the surface area of the bollard
- **c** the volume of the bollard.

$P = 3\pi r^2 + 2\pi rh$

$Q = \pi r^2 + h$

$R = r + h$

$S = \pi r^2 h + \frac{2}{3}\pi r^3$

$T = \pi rh + \frac{4}{3}\pi r^3$

**a** The overall height is given by $R = r + h$.

💡 Height is one-dimensional so the formula must have single letter terms.

**b** The surface area is given by $P = 3\pi r^2 + 2\pi rh$.

💡 Area is two-dimensional so each term in the formula must have a product of two letter terms.

**c** The volume is given by $S = \pi r^2 h + \frac{2}{3}\pi r^3$.

💡 Volume is three-dimensional so each term in the formula must have a product of three letter terms.

**8** From the list of expressions given below, choose the correct one for
- **a** the perimeter of the shaded region
- **b** the area of the unshaded region.

$2r + \pi r^2 \quad 2r + \pi r \quad \pi r^2 - \frac{1}{4}\pi r^2 \quad \pi r^2 - 2\pi r$

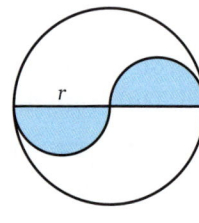

**9**

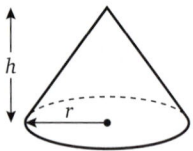

Which of the following formulas could give the volume of this cone?

$V = \pi rh \quad V = \frac{1}{3}\pi r^2 h \quad V = \pi r^2 + \frac{1}{3}h \quad V = \pi r\sqrt{r^2 + h^2}$

Give reasons for your answer.

**13 Areas and volumes**

10 The mathematical name for this ring doughnut is a *torus*.
Which of the following formulas could give

a its surface area      b its volume?

$\frac{1}{4}\pi(R+r)(R-r)^2$

$\frac{3}{4}(\pi^2 R^2 - 3\pi Rr + R^3)$

$\pi^2(R^3 - r^2)$

$\pi^2(R^2 - r^2)$

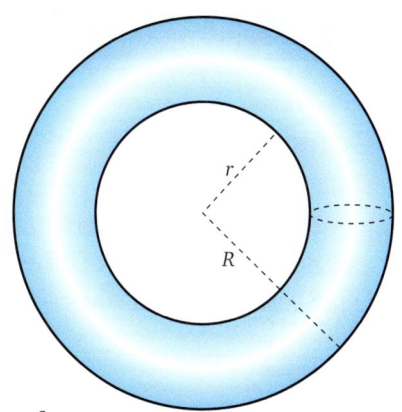

11 Alice wasn't sure of the formula for the surface area of a closed cylinder of radius $r$ and height $h$. Her elder brother thought that the formula was $A = 2\pi r^2(r+h)$, whereas her younger brother thought it was $A = \pi(2r+h)$. Her father suggested $A = 2\pi r(r+h)$. One of these formulas is correct. Which one is it? Justify your choice.

12 This solid is a cylinder of radius $a$ cm on a square base of side $2a$ cm.

Find a formula for

a the volume

b the surface area.

Check that the dimensions of your answers are correct.

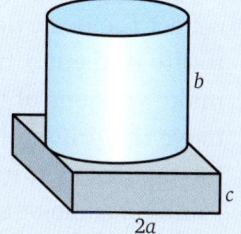

## Consider again

These are the drawings for a conduit for cables, which has a uniform cross-section.

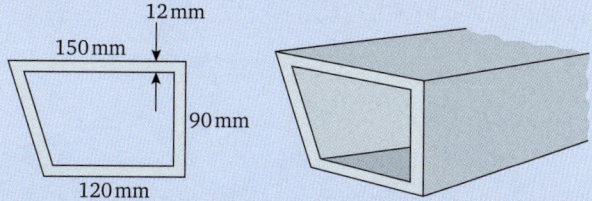

The conduit is to be manufactured in lengths of 2.5 metres. The material used to make this conduit has a density of 0.75 g/cm³.

Anna is ordering shelving for storing 5000 lengths of this conduit.

She needs to know how much space they take up, and their mass.

Now can you find the volume of these 5000 lengths of conduit and their mass?

# 14 Transformations

## Consider

Helen designs cushion covers. She spends a great deal of time on the original design of, say, a petal because she needs to be able to manipulate it into different sizes and positions. If she can do lots of different things with the basic design, she has much more scope to produce an interesting cushion cover. Normally she uses her computer to work on the design.

To investigate as many different designs as possible, she would like to
- reflect the basic design about a line
- rotate it about different points
- make versions of it that are both larger and smaller
- move the original design around so that it looks exactly the same except for the fact that it is in a different position
- combine some of these transformations.

For example

Describe the transformations needed to produce each of these designs.

*You should be able to answer this question after you have worked through this chapter.*

## Class discussion

1. Discuss how you would paint the word AMBULANCE on the front of an ambulance so that it can be read in the mirror of a car travelling in front of it.

2. Give examples of common everyday objects that have been designed using reflections and/or rotations. Describe how each transformation has been used.

3.

   Discuss how to draw this accurate machine part that is to be made from a circular metal disc and has five axes of symmetry. How would you get over the design problem if the finished part is to be cut from a circle of diameter

   a  1 cm  b  4 m?

## Negative scale factors

In Book 8 we considered enlarging various shapes using positive and fractional **scale factors**. This work is revised in Revision exercise 7 at the beginning of this book. Now we extend that work to include negative scale factors.

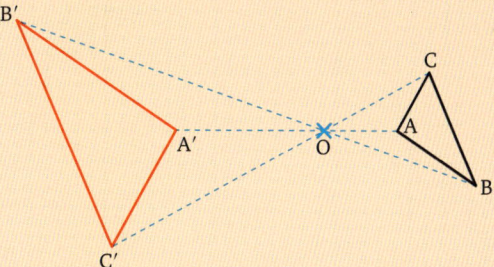

As you can see in the diagram above, it is possible to produce an image twice the size of the object by drawing the guidelines backwards rather than forwards from the centre O so that OA′ = 2OA, and so on.

To show that we are going the opposite way, we say that the scale factor is −2.

The image is the same shape, but it has been rotated through a half turn compared with the image produced by a scale factor of +2.

The following diagrams show **enlargements** which have scale factors of −3 and +3.

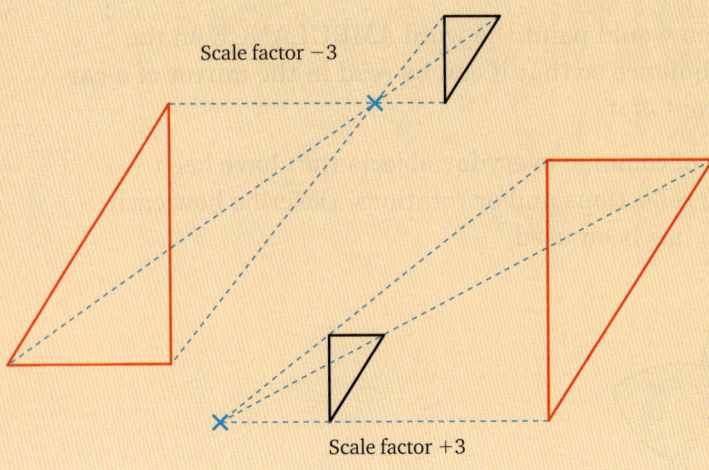

## Exercise 14a

In questions **1** and **2**, give the centre of enlargement and the scale factor.

**1**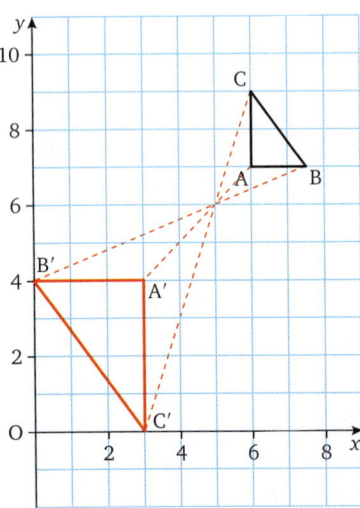

**2**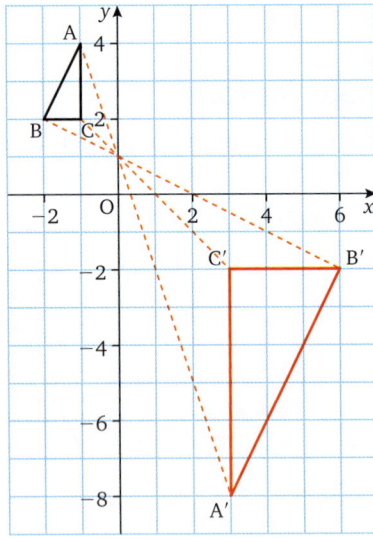

Copy the diagrams in questions **3** to **6**, using 1 cm to 1 unit. Find the centre of enlargement and the scale factor. *Remember:* to find the centre of enlargement, draw lines similar to those in the diagrams in questions **1** and **2**.

**3**

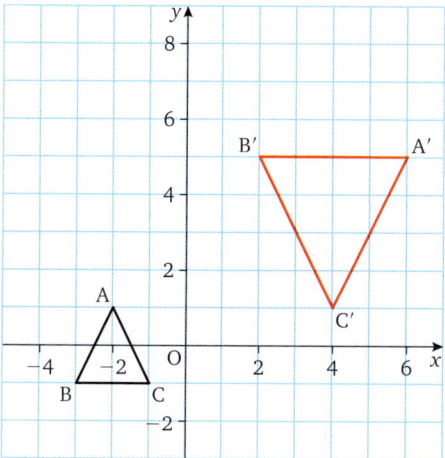

**5**

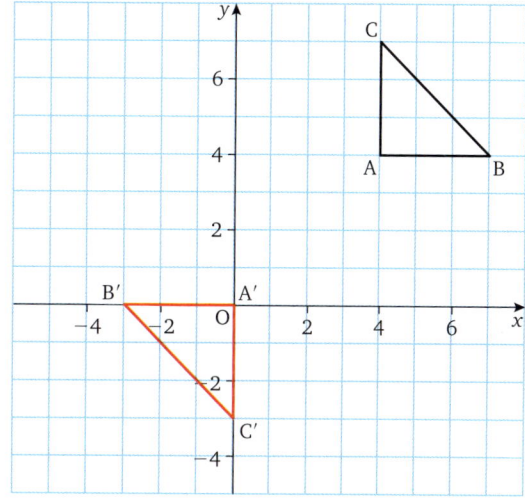

**4**

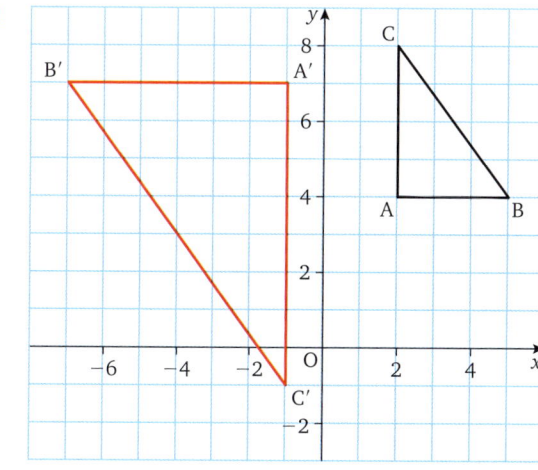

**6**

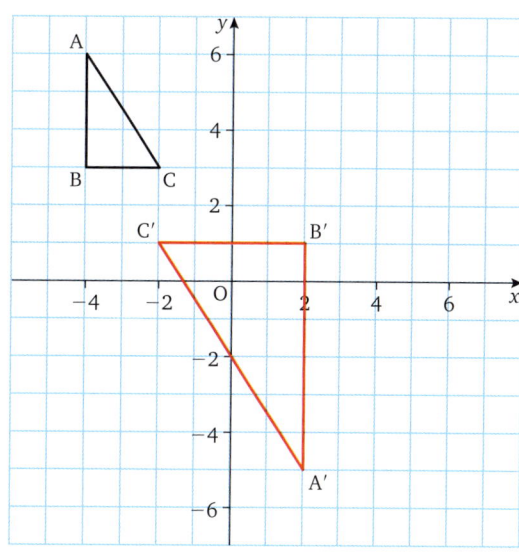

In questions **7** and **8**, draw axes for $x$ and $y$ from $-6$ to $6$.
Draw the object and the image and find the centre of enlargement and the scale factor.

**7**  Object: △ABC with A (6, −1), B (4, −3), C (4, −1)
Image: △A′B′C′ with A′ (−3, 2), B′ (1, 6), C′ (1, 2)

**8**  Object: △ABC with A (2, 3), B (4, 3), C (2, 6)
Image: △A′B′C′ with A′ (2, 3), B′ (−4, 3), C′ (2, −6)

**9**
a  If A'B'C'D' is the image of ABCD under enlargement, give the centre and the scale factor.
b  What other transformation would map ABCD to A'B'C'D'?

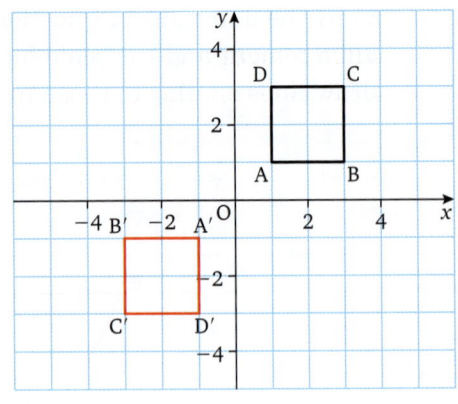

**10** On plain paper, draw an object such as a pin man in the top left-hand corner. Mark the centre of enlargement somewhere between the object and the centre of the page. By drawing guidelines, draw the image with a scale factor of −2.

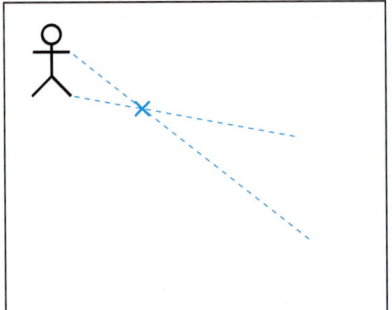

In questions **11** and **12**, copy the diagrams and find the images of the triangles, using P as the centre of enlargement and a scale factor of −2.

**11**

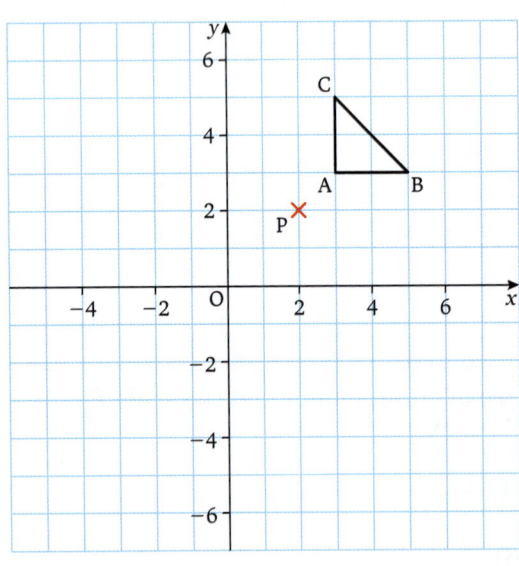

**12**

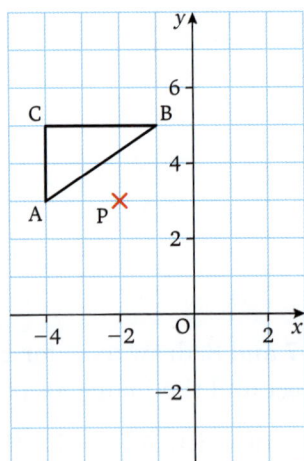

**13** Draw axes for $x$ from −10 to 4 and for $y$ from −2 to 2.
Draw △ABC with A (2, 1), B (4, 1) and C (2, 2).

If the centre of enlargement is (1, 1) and the scale factor is −3, find the image of △ABC.

## 14 Transformations

### Invariant points

A point which is its own image, that is, such that the object point and its image are in the same place, is called an **invariant point**. With reflection, the invariant points lie on the mirror line. The mirror line is an **invariant line**.

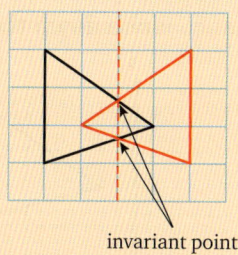
invariant points

### Reflections: finding the mirror line

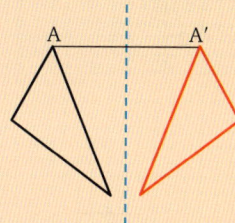
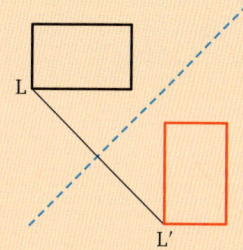

We saw in the work on **reflection** in Book 8 that the object points and the corresponding image points are at equal distances from the mirror line, and that the lines joining them (for example, AA′ and LL′) are perpendicular (at right angles) to the **mirror line**.

## Exercise 14b

### Worked example

→ Find the mirror line if △A′B′C′ is the image of △ABC.

 The mirror line is halfway between A and A′ and perpendicular to AA′.

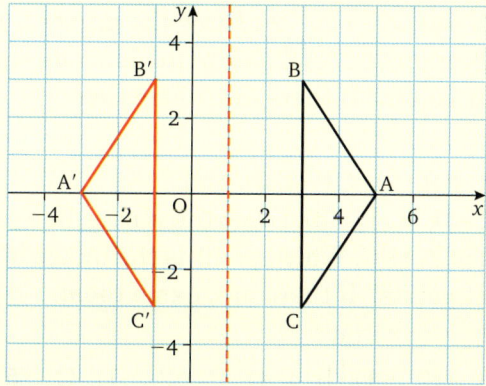

The mirror line is the line $x = 1$.

289

# STP Maths 9

In each of questions **1** to **4**, copy the diagram and draw the mirror line. Hence write down the equation of the mirror line.

**1**
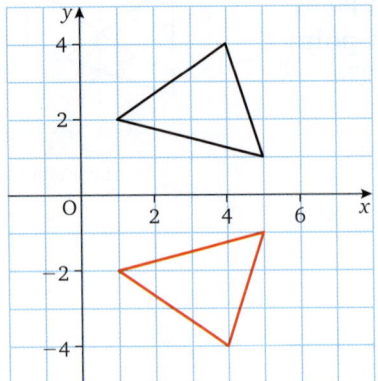

**3**
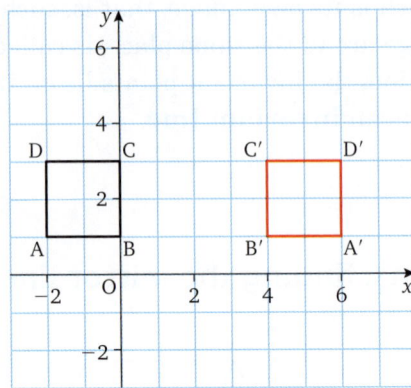

**2**
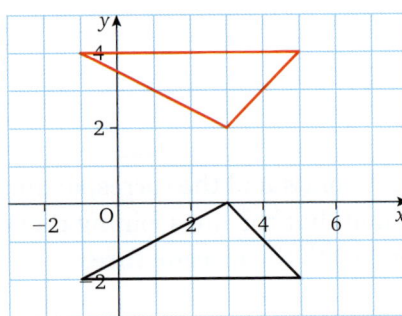

**4**
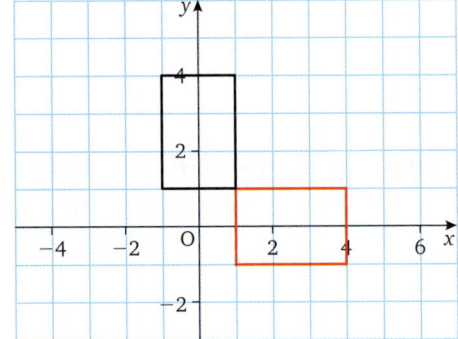

For each of questions **5** to **8**, draw axes for $x$ and $y$ from $-5$ to $5$.

**5** Draw square PQRS: $P(1, 1)$, $Q(4, 1)$, $R(4, 4)$, $S(1, 4)$.
Draw square P'Q'R'S': $P'(-2, 1)$, $Q'(-5, 1)$, $R'(-5, 4)$, $S'(-2, 4)$.
Draw the mirror line so that P'Q'R'S' is the reflection of PQRS and write down its equation.

**6** Draw △XYZ: $X(2, 1)$, $Y(4, 4)$, $Z(-2, 4)$, and
△X'Y'Z': $X'(2, 1)$, $Y'(4, -2)$, $Z'(-2, -2)$. Draw the mirror line so that △X'Y'Z' is the reflection of △XYZ and write down its equation. Are there any invariant points? If there are, name them.

**7** Draw △ABC: $A(-2, 0)$, $B(0, 2)$, $C(-3, 3)$, and
△PQR: $P(3, -1)$, $Q(4, -4)$, $R(1, -3)$. Draw the mirror line so that △PQR is the reflection of △ABC. Which point is the image of A? Are there any invariant points? If there are, name them.

**8** Draw lines AB and PQ: $A(2, -1)$, $B(4, 4)$, $P(-2, -1)$, $Q(-5, 4)$.
Is PQ a reflection of AB? If it is, draw the mirror line. If not, give a reason.

## 14 Transformations

**Worked example**

→ If A'B'C' is the reflection of ABC, draw the mirror line.

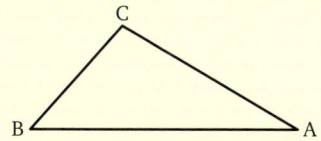
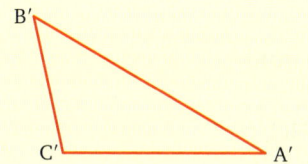

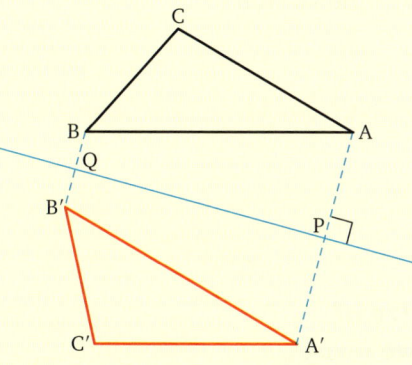

 Join AA' and BB' and find the midpoints P and Q. Then PQ is the mirror line.

 Whenever you attempt to draw a mirror line in this way, always check that it is at right angle to AA' and BB'. If it is not, then A'B'C' cannot be a reflection of ABC.

**9** Trace the diagrams and draw the mirror lines.

**a**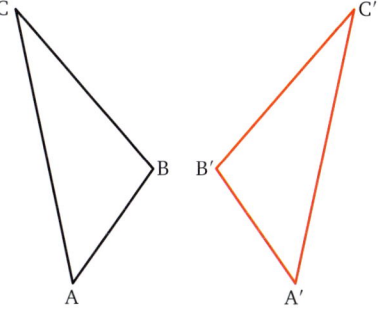

**b**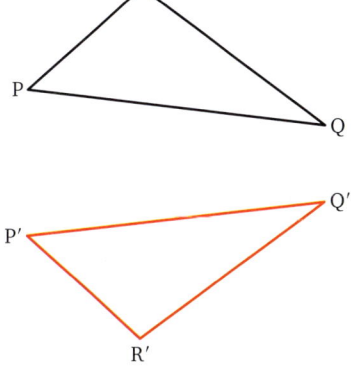

291

c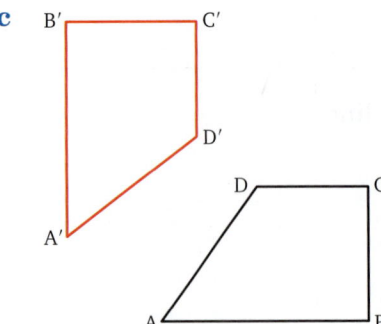

10 On squared paper, draw △ABC: A (3, 1), B (4, 5), C (1, 4), and △A'B'C': A' (0, −2), B' (−4, −3), C' (−3, 0).
Draw the mirror line so that △A'B'C' is the image of △ABC.

11 Draw axes for $x$ and $y$ from −2 to 8, using 1 cm to 1 unit. B is the point (−2, 0) and B' is the point (6, 2). Draw the mirror line so that B' is the reflection of B.

12 Find the gradient and $y$-intercept of the mirror line in question **11**. Hence find the equation of the mirror line.

# Rotations

In Book 8, we saw that it is often easy to spot both the **centre of rotation** and the **angle of rotation**. In the next exercise we revise this work.

## Exercise 14c

For each diagram state the centre of rotation and the angle of rotation. △ABC is the object in each case.

1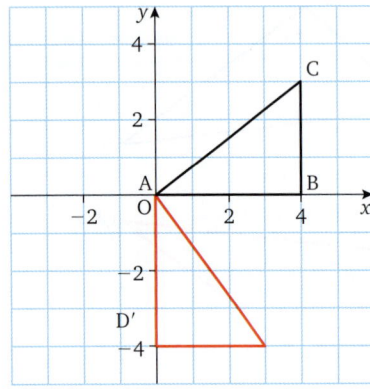

2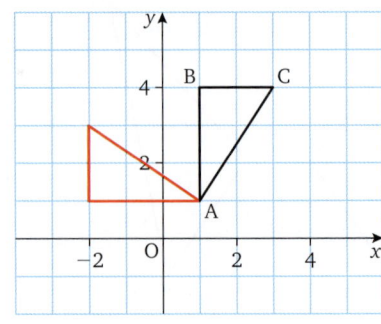

3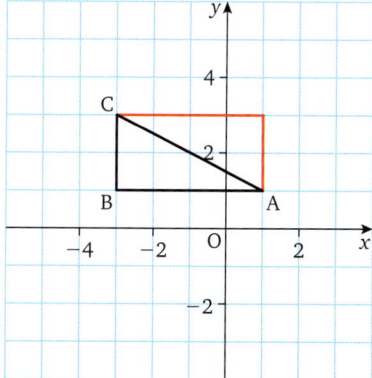

5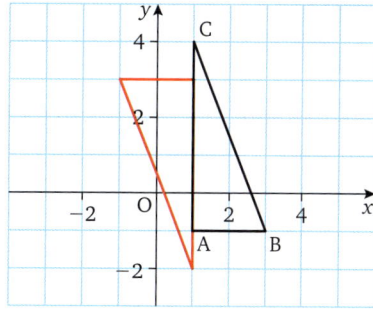

4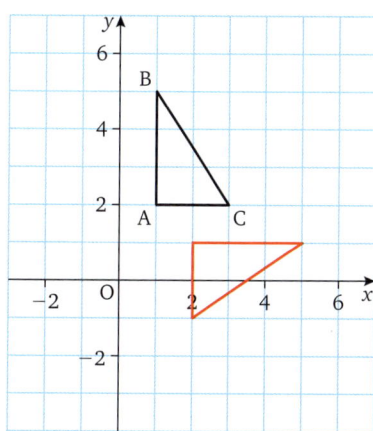

6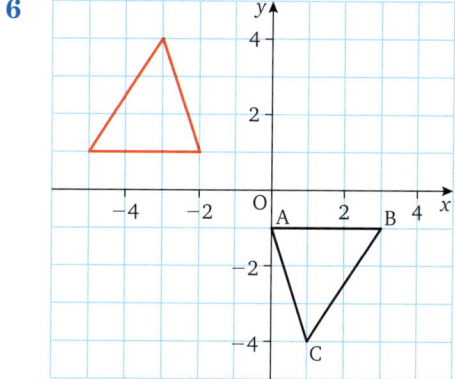

### Finding the angle of rotation

When we know the centre of rotation, we can find the angle of rotation by joining both an object point and its image to the centre.

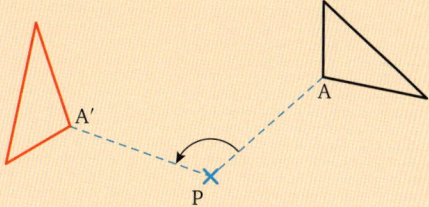

In the diagram above, A′ is the image of A, and P is the centre of rotation.

Join both A and A′ to P. AP̂A′ is the angle of rotation.

In this case the angle of rotation is 120° anticlockwise.

293

# STP Maths 9

## Exercise 14d

Trace each of the diagrams and, by drawing in the necessary lines, find the angle of rotation when △ABC is rotated about the centre P to give △A'B'C'.

**1**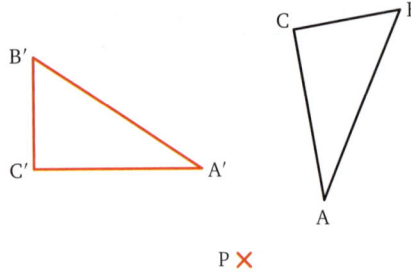

**2**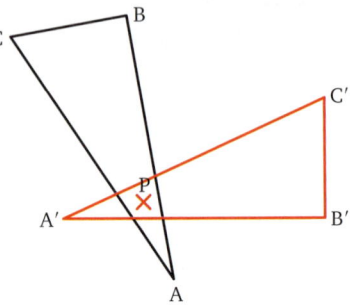

## Compound transformations

If we reflect an object in the $x$-axis and then reflect the resulting image in the $y$-axis, we are carrying out a **compound transformation**.

## Exercise 14e

In each of questions **1** to **4**, copy the diagram and carry out the given compound transformation. Label the final image P.

**1**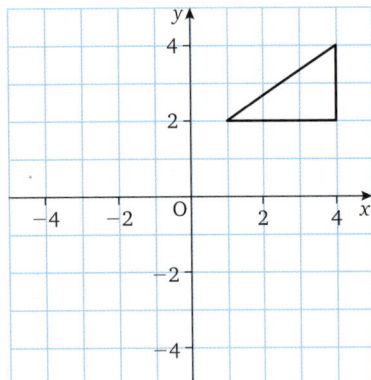

A reflection in the $x$-axis, followed by a reflection the $y$-axis.

**2**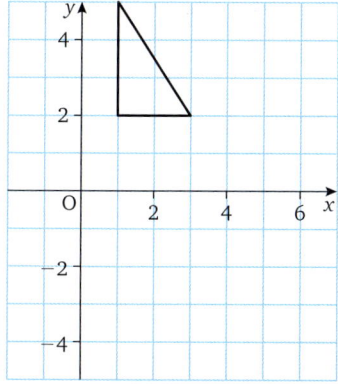

A rotation of 90° clockwise about the point (3, 2), followed by a reflection in the $x$-axis.

**3**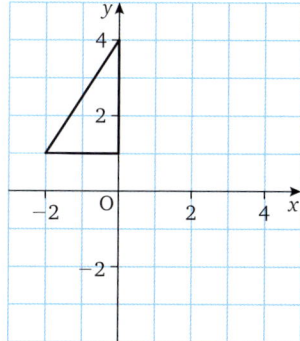

A rotation of 180° about the point (0, 2), followed by a reflection in the line $x = 2$.

**4**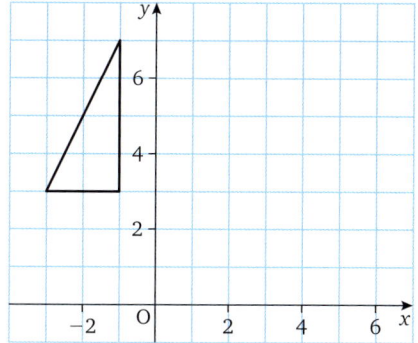

A reflection in the $y$-axis, followed by a translation which is 2 units parallel to the $x$-axis to the right and 3 units parallel to the $y$-axis down.

In each of questions **5** to **8**, describe a compound transformation that maps △ABC onto △A'B'C'.

**5**
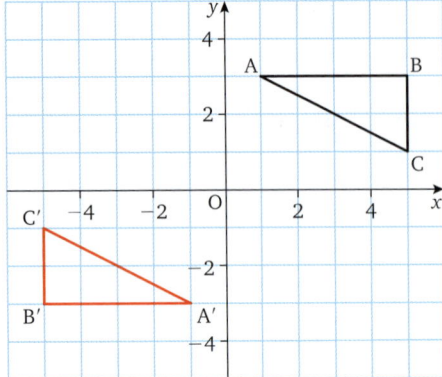

**7**
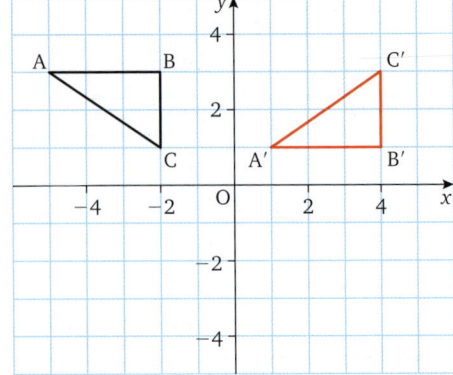

**6**
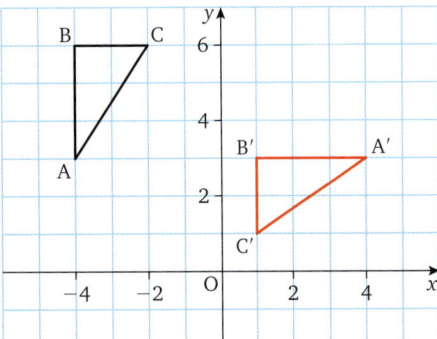

**8**
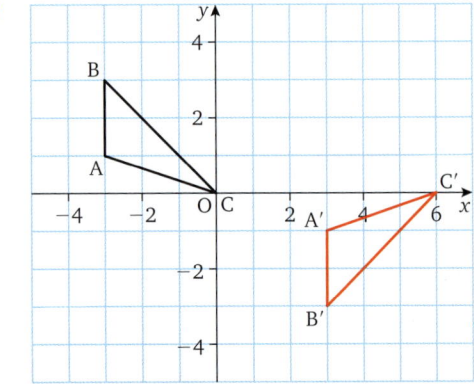

## Vectors

If you arranged to meet your friend 3 km from your home, this information would not be enough to ensure that you both went to the same place. You would also need to know which way to go.

Two pieces of information are required to describe where one place is in relation to another: the distance and the direction. Quantities which have both *magnitude* (size) and *direction* are called **vectors**.

A quantity which has magnitude but not direction is called a **scalar**. For example, the amount of money in your pocket and the number of students in your school are scalar quantities.

### Exercise 14f

State whether the following sentences refer to vector or scalar quantities.

1. There are 24 students in my class.
2. To get to school I walk 2 km due North.

**3** There are 11 players in a cricket team.

**4** John walked at 6 km per hour.

**5** The vertical cliff face is 50 m high.

**6** Flight BA101 is flying due West at 350 mph.

**7** Castle Millington is 50 miles due East of Junction 48 on the M7.

**8** Give other examples of
    **a** vector quantities
    **b** scalar quantities.

## Representing vectors

We can represent a vector by a line segment (i.e. a straight line whose length represents the size of the vector) and indicate its direction with an arrow.

For example

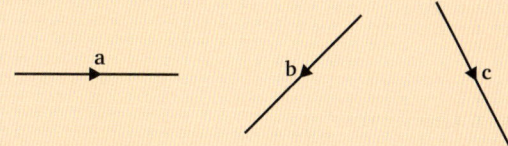

We use **a**, **b**, **c**, ... to name the vectors.

When writing by hand, it is difficult to write **a**, which is in bold type, so we use a̲.

## Vectors in the form $\begin{pmatrix} a \\ b \end{pmatrix}$

When vectors are drawn on squared paper, we can describe them in terms of the number of squares we need to go across and up. For example, in the diagram below, the vector **a** corresponds to a movement of 4 across and 2 up and we write $\mathbf{a} = \begin{pmatrix} 4 \\ 2 \end{pmatrix}$.

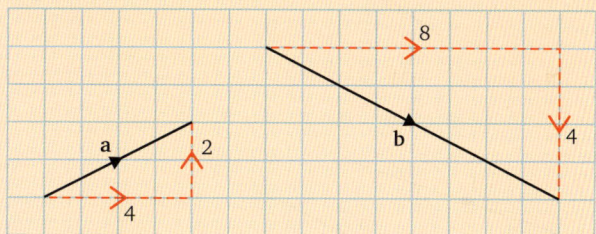

The vector **b** can be described as 8 across and 4 down. As with coordinates, we use negative numbers to indicate movement down or movement to the left.

Therefore, $\mathbf{b} = \begin{pmatrix} 8 \\ -4 \end{pmatrix}$.

Notice that the top number represents movement across and that the bottom number represents movement up or down.

# Exercise 14g

## Worked example

→ Write the following vectors in the form $\begin{pmatrix} p \\ q \end{pmatrix}$.

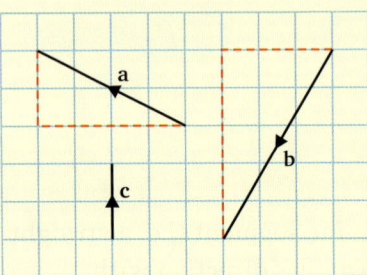

$\mathbf{a} = \begin{pmatrix} -4 \\ 2 \end{pmatrix}$, $\mathbf{b} = \begin{pmatrix} -3 \\ -5 \end{pmatrix}$, $\mathbf{c} = \begin{pmatrix} 0 \\ 2 \end{pmatrix}$

Write the following vectors in the form $\begin{pmatrix} p \\ q \end{pmatrix}$.

1
2
3

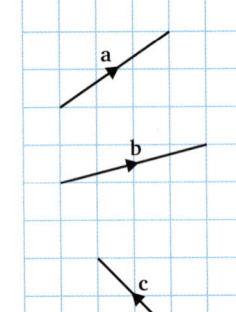

4
5
6

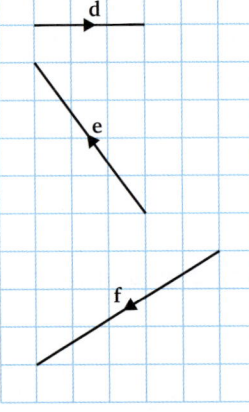

7

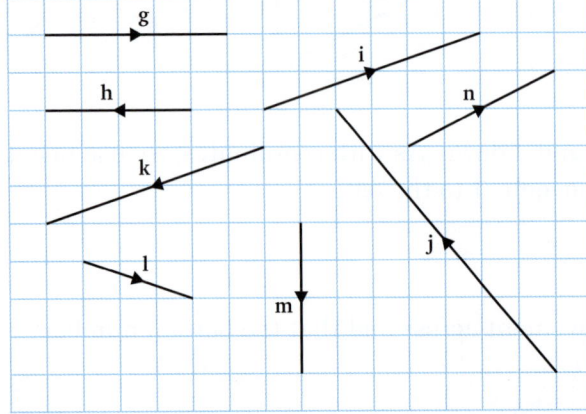

On squared paper, draw the following vectors. Label each vector with its letter and an arrow.

8  $\mathbf{a} = \begin{pmatrix} 3 \\ 5 \end{pmatrix}$    11  $\mathbf{d} = \begin{pmatrix} 6 \\ -12 \end{pmatrix}$    14  $\mathbf{g} = \begin{pmatrix} -4 \\ 3 \end{pmatrix}$

9  $\mathbf{b} = \begin{pmatrix} -4 \\ -3 \end{pmatrix}$    12  $\mathbf{e} = \begin{pmatrix} 6 \\ 10 \end{pmatrix}$    15  $\mathbf{h} = \begin{pmatrix} -1 \\ -5 \end{pmatrix}$

10  $\mathbf{c} = \begin{pmatrix} 2 \\ -4 \end{pmatrix}$    13  $\mathbf{f} = \begin{pmatrix} -2 \\ 5 \end{pmatrix}$    16  $\mathbf{i} = \begin{pmatrix} -6 \\ 2 \end{pmatrix}$

17  What do you notice about the vectors in questions **8** and **12**, and in questions **10** and **11**?

## Worked example

→ The coordinates of the starting point of the vector $\begin{pmatrix} 5 \\ 2 \end{pmatrix}$ are (2, 1). Find the coordinates of its other end.

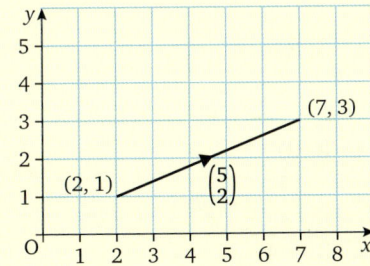

💡 Starting at (2, 1) we go 5 units to the right followed by 2 units up.

The coordinates of its other end are (7, 3).

In each of the following questions you are given a vector, followed by the coordinates of its starting point. Find the coordinates of its other end.

18  $\begin{pmatrix} 3 \\ 3 \end{pmatrix}$, (4, 1)    22  $\begin{pmatrix} 5 \\ 2 \end{pmatrix}$, (3, −1)    26  $\begin{pmatrix} 4 \\ 3 \end{pmatrix}$, (−2, −3)

19  $\begin{pmatrix} 3 \\ 1 \end{pmatrix}$, (−2, −3)    23  $\begin{pmatrix} 4 \\ -2 \end{pmatrix}$, (4, 2)    27  $\begin{pmatrix} 5 \\ -3 \end{pmatrix}$, (2, −1)

20  $\begin{pmatrix} -6 \\ 2 \end{pmatrix}$, (3, 5)    24  $\begin{pmatrix} -3 \\ 4 \end{pmatrix}$, (2, −4)    28  $\begin{pmatrix} -5 \\ 2 \end{pmatrix}$, (−4, −3)

21  $\begin{pmatrix} -4 \\ -3 \end{pmatrix}$, (5, −2)    25  $\begin{pmatrix} -6 \\ -6 \end{pmatrix}$, (−3, −2)    29  $\begin{pmatrix} -4 \\ -2 \end{pmatrix}$, (−3, −1)

# STP Maths 9

> **Worked example**
>
> → The coordinates of the other end of the vector $\binom{6}{4}$ are (8, 6). Find the coordinates of its starting point.
>
>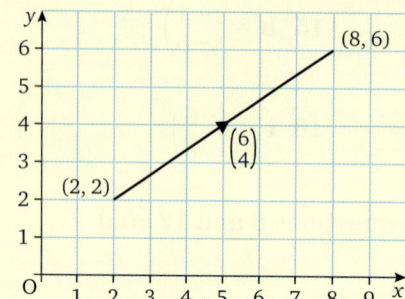
>
> 💡 Starting at (8, 6), we have to go 6 units to the left and 4 units down.
>
> The coordinates of the starting point of the vector are (2, 2).

In each of the following questions, a vector is given, followed by the coordinates of its other end. Find the coordinates of its starting point.

30  $\binom{10}{2}$, (4, 1)

31  $\binom{5}{-1}$, (3, −4)

32  $\binom{-5}{-2}$, (−2, −4)

33  $\binom{8}{6}$, (6, 3)

34  $\binom{-3}{4}$, (−2, 1)

35  $\binom{-6}{-3}$, (−5, 2)

36  $\binom{4}{-2}$, (−3, 2)

37  $\binom{-2}{6}$, (−3, −4)

38  $\binom{1}{4}$, (−5, −2)

## Translations

We can describe a **translation** only by stating what the movement, or **displacement**, is. The easiest way to do this is to give the vector that describes the displacement.

# 14 Transformations

## Exercise 14h

### Worked example

→ Describe the transformation that maps △ABC onto △A′B′C′.

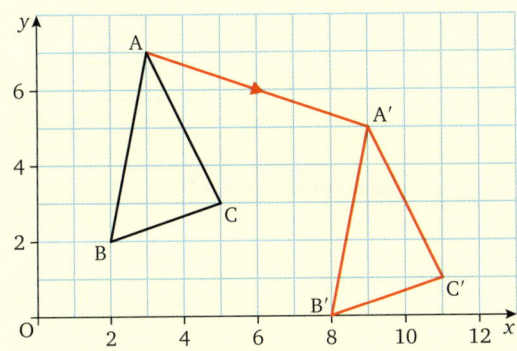

Consider the movement, or displacement, from A to A′.

The transformation is given by the vector $\begin{pmatrix} 6 \\ -2 \end{pmatrix}$.

In questions **1** to **4**, use a vector to describe the transformation that maps the object onto its red image.

**1**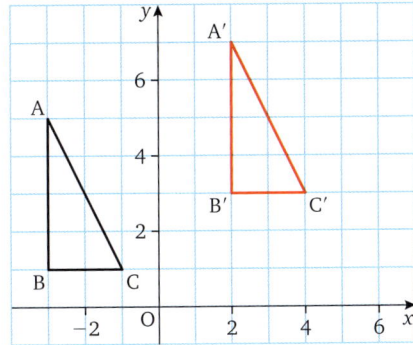

**3**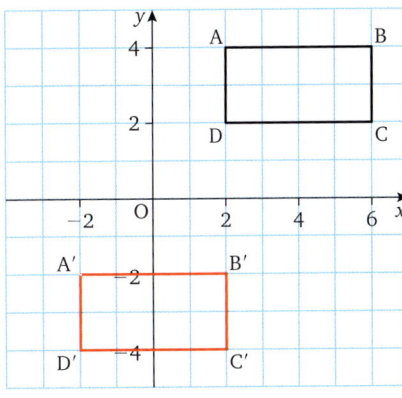

**2**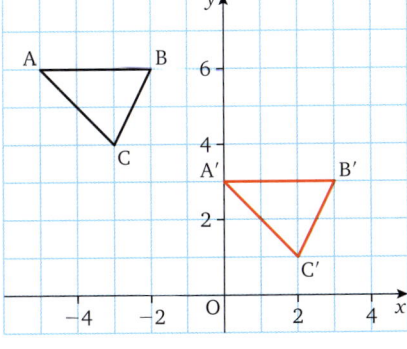

**4**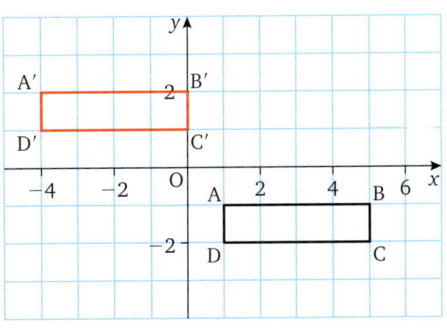

# STP Maths 9

In each of questions **5** to **8**, copy the diagram and draw the image of the given object under a translation defined by the given vector.

**5**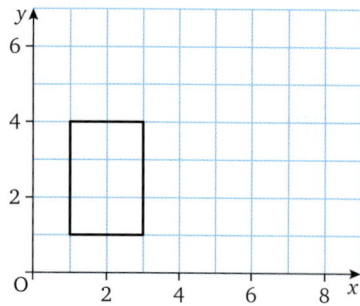

The vector is $\begin{pmatrix} 5 \\ 1 \end{pmatrix}$.

**7**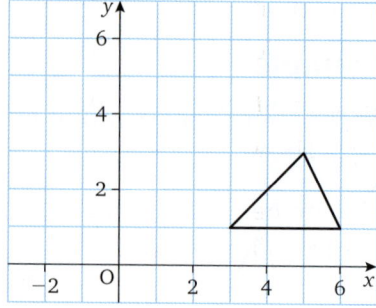

The vector is $\begin{pmatrix} -4 \\ 2 \end{pmatrix}$.

**6**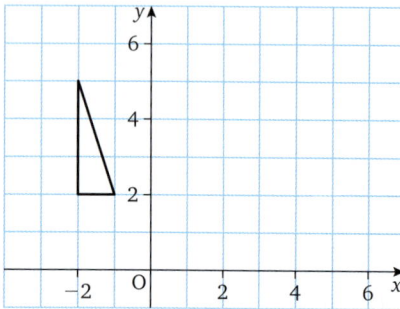

The vector is $\begin{pmatrix} 5 \\ -3 \end{pmatrix}$.

**8**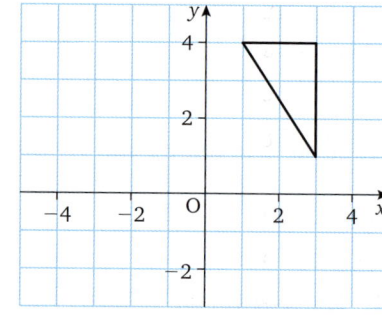

The vector is $\begin{pmatrix} -4 \\ -3 \end{pmatrix}$.

**9** In each of questions **5** to **8**, use a vector to describe the transformation that maps the image onto the given object.

## Mixed exercise

### Exercise 14i

**1 a** Copy the diagram and find the centre of enlargement and the scale factor when △ABC is enlarged to give △A′B′C′.
**b** Draw the reflection of △A′B′C′ in the line $x = 3$, followed by a translation defined by the vector $\begin{pmatrix} 4 \\ -4 \end{pmatrix}$. Label it Q.

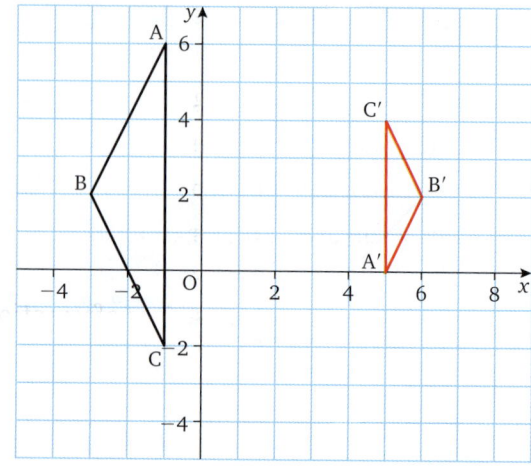

**2 a** On $x$- and $y$-axes scaled from 0 to 12, plot the points A (6, 3), B (9, 9), C (12, 3).

**b** An enlargement, scale factor $\frac{1}{3}$ and centre (0, 0), transforms $\triangle ABC$ onto $\triangle A_1B_1C_1$. Draw $\triangle A_1B_1C_1$.

**c** A translation of $\triangle A_1B_1C_1$ onto $\triangle A_2B_2C_2$ is defined by the vector $\binom{8}{2}$. Draw $\triangle A_2B_2C_2$. What are the coordinates of $B_2$?

**d** $\triangle A_2B_2C_2$ can be transformed onto $\triangle ABC$ by an enlargement. Give the scale factor and the centre of enlargement.

**3 a** Copy the diagram onto squared paper. Draw the mirror line so that $\triangle A_1B_1C_1$ is the image of $\triangle ABC$. Write down the coordinates of the points where this line crosses the axes.

**b** $\triangle ABC$ is mapped onto $\triangle A_2B_2C_2$ by a rotation. Give the angle of rotation.

**c** $\triangle A_2B_2C_2$ is reflected in the line $y = -3$ to give $\triangle A_3B_3C_3$. Mark $\triangle A_3B_3C_3$ on your diagram and write down the coordinates of $C_3$.

**d** Describe the transformation that maps $\triangle A_1B_1C_1$ onto $\triangle A_3B_3C_3$.

**e** Write down the vector that defines the translation that maps
  **i** $C_3$ onto $C_1$
  **ii** $C_1$ onto C
  **iii** C onto $C_2$.

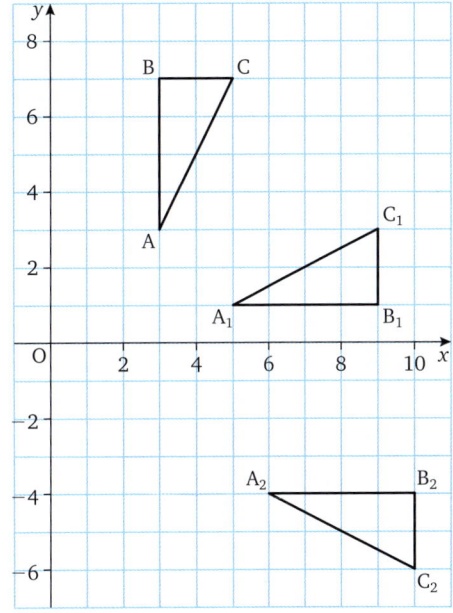

### Consider again

Helen designs cushion covers. She needs to be able to manipulate each shape into different sizes and positions. To investigate as many different designs as possible, she would like to

- reflect the basic design about a line
- rotate it about different points
- make versions of it that are both larger and smaller
- move the original design around so that it looks exactly the same except for the fact that it is in a different position
- combine some of these transformations.

For example

Now can you describe the transformations needed to produce each of these designs?

##  Investigation

This diagram shows the vectors $\mathbf{a} = \begin{pmatrix} 2 \\ 3 \end{pmatrix}$ and $\mathbf{b} = \begin{pmatrix} 2 \\ 1 \end{pmatrix}$.

The point (6, 7) can be reached from the origin by adding **a** and **b**, and then adding **a** to the result. This is shown in the diagram below.

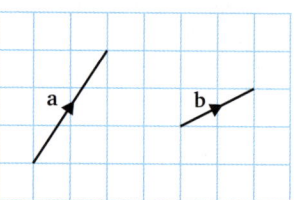

**a** In how many different ways can these vectors be combined to get from the origin to the point (6, 7)? Show them on a diagram.

**b** The vector in the direction opposite to **a** is called $-\mathbf{a}$, so as $\mathbf{a} = \begin{pmatrix} 2 \\ 3 \end{pmatrix}$, $-\mathbf{a} = \begin{pmatrix} -2 \\ -3 \end{pmatrix}$.

Similarly, $-\mathbf{b} = \begin{pmatrix} -2 \\ -1 \end{pmatrix}$.

How many more ways of getting from the origin to point (6, 7) are possible, if you must always remain on the given grid but you are allowed to combine the vectors **a**, **b**, $-\mathbf{a}$ and $-\mathbf{b}$?

**c** How many of the points on this grid is it possible to get to from the origin if any combinations of **a**, **b**, $-\mathbf{a}$ and $-\mathbf{b}$ are allowed?

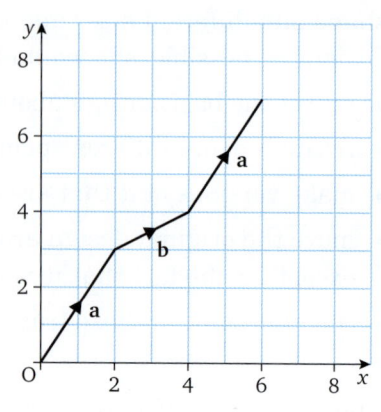

# 15 Similar figures

## Consider

The diagram shows the cross-section of a swimming pool which is 3.5 m deep at one end and 1 m deep at the other. The pool is 30 m long and 10 m wide. Water flows into the pool to a depth of 1.5 m at the deep end. What are the dimensions of the surface of the water?

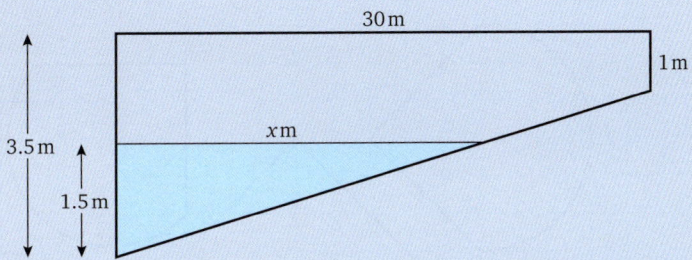

*You should be able to answer this question after you have worked through this chapter.*

## Similar figures

In Chapter 14 we studied different types of transformation. One type of transformation is an enlargement. In the diagram below, rectangle **A** is enlarged by a scale factor of 2 with P as the centre of enlargement. This transformation gives rectangle **B**. Rectangle **A** is also enlarged by a scale factor of $\frac{1}{3}$ with Q as the centre of enlargement. This gives rectangle **C**, which is smaller than **A** because the scale factor is a fraction less than 1.

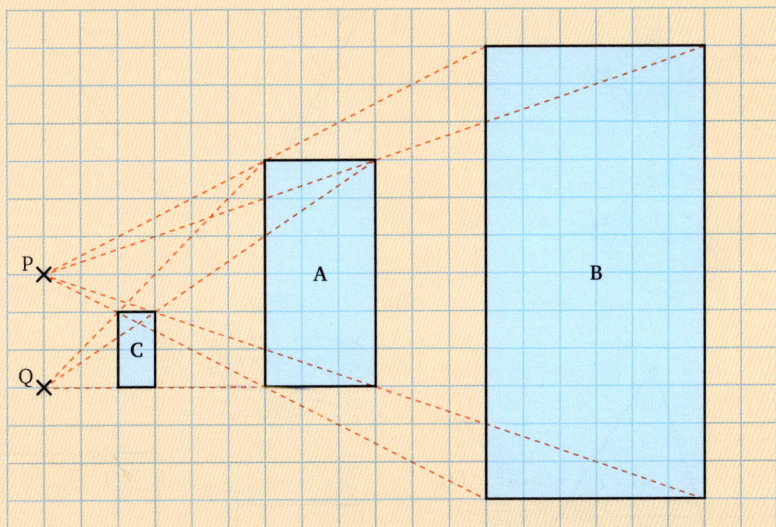

The three rectangles **A**, **B** and **C** are all *exactly the same shape* but are *different in size*.

305

When we say they are the same shape, we mean that the corresponding sides are in proportion. For example, the long and short sides in **B** are twice the length of the long and short sides in **A**. Likewise, the long and short sides in **C** are $\frac{1}{3}$ of the length of the long and short sides in **A**. When one figure is an enlargement of another (it can be larger or smaller), we say that the two shapes are **similar**.

It doesn't matter whether one shape is then rotated or reflected or translated – the two shapes are still similar. In the diagram given below, any one shape is similar to each of the other four.

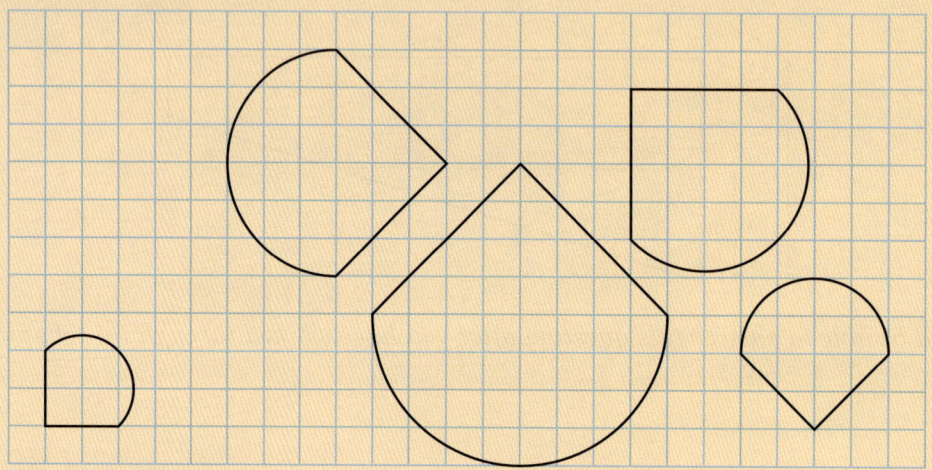

## Class discussion

1  Discuss whether or not the pairs of figures in parts **a** to **j** are similar.

   **a**

   **b**

   **c**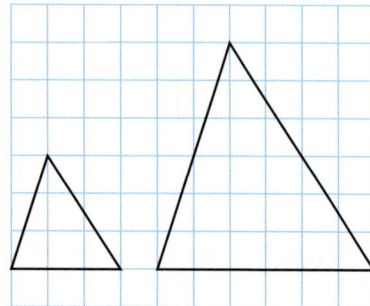

   **d**

   **e**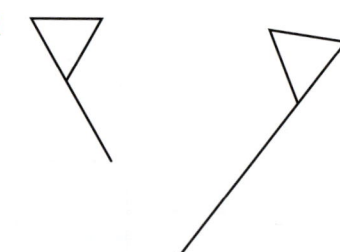

   **f**

**15 Similar figures**

g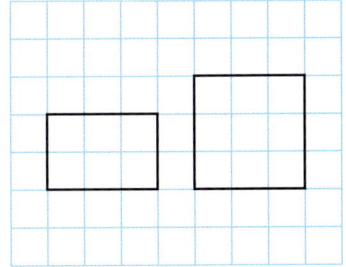

i

h

j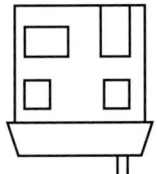

**2** Discuss which two rectangles are similar. How can you tell?

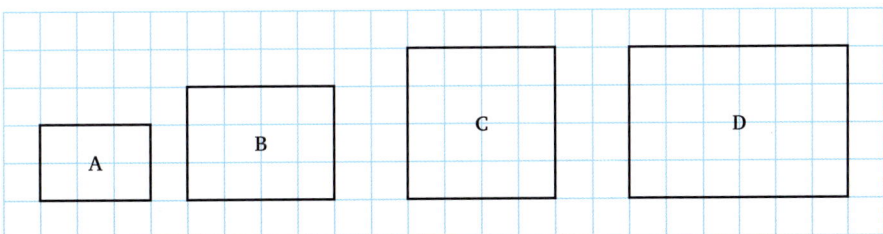

**3** Discuss which of the other five shapes are similar to **X**. How can you tell?

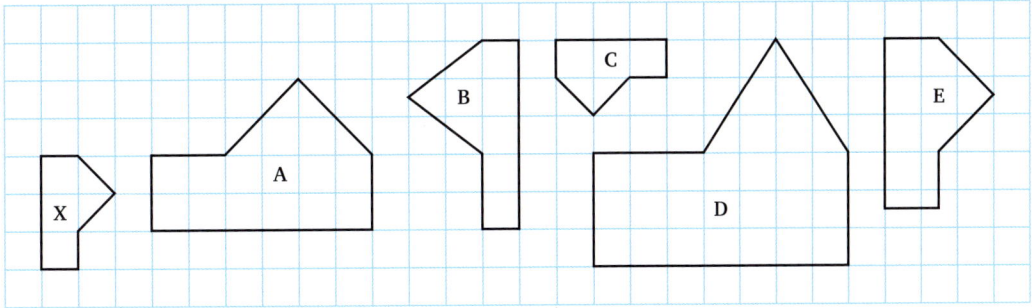

**4** Discuss which of the following statements are true.
  a  All circles are similar.
  b  All rectangles are similar.
  c  All squares are similar.
  d  All triangles are similar.
  e  All equilateral triangles are similar.
  f  All isosceles triangles are similar.

307

# STP Maths 9

## Similar triangles

Some of the easiest similar figures to deal with are triangles. This is because only a small amount of information is needed to prove them to be similar.

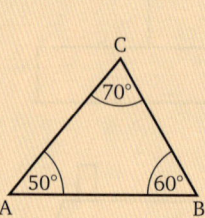
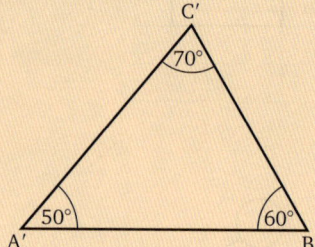

In these triangles the corresponding angles are equal and so the triangles are the same shape. One triangle is an enlargement of the other. These triangles are *similar*.

**We can prove that two triangles are similar if we can show that they have the same angles.**

## Exercise 15a

**1**

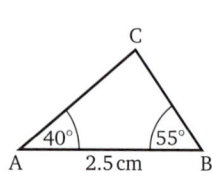
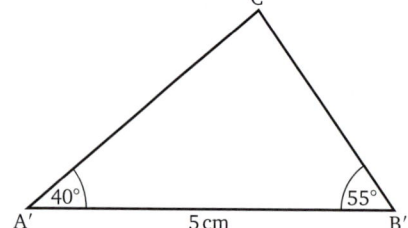

  **a** Are the triangles similar?

  **b** Measure the remaining sides.

  **c** Find $\dfrac{A'B'}{AB}$, $\dfrac{B'C'}{BC}$ and $\dfrac{C'A'}{CA}$

  (as decimals if necessary, correct to 1 d.p.)

  **d** What do you notice about the answers to part **c**?

Repeat question **1** for the pairs of triangles in questions **2** to **4**.

**2**

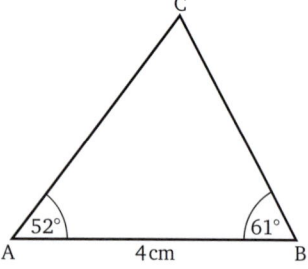
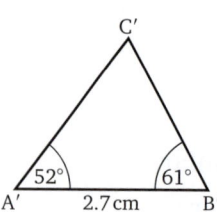

3

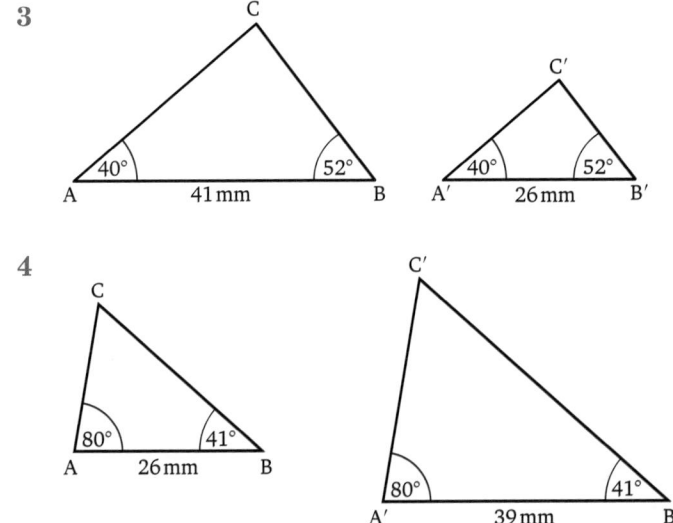

4

In questions **5** to **7**, sketch the pairs of triangles and find the size of each of the missing angles. In each question, state whether the two triangles are similar. (One triangle may be turned round or over compared with the other.)

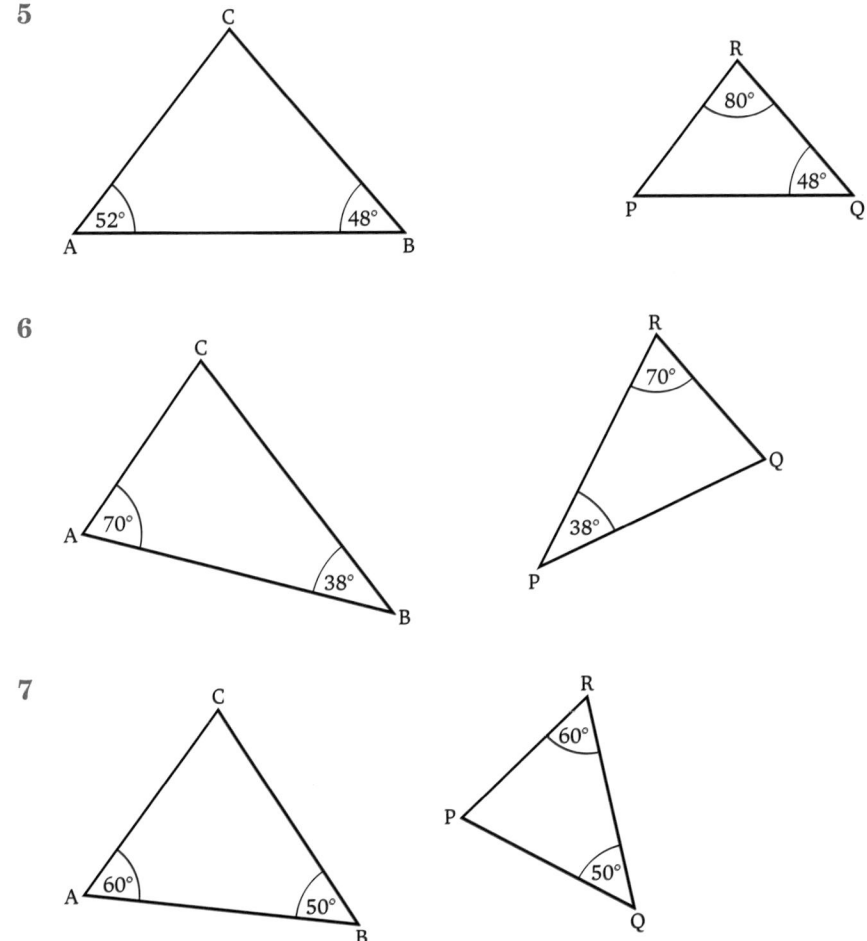

# STP Maths 9

## Corresponding vertices

These two triangles are similar and we can see that X corresponds to A, Y to B and Z to C.

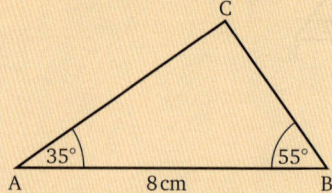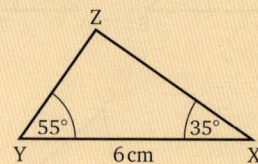

We can write  $\triangle$s $\dfrac{ABC}{XYZ}$ are similar.

**Make sure that X is written below A, Y below B and Z below C, i.e. that the angles of one triangle are below the corresponding angles of the other.**

The pairs of corresponding sides are in the same ratio and we can read these from $\dfrac{ABC}{XYZ}$, that is $\dfrac{AB}{XY} = \dfrac{BC}{YZ} = \dfrac{CA}{ZX}$

## Exercise 15b

### Worked example

→ State whether triangles ABC and PQR are similar and, if they are, give the ratios of the sides.

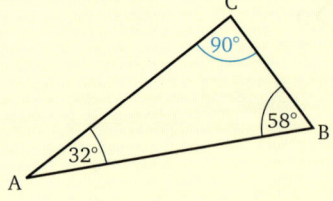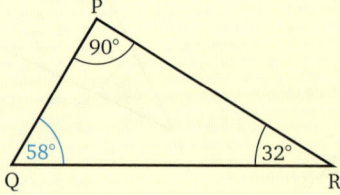

$\widehat{Q} = 58°$  (angle sum of a triangle)

and $\widehat{C} = 90°$

so $\triangle$s $\dfrac{ABC}{RQP}$ are similar

 The angles in each triangle are 32°, 58° and 90°.

and $\dfrac{AB}{RQ} = \dfrac{BC}{QP} = \dfrac{CA}{PR}$

310

# 15 Similar figures

In questions **1** to **3**, state whether the two triangles are similar and, if they are, give the ratios of the sides.

**1** 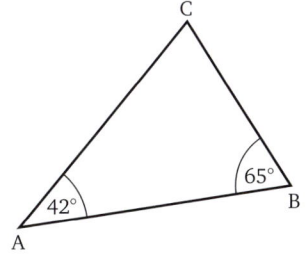   **2** 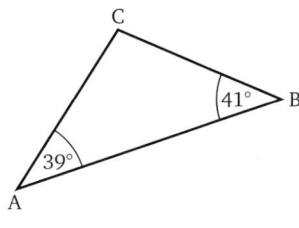   **3**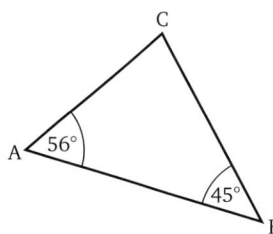

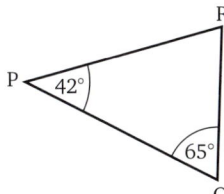   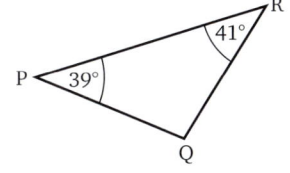   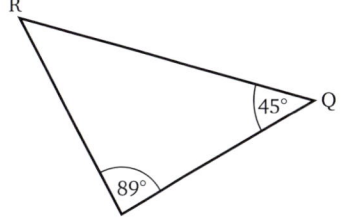

## Finding a missing length

### Exercise 15c

**Worked example**

→ State whether the two triangles are similar. If they are, find AB.

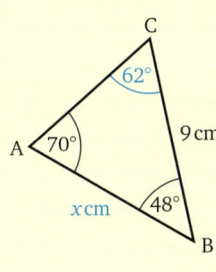   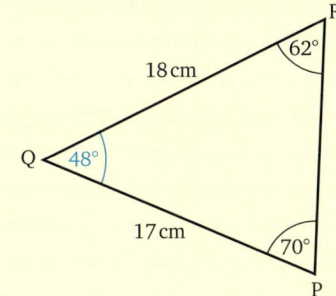   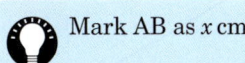

Mark AB as $x$ cm.

$\hat{C} = 62°$ and $\hat{Q} = 48°$  (angle sum of a triangle)

so  $\triangle$s $\dfrac{ABC}{PQR}$ are similar and $\dfrac{AB}{PQ} = \dfrac{BC}{QR} = \dfrac{CA}{RP}$

i.e.  $\dfrac{x}{17} = \dfrac{9}{18}$

$\cancel{17}^{1} \times \dfrac{x}{\cancel{17}_{1}} = \dfrac{\cancel{9}^{1}}{\cancel{18}_{2}} \times 17$

$x = \dfrac{17}{2} = 8.5$  so  AB = 8.5 cm

In questions **1** to **5**, state whether the pairs of triangles are similar. If they are, find the required side.

**1** Find PR.

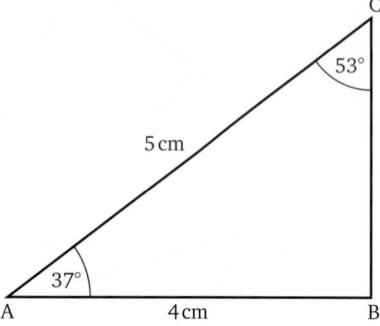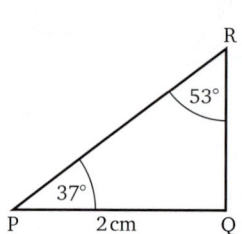

**2** Find QR.

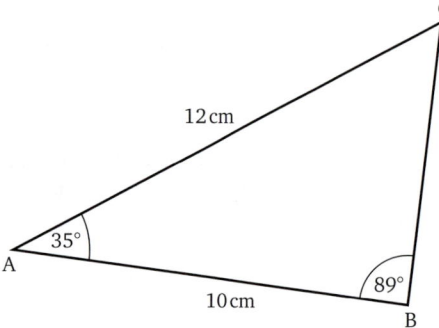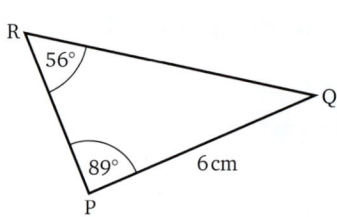

**3** Find BC.

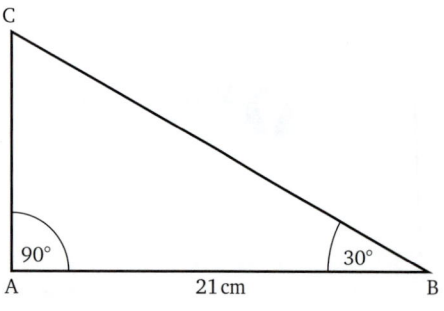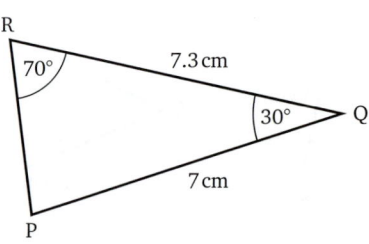

**4** Find PR.

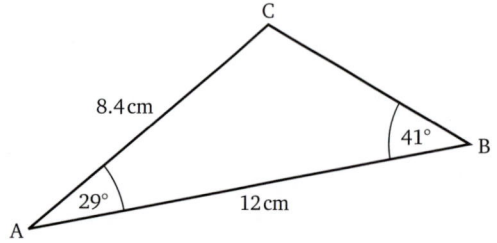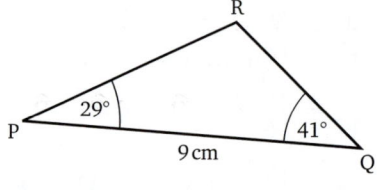

## 15 Similar figures

**5** Find QR.

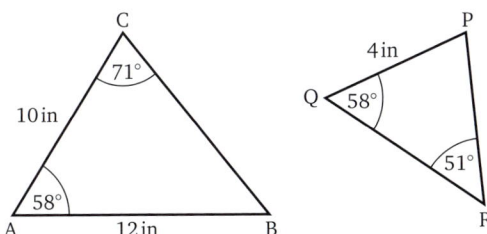

> In some cases we do not need to know the sizes of the angles, as long as we know that pairs of angles are equal. (Two pairs only are needed, as the third pair must then be equal.)

### Worked example

→ In △s ABC and DEF, $\hat{A} = \hat{E}$ and $\hat{B} = \hat{D}$. AB = 4 cm, DE = 3 cm and AC = 6 cm. Find EF.

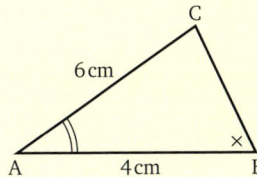 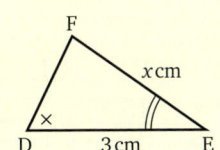

> Draw the triangles and mark the equal angles in the diagram. This helps to identify the corresponding sides.

△s $\dfrac{\text{EDF}}{\text{ABC}}$ are similar

> We put the triangle with the unknown side on the top; this makes sure that the equation has $x$ as the numerator of a fraction.

so $\quad \dfrac{FE}{CA} = \dfrac{ED}{AB} = \dfrac{DF}{BC}$

> We need $\dfrac{FE}{CA}$ and $\dfrac{ED}{AB}$.

$$\dfrac{x}{6} = \dfrac{3}{4}$$

$$\cancel{6}^1 \times \dfrac{x}{\cancel{6}_1} = \dfrac{3}{\cancel{4}_2} \times \cancel{6}^3$$

$$x = \dfrac{9}{2} = 4.5$$

so $\quad$ EF = 4.5 cm

**6** In △s ABC and XYZ, $\hat{A} = \hat{X}$ and $\hat{B} = \hat{Y}$.
AB = 6 cm, BC = 5 cm and XY = 9 cm. Find YZ.

**7** In △s ABC and PQR, $\hat{A} = \hat{P}$ and $\hat{C} = \hat{R}$.
AB = 10 cm, PQ = 12 cm and QR = 9 cm. Find BC.

**8** In △s ABC and DEF, $\hat{A} = \hat{E}$ and $\hat{B} = \hat{F}$.
AB = 3 cm, EF = 5 cm and AC = 5 cm. Find DE.

**9** In △s ABC and PQR, $\hat{A} = \hat{Q}$ and $\hat{C} = \hat{R}$.
AC = 8 cm, BC = 4 cm and QR = 9 cm. Find PR.

## Worked example

→ Show that triangles ABC and CDE are similar.

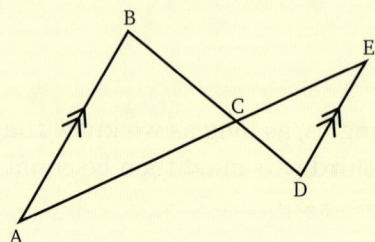

$\hat{A} = \hat{E}$ (alternate angles, AB∥DE)
$\hat{B} = \hat{D}$ (alternate angles, AB∥DE)

so △s $\genfrac{}{}{0pt}{}{ABC}{EDC}$ are similar.

 Or we could use $B\hat{C}A = E\hat{C}D$, as these are vertically opposite angles.

→ Given that AC = 15 cm, CE = 9 cm and DE = 8 cm, find AB.

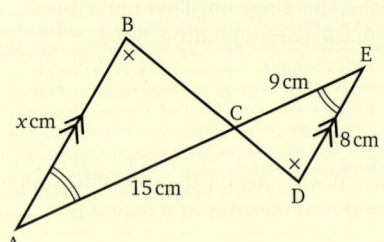

$$\frac{AB}{ED} = \frac{BC}{DC} = \frac{CA}{CE}$$

$$\frac{x}{8} = \frac{15}{9}$$

$$\overset{1}{\cancel{8}} \times \frac{x}{\cancel{8}_1} = \frac{\overset{5}{\cancel{15}}}{\cancel{9}_3} \times 8$$

$$x = \frac{40}{3}$$

$$= 13\tfrac{1}{3}$$

AB = $13\tfrac{1}{3}$ cm, or 13.3 cm correct to 3 s.f.

 **10 a** Show that △s ABC and BDE are similar.
   **b** If AB = 6 cm, BD = 3 cm and DE = 2 cm, find BC.

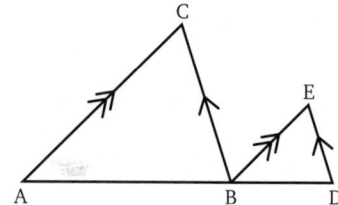

 **11 a** Show that △s ABC and CDE are similar.
   **b** If AB = 7 cm, BC = 6 cm, AC = 4 cm and CE = 6 cm, find CD and DE.

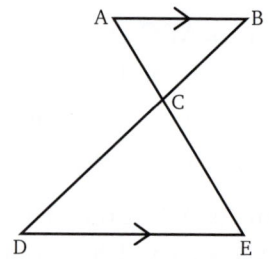

## 15 Similar figures

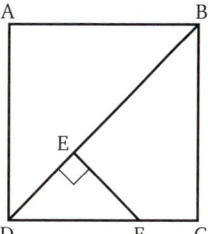

**12 a** ABCD is a square.
 EF is at right angles to BD.
 Show that △s ABD and DEF are similar.

 **b** If AB = 10 cm, DB = 14.2 cm and DF = 7.1 cm, find EF.

**13**

 **a** Show that △s ABC and ADE are similar. (Notice that $\hat{A}$ is *common* to both triangles.)
 **b** If AB = 10 cm, AD = 15 cm, BC = 12 cm and AC = 9 cm, find DE, AE and CE.

### Using the scale factor to find the missing length

Sometimes the scale factor for enlarging one triangle to the other is very obvious and we can make use of this to save ourselves some work.

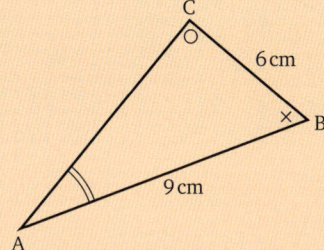 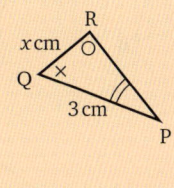

The two triangles above are similar and we can see that the scale factor for 'enlarging' the first triangle to the second is $\frac{1}{3}$.

We can say straight away that $x$ is $\frac{1}{3}$ of 6, which is 2 cm.

If we want to find a length in the first triangle, we use the scale factor that enlarges the second triangle to the first.

The scale factor is 4, so $x = 4 \times 2\frac{1}{2} = 10$

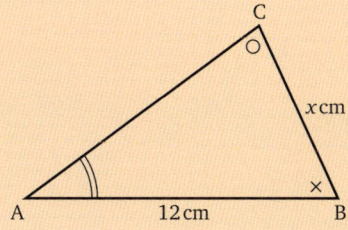 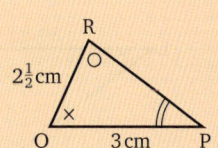

# STP Maths 9

## Exercise 15d

### Worked example

→ Find QR.

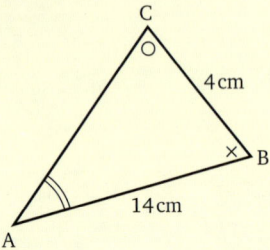 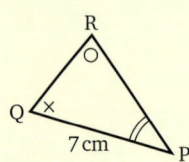

△s $\frac{PQR}{ABC}$ are similar

The scale factor is $\frac{1}{2}$

∴ QR = $\frac{1}{2}$ × 4 cm = 2 cm

---

**1** Find BC.

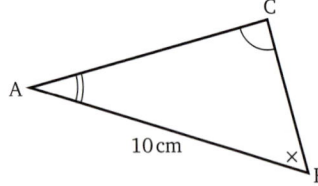 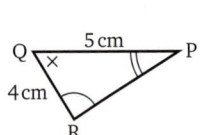

**2** Find PR.

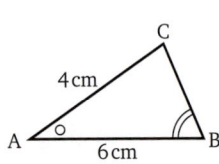 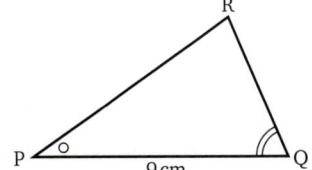

**3** Find PR.

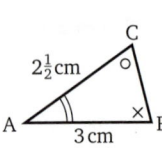 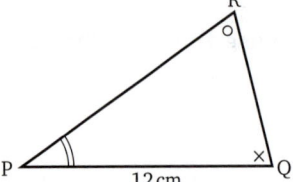

**4** Find LN.

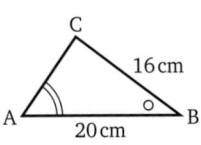 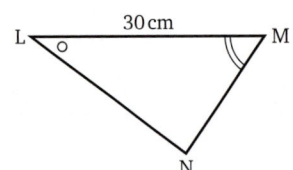

**5** Find PQ.

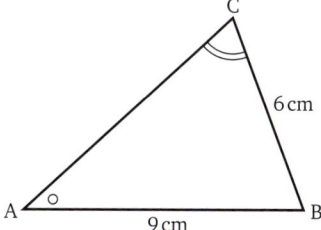
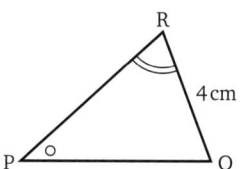

## Corresponding sides

**If all three pairs of corresponding sides of two triangles are in the same ratio, then the triangles are similar.**

We can use this fact to prove that two triangles are similar. It then follows that their corresponding angles are equal.

When finding the three ratios of pairs of sides, give each ratio as a whole number or as a fraction in its lowest terms.

## Exercise 15e

### Worked example

→ State whether triangles ABC and PQR are similar. Say which angle, if any, is equal to $\hat{A}$.

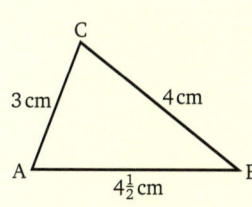
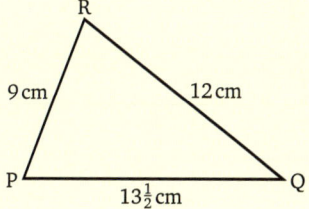

$\dfrac{PR}{AC} = \dfrac{9}{3} = 3$

$\dfrac{QR}{BC} = \dfrac{12}{4} = 3$

$\dfrac{PQ}{AB} = \dfrac{13\frac{1}{2}}{4\frac{1}{2}} = \dfrac{27}{9} = 3$

💡 Start with the shortest side of each triangle.

i.e. $\dfrac{PR}{AC} = \dfrac{QR}{BC} = \dfrac{PQ}{AB}$

so $\triangle$s $\dfrac{PQR}{ABC}$ are similar

∴ $\hat{P} = \hat{A}$

317

STP Maths 9

State whether the following pairs of triangles are similar. In each case say which angle, if any, is equal to $\hat{A}$.

**1**

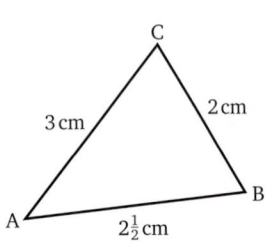 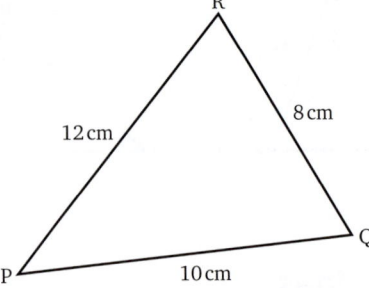

**2**

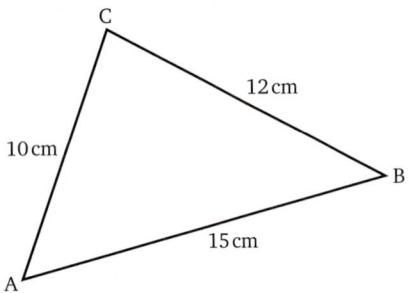 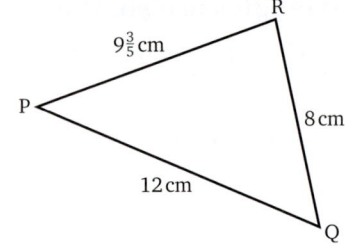

**3**

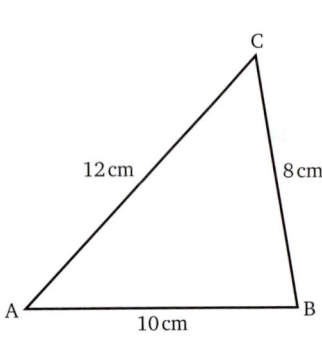 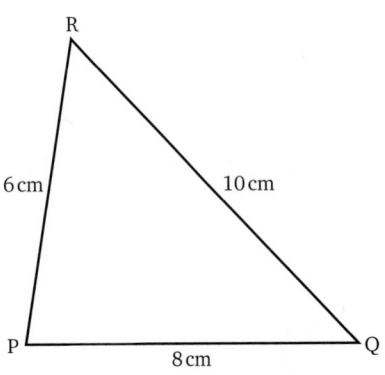

**4**

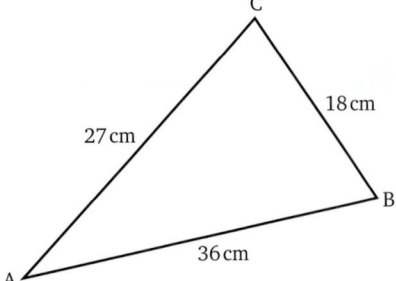 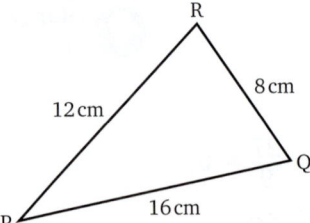

**15** Similar figures

**5**

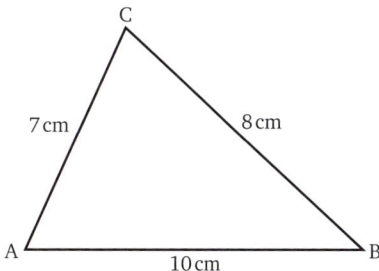
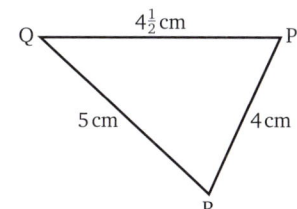

## Other shapes

So far in this chapter we have concentrated on triangles that are similar. However, relationships we have found for similar triangles apply equally well to other shapes.

Enlarging a figure does not alter the angles. It does change the lengths of the lines, but all the lengths change in the same ratio. For example, if one line is trebled in length, all lines are trebled in length.

**Exercise 15f**

In questions **1** to **5**, you are given two similar figures. Find each length marked with a letter.

**1**

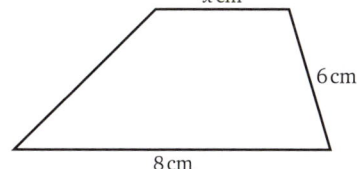

**2**

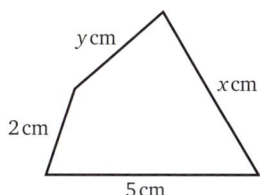
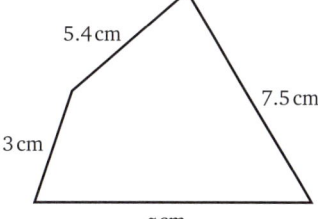

**3**

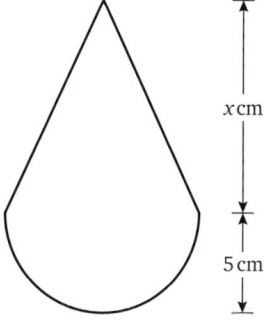
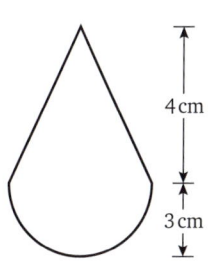

319

**4**

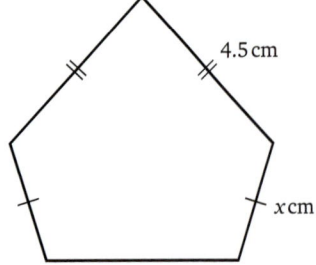

**5**

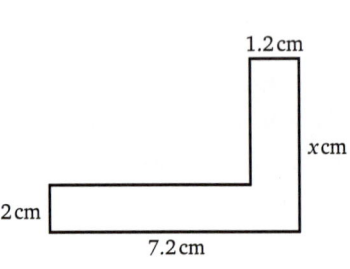
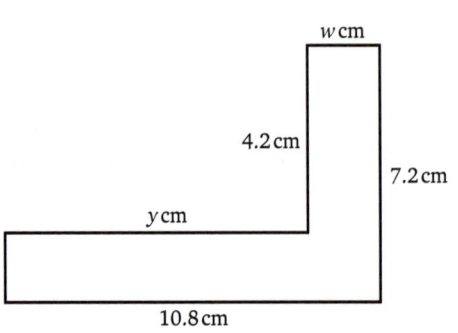

**6** The sketches show the length and width of a lake on a map and the corresponding length of the actual lake. What is the width of the lake at its widest point?

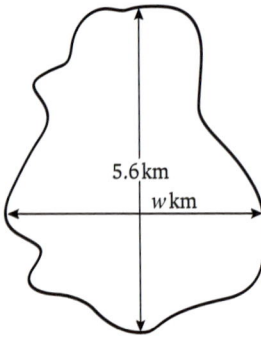
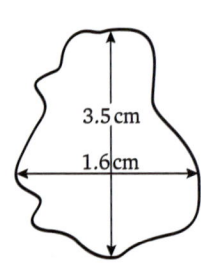

**7** The sketch shows a field which is planted with two different crops of cereal. Find the length of the common border between the two crops.

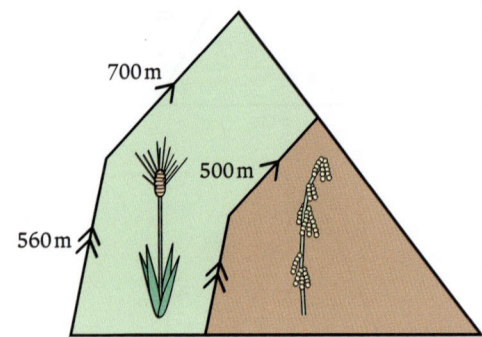

320

# 15 Similar figures

## Mixed exercise

### Exercise 15g

1 Which of the following shapes are similar to **X**?

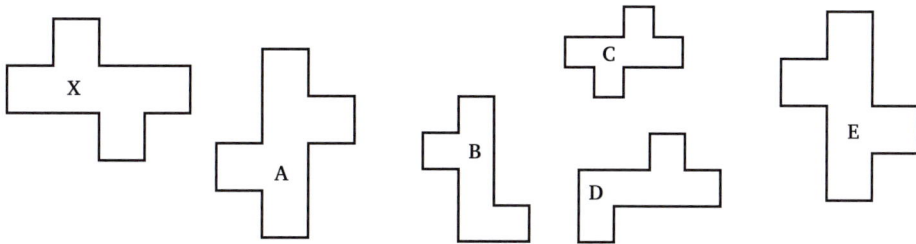

2 State whether or not triangles ABC and DEF are similar. If they are, give the ratio of the sides.

a
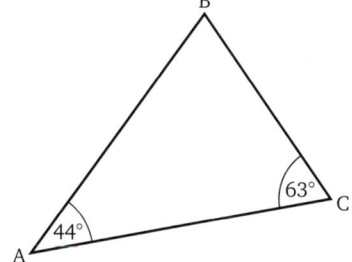
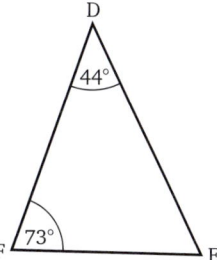

b
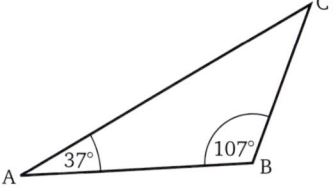
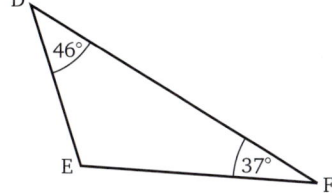

3 State whether triangles ABC and DEF are similar. If they are, find the length of DE.

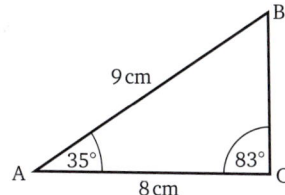
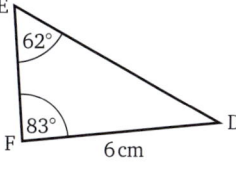

**4** Find the length of XY.

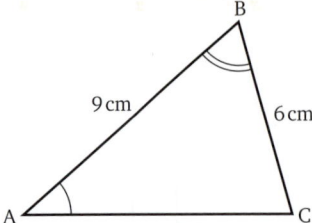 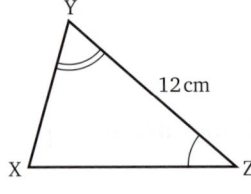

**5** State whether or not triangles ABC and XYZ are similar.

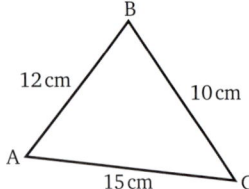 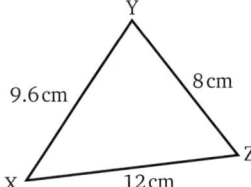

**6** The two given shapes are similar. Find the lengths marked with letters.

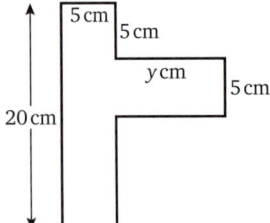 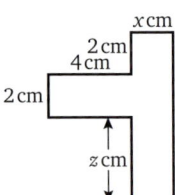

## Consider again

The diagram shows the cross-section of a swimming pool which is 3.5 m deep at one end and 1 m deep at the other. The pool is 30 m long and 10 m wide. Water flows into the pool to a depth of 1.5 m at the deep end. Now can you work out the dimensions of the surface of the water?

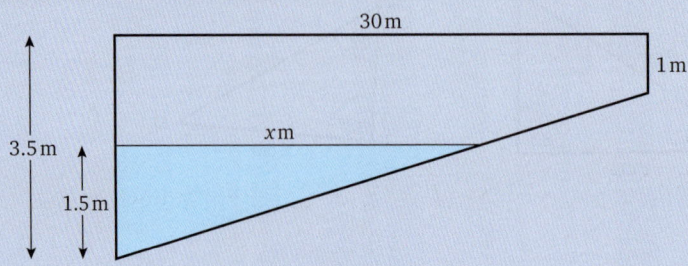

# Summary 3

## Quadratic equations

A *quadratic equation* such as $x^2 - 3x + 2 = 0$ can be solved algebraically if the left-hand side can be factorised.

$$x^2 - 3x + 2 = 0 \text{ becomes } (x - 2)(x - 1) = 0$$

Now we use the fact that if the product of two numbers is zero, then one or both of the numbers must be zero.

Therefore, as $(x - 2)(x - 1) = 0$

then either $\quad x - 2 = 0$, in which case $x = 2$

or $\quad\quad\quad\quad x - 1 = 0$, in which case $x = 1$

Hence the equation $x^2 - 3x + 2 = 0$ has two solutions:

$\quad x = 2$ and $x = 1$.

If necessary, we must rearrange a quadratic equation in the order $ax^2 + bx + c = 0$ before we attempt factorisation.

## Graphs

### Straight lines and parabolas

An equation of the form $y = mx + c$ gives a straight line, where $m$ is the gradient of the line and $c$ is the $y$-intercept.

An equation of the form $y = ax^2 + bx + c$ gives a curve whose shape is called a *parabola* and looks like this when $a$ is positive.

When the $x^2$ term is negative, the curve is 'upside down'.

### Cubic curves

When the equation of a curve contains $x^3$ (and maybe terms involving $x^2$, $x$ and a number), the curve is called a *cubic curve*.

These equations give cubic curves:

$$y = x^3,\ y = 2x^3 - x + 5,\ y = x^3 - 2x^2 + 6$$

A cubic curve looks like ⟋ or ⟿ when the $x^3$ term is positive

and ⟍ or ⟾ when the $x^3$ term is negative.

323

# STP Maths 9

## Reciprocal curves

The equation $y = \frac{a}{x}$, where $a$ is a number, is called a *reciprocal equation*.

An equation of the form $y = \frac{a}{x}$, where $a$ is a constant (that is, a number), gives a two-part curve called a *reciprocal curve* or a *hyperbola*. There are no values for $y$ when $x = 0$, as division by zero is not possible.

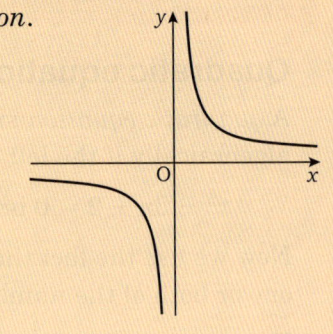

Any two quantities, $x$ and $y$, that are *inversely proportional*, are related by the equation $y = \frac{k}{x}$ and the graph representing them is a hyperbola.

## Gradient

The *gradient*, or slope, of a curve changes from point to point.

Gradient gives the rate at which the quantity on the vertical axis is changing as the quantity on the horizontal axis increases. The steeper the graph, and therefore the greater the gradient, the faster $y$ is increasing compared with $x$.

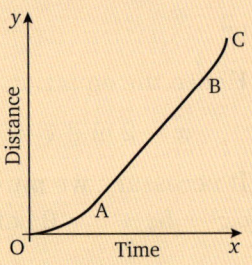

For graphs of distance plotted against time the steeper the graph is, the greater is the speed. In this graph, the greatest speed is at C; for the section AB, the speed is steady.

## Area

The *area of a trapezium* is equal to
$\frac{1}{2}$(sum of the parallel sides) × (distance between them)

$\frac{1}{2}(a + b) \times h$

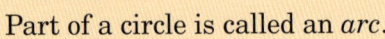

Part of a circle is called an *arc*.

Length of the arc AB $= \frac{A\hat{O}B}{360°} \times 2\pi r$

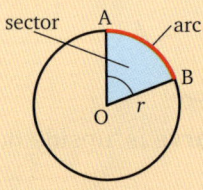
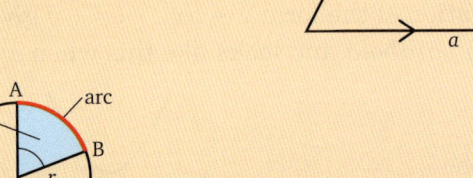

The slice of the circle enclosed by the arc and the radii is called a *sector*.

The area of the sector AOB $= \frac{A\hat{O}B}{360°} \times \pi r^2$

## Units of area

1 hectare = 10 000 m²     1 acre = 4840 sq yd

1 hectare = 2.5 acres correct to 1 d.p.

## Vectors

A *vector* is any quantity that needs to be described by giving both its magnitude (size) and its direction. For example, to explain where one village is in relation to another, we need to give both distance and direction, that is, a vector. A quantity that needs only size to describe it is called a *scalar*, for example, time.

# Summary 3

A vector can be represented by a straight line with an arrow to show direction, e.g.

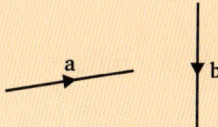

When vectors are drawn on squared paper, they can be described in terms of the number of squares needed to go across to the right and up. They are written in the form $\binom{a}{b}$, where $a$ is the number of squares across and $b$ is the number of squares up.

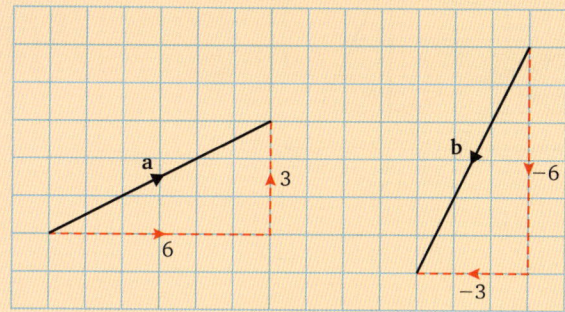

In the diagram, $\mathbf{a} = \binom{6}{3}$ and $\mathbf{b} = \binom{-3}{-6}$.

## Transformations

### Enlargement by a negative scale factor
When an object is enlarged by a scale factor of $-2$, we draw the guidelines *backwards* through the centre of enlargement, O, so that $OA' = 2OA$.

This produces an image each of whose lines is twice as long as the corresponding line on the object. The image is also rotated by 180° with respect to the object.

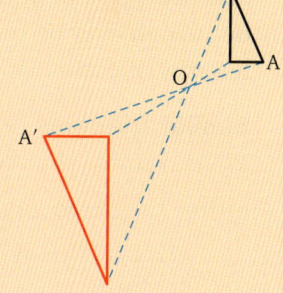

### Finding the mirror line
When an object has been reflected, corresponding points on the object and the image are the same distance from the mirror line. Therefore, we can find the mirror line by joining a pair of corresponding points on the object and the image, such as AA′, and then finding the line that goes through the midpoint of AA′ and is perpendicular to it.

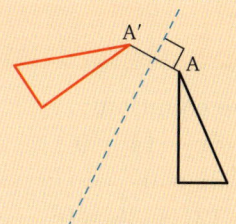

## Translation

A *translation* can be described by the vector that gives the movement from the object to the image.

In the diagram, the movement of the point A to A′ is 9 units to the left and 3 units up, so the translation is described by the vector $\begin{pmatrix} -9 \\ 3 \end{pmatrix}$.

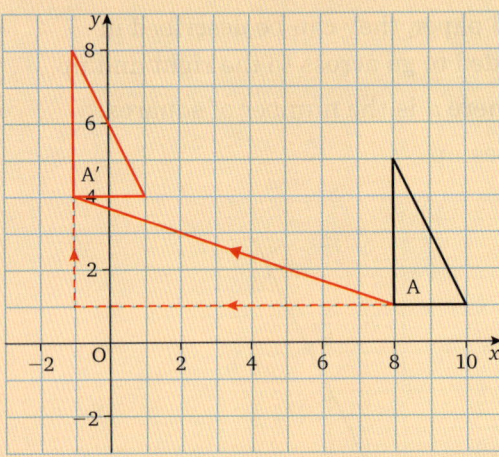

## Compound transformation

A compound transformation is the result of one transformation followed by another; for example, the result of reflecting an object in the *y*-axis and then rotating the image obtained by 30° about the origin.

# Similar figures

Two figures are similar if they are the same shape but are different in size, that is, one figure is an enlargement of the other. (One figure may be turned over or round with respect to the other.) It follows that the lengths of corresponding sides are all in the same ratio.

## Similar triangles

We can prove that two triangles are similar if we can show that

either the three angles of one triangle are equal to the three angles of the other

> (In practice, we only need to show that two pairs of angles are equal because, since the sum of the three angles in any triangle is 180°, it follows that the third pair must be equal.)

or the three pairs of corresponding sides are in the same ratio.

> (Two pairs of sides in the same ratio is not enough to prove that the triangles are similar.)

## REVISION EXERCISE 3.1 (Chapters 11 to 13)

**1** Solve the equations.
  **a** $(x + 3)(x - 4) = 0$
  **b** $x(x - 5) = 0$

**2** Solve the equations.
  **a** $x^2 - 5x - 14 = 0$
  **b** $x^2 - 3x = 0$
  **c** $x^2 - 12x + 36 = 0$

**3 a** Solve the equations.
  **i** $x^2 - 8x + 15 = 0$
  **ii** $3x^2 - x = 0$
  **iii** $x^2 = 8x + 20$
  **iv** $(x + 4)(x - 4) + 12 = 0$

  **b** A rectangle is $x$ cm wide and is 5 cm longer than it is wide. Its area is 24 cm². Form an equation in $x$ and solve it to find the dimensions of the rectangle.

**4** Suppose that you have drawn the graph of $y = x^2 - 3x - 5$. What values of $y$ on this graph give the values of $x$ that are solutions to these equations?
  **a** $x^2 - 3x - 7 = 0$
  **b** $x^2 - 3x = 0$
  **c** $x^2 - 3x + 1 = 0$
  **d** $2x^2 - 6x - 11 = 0$

**5** Which of the following graphs could represent the equation $y = x^2 - x$?

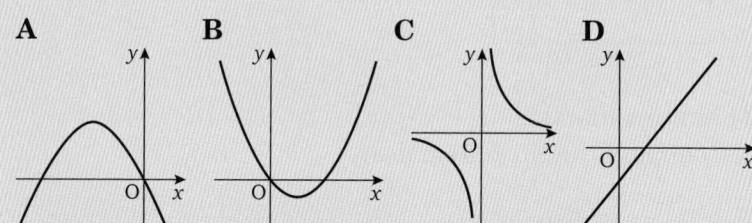

**6** Liquid is poured at a constant rate into each of the containers whose cross-sections are shown below. For each shape, sketch the graph showing how the depth of liquid in the container increases as the liquid is poured in.

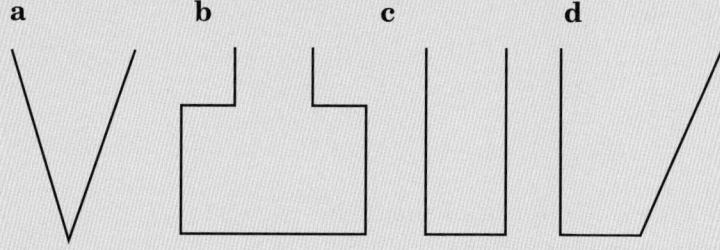

**7** Find the area of each shape.

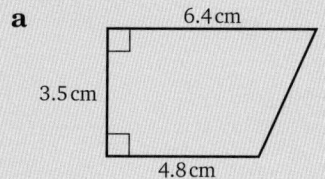
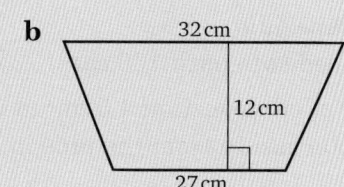

**8** Find  **a**  the length of the minor arc AB
   **b**  the area of the blue sector

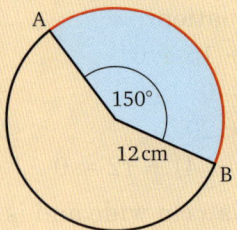

**9 a**  The volume of this solid is 234 cm³.
   Find  **i**  the area of cross-section
       **ii**  the length of the solid.

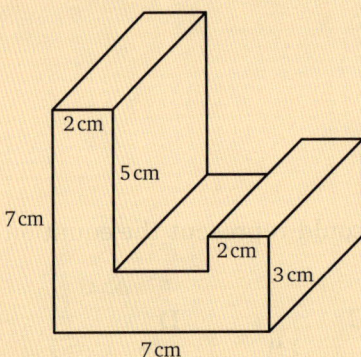

**b**  Chicken liver pâté is sold in tins like this.
The top has an area of 32 cm² and the tin is 3 cm deep.

The manufacturer decides to change to cylindrical tins that are 4 cm deep but hold the same quantity of pâté. Find the diameter of the top of the cylindrical tin.

**10 a**  State whether each of the following quantities should be measured in units of length or area or volume
  **i**  the space inside a cubical box
  **ii**  the distance round the edge of lake
  **iii**  the region inside a trapezium
  **iv**  the amount of land needed to park 100 cars.

**b**  Water issues from a pipe of cross-section 4.3 cm² at a speed of 0.6 m/s. How much water issues from the pipe in one second?

## REVISION EXERCISE 3.2 (Chapters 14 and 15)

1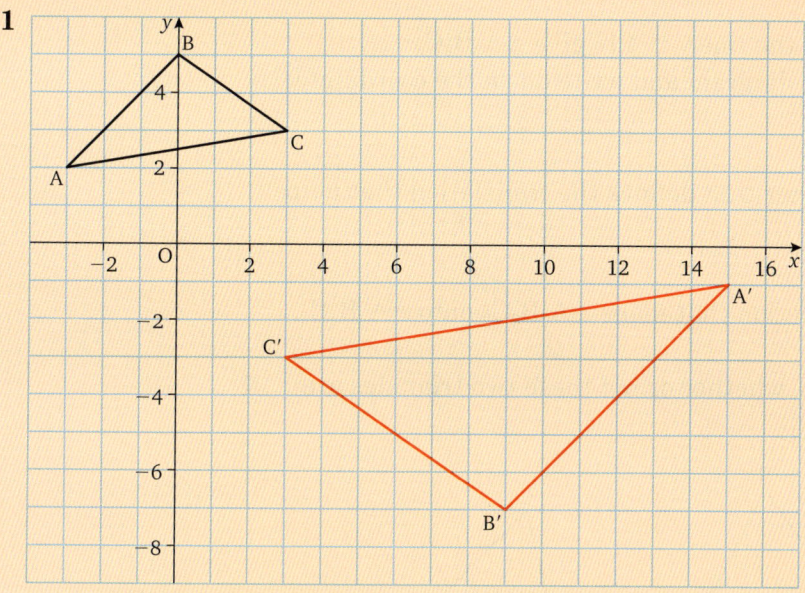

   **a** Copy the diagram and find the centre of enlargement and the scale factor when △ABC is enlarged to give △A′B′C′.
   
   **b** Draw the reflection of △ABC in the line $x = 5$.

Copy the following diagram onto 5 mm squared paper using 1 square as 1 unit on both axes, and use it to answer questions **2** to **5**.

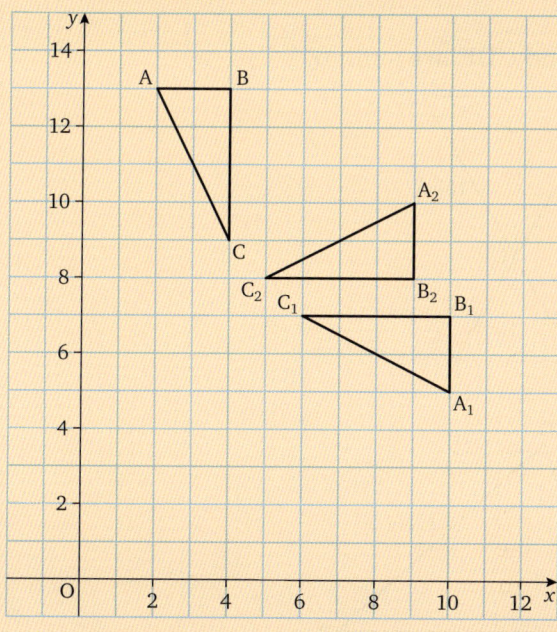

**2** Draw the mirror line so that △$A_1B_1C_1$ is the image of △ABC.

Write down its gradient and $y$-intercept.

**3** $\triangle ABC$ is mapped onto $\triangle A_2B_2C_2$ by a rotation. Find the centre and angle of rotation.

**4** $\triangle A_1B_1C_1$ is reflected in the line $y = 4$ to give $\triangle A_3B_3C_3$.
Mark $\triangle A_3B_3C_3$ on your diagram and write down the coordinates of $C_3$.

**5 a** Write down the vector that defines the translation that maps
    **i** $C_3$ onto $C_2$
    **ii** $C_2$ onto $C_3$.
 **b** Describe the translation that maps $\triangle A_2B_2C_2$ onto $\triangle A_3B_3C_3$.

**6 a** State, with reasons, whether or not these two triangles are similar.

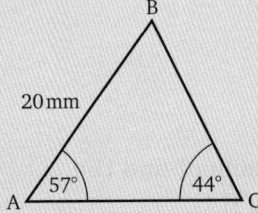
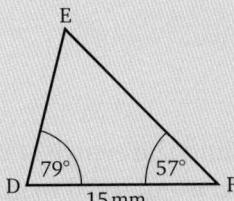

 **b** If they are, give the ratio of the corresponding sides.

**7** In triangles ABC and PQR, $\hat{A} = \hat{P}$ and $\hat{C} = \hat{R}$.
AB = 12 cm, PQ = 15 cm and QR = 10 cm. Find BC.

**8 a** Show that triangles ABC and CDE are similar.
 **b** If AB = 12 cm, BC = 10 cm, AC = 9 cm and CE = 6 cm, find DE and CD.

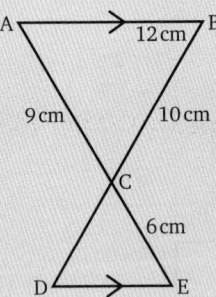

**9**

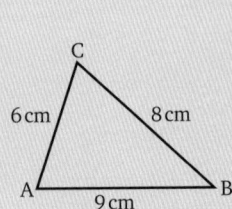
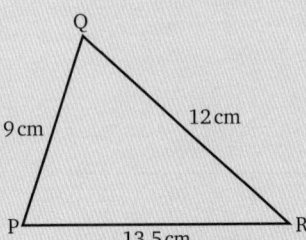

State whether or not triangles ABC and PQR are similar. Say which angle, if any, is equal to $\hat{A}$.

**10**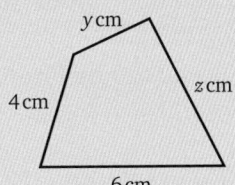

These two quadrilaterals are similar. Find the values of $x$, $y$ and $z$.

## REVISION EXERCISE 3.3 (Chapters 11 to 15)

**1** Solve the equations.
  **a** $x^2 + 7x = 0$
  **b** $x^2 - 49 = 0$
  **c** $x^2 + 3x - 28 = 0$

**2** Solve the equations.
  **a** $x^2 - 2x + 1 = 0$
  **b** $18 = 9x - x^2$
  **c** $3x^2 + 12x + 9 = 0$

**3** A curve has the equation $y = (x + 2)(x - 1)(x - 5)$.
  **a** What name do we give to this type of equation?
  **b** How many values of $x$ satisfy the equation $(x + 2)(x - 1)(x - 5) = 0$? What are they?

**4** Which of the following could be the equation of this curve?

  **A** $y = \dfrac{12}{x}$

  **B** $y = x^3 - 4x$

  **C** $y = 4x - x^3$

  **D** $y = x^2 + 4x - 4$

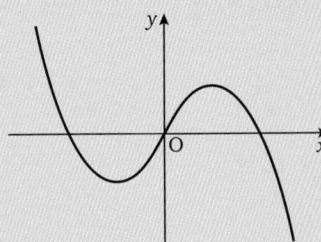

**5**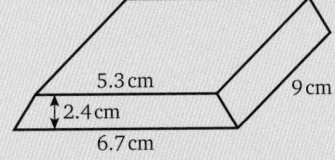

The cross-section of this metal ingot is a trapezium,
  **a** Find
    **i** the area of the cross-section
    **ii** the volume of the ingot.
  **b** The ingot is melted down and recast, without any change in volume, into a cube. What is the length of an edge of this cube? Give your answer correct to 3 significant figures.

**6** On squared paper, draw axes for $x$ and $y$ using the ranges $-6 \leqslant x \leqslant 6$ and $-6 \leqslant y \leqslant 6$ and a scale of one square to 1 unit. Plot the points A $(-5, 5)$, B $(4, 5)$, C $(2, -3)$, D $(-1, -3)$ and join them in alphabetical order to form a closed quadrilateral. Find, in square units, the area of the resulting shape.

**7** Copy the diagram.

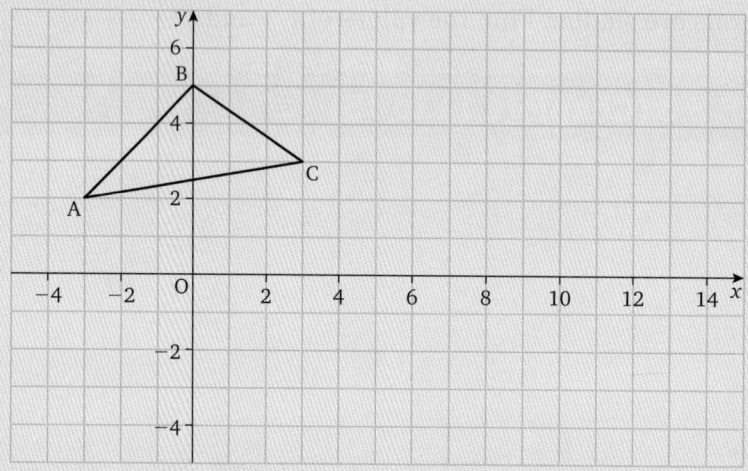

  **a** Draw the image of $\triangle ABC$ when it is rotated through 90° clockwise about the point $(-2, 1)$. Mark it $\triangle A_1B_1C_1$.

  **b** Draw the reflection of $\triangle A_1B_1C_1$ in the line $x = 3$. Mark it $\triangle A_2B_2C_2$.

  **c** Draw the image of $\triangle ABC$ when it undergoes a translation defined by the vector $\begin{pmatrix} 10 \\ 0 \end{pmatrix}$. Mark it $\triangle A_3B_3C_3$.

  **d** Describe the single transformation that maps $\triangle A_3B_3C_3$ onto $\triangle A_2B_2C_2$.

**8 a** One rectangle measures 12 cm by 8 cm and another rectangle measures 10 cm by 6 cm. Are these rectangles similar? Justify your answer.

  **b** Which of the following statements are true and which are false?
    **i** Any two isosceles triangles are similar.
    **ii** Two regular octagons are always similar.
    **iii** Any two parallelograms are similar.
    **iv** All squares are similar.

**9 a** Show that triangles ABC and CDE are similar.
  **b** Given that AC = 4 cm, CE = 10 cm and AB = 5 cm, find DE.

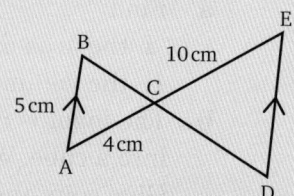

# REVISION EXERCISE 3.4 (Chapters 1 to 15)

**1 a** Express each fraction as a decimal.

    **i** $\frac{1}{200}$     **ii** $5\frac{5}{8}$     **iii** $\frac{8}{25}$     **iv** $\frac{39}{50}$     **v** $\frac{27}{500}$     **vi** $\frac{9}{8}$

**b** The mass, $w$ kg, of a bag of apples is given as 5 kg correct to the nearest whole number. Find the range of values in which $w$ lies.

**2 a** Express 9 : 4 in the form $n$ : 1.

**b** If 15 machines cost £9465, what is the cost of 18 similar machines?

**3** The following marks were obtained by the 80 candidates in a history test, which was marked out of 50.

| 48 | 43 | 32 | 8  | 47 | 15 | 36 | 46 | 43 | 30 |
|----|----|----|----|----|----|----|----|----|----|
| 41 | 25 | 36 | 29 | 17 | 20 | 22 | 35 | 23 | 9  |
| 30 | 7  | 38 | 44 | 35 | 31 | 36 | 44 | 10 | 34 |
| 46 | 19 | 31 | 28 | 39 | 37 | 48 | 17 | 28 | 33 |
| 23 | 45 | 39 | 7  | 22 | 33 | 24 | 40 | 36 | 23 |
| 35 | 31 | 36 | 12 | 27 | 12 | 37 | 33 | 33 | 46 |
| 27 | 26 | 37 | 32 | 33 | 26 | 24 | 42 | 23 | 10 |
| 14 | 30 | 38 | 34 | 34 | 31 | 26 | 11 | 26 | 48 |

Use this data to complete the following table.

| Interval | Tally | Frequency | Mark | Cumulative frequency |
|----------|-------|-----------|------|----------------------|
| 0–9      |       |           | ⩽9   |                      |
| 10–19    |       |           | ⩽19  |                      |
| 20–29    |       |           | ⩽29  |                      |
| 30–39    |       |           |      |                      |
| 40–50    |       |           |      |                      |

Use the information in your table to draw a cumulative frequency curve and from it estimate

**a** the median mark

**b** the upper and lower quartiles

**c** the number of candidates who passed if the pass mark was 30

**d** the pass mark if 75% of the candidates passed

**e** the probability that a candidate selected at random scored less than 20.

**4 a** Make the letter in brackets the subject of the formula.

    **i** $p = 5 + q$    $(q)$

    **ii** $x = \frac{1}{3}t$    $(t)$

    **iii** $z = \frac{x}{y}$    $(x)$

**b** Given that $V = 3b + \frac{8a}{b}$ and that $a = \frac{b^2}{2}$, find an expression for $V$ in terms of $b$.

**5**

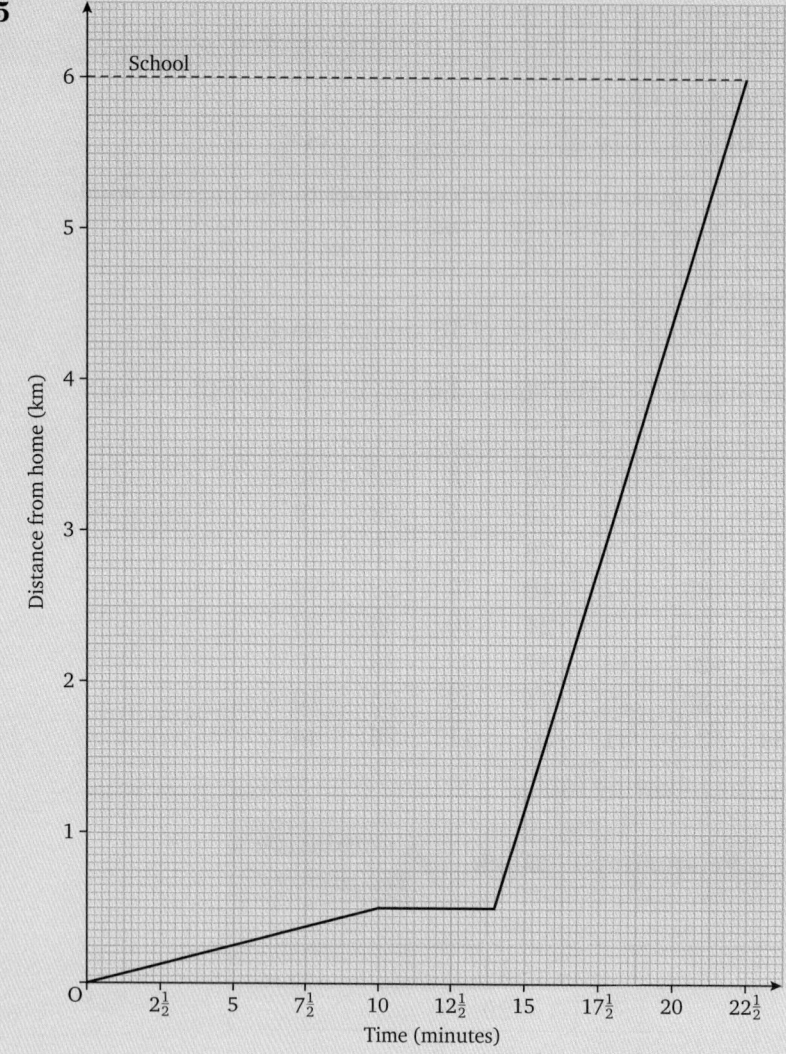

The graph shows Emily's journey from her home to school.
She walked to the bus stop and had to wait until the bus arrived.

Use the graph to answer the following questions:.

- **a** How far is it from Emily's home to school?
- **b** How long did it take her to walk to the bus stop?
- **c** How far did Emily walk?
- **d** Work out Emily's walking speed in km/h.
- **e** How long did she have to wait for the bus?
- **f** How far did she travel by bus?
- **g** How long did the journey by bus take?
- **h** At what speed, in km/h, did the bus travel?
- **i** What was her average speed in km/h, for the whole journey from home to school?

**6** Solve the simultaneous equations.

a   $4x + 3y = 3$
    $2x - y = 9$

b   $2x + 5y = -14$
    $5x + 4y = -1$

**7** Solve the equations.

a   $x^2 + 10x + 24 = 0$

b   $x^2 = 3x + 28$

c   $x(x + 2) = 24$

**8** Which of the following could be the equation of this curve?

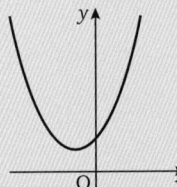

A   $y = x^2$

B   $y = 9 - x^2$

C   $y = \dfrac{5}{x} + 4$

D   $y = x^2 + 3x + 3$

**9 a** On $x$- and $y$-axes scaled from 0 to 12, plot the points A (6, 3), B (12, 12), C (12, 3).

**b** An enlargement, centre the origin, with scale factor $\tfrac{1}{3}$, maps $\triangle ABC$ onto $\triangle A_1B_1C_1$. Draw $\triangle A_1B_1C_1$.

**c** A translation of $\triangle A_1B_1C_1$ onto $\triangle A_2B_2C_2$ is defined by the vector $\begin{pmatrix}6\\2\end{pmatrix}$. Draw $\triangle A_2B_2C_2$. What are the coordinates of $C_2$?

**d** $\triangle ABC$ can be transformed onto $\triangle A_2B_2C_2$ by an enlargement. Give the centre of enlargement and the scale factor.

**10**

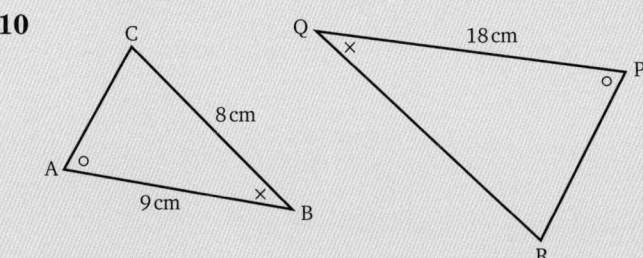

In triangles ABC and PQR, $\widehat{A} = \widehat{P}$ and $\widehat{B} = \widehat{Q}$, AB = 9 cm, BC = 8 cm and PQ = 18 cm. Find QR.

# REVISION EXERCISE 3.5 (Chapters 1 to 15)

**1 a** Abe walks 600 m in 5 minutes. Find his average speed in
    **i** metres per minute
    **ii** metres per second
    **iii** km/h.
 **b** Express the following decimals as fractions.
    **i** 0.55
    **ii** 0.56
    **iii** 0.056
    **iv** 0.0055
 **c** Simplify $\dfrac{2\frac{1}{9} - \frac{2}{3} \times 1\frac{5}{6}}{\frac{1}{5} \times 3\frac{1}{3} + 1\frac{5}{9}}$

**2** The school bus never leaves early in the morning. The probability that it leaves on time is $\frac{2}{5}$.
 **a** What is the probability that it leaves late?
 **b** Copy and complete the tree diagram by writing the probabilities on the branches.

What is the probability that
 **c** the bus is late on both days
 **d** the bus is late on one day and on time on the other day?

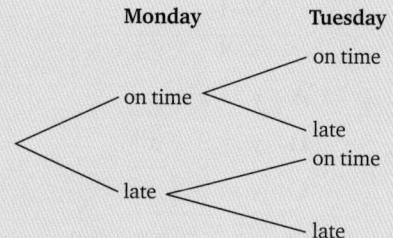

**3 a** George Sharp buys a greyhound for £360 and sells it at a loss of $12\frac{1}{2}\%$. How much does he lose?
 **b** A small business made a profit of £8909 this year. This is an increase of 18% on its profit last year. How much profit did it make last year?

**4 a** Divide 88 cm in the ratio 4:7.
 **b** A ball of string can be cut into 45 pieces if each piece is 28 cm long. How many pieces, each 63 cm long, could be cut from the same ball?

**5** Expand
 **a** $(a - b)(a - 2b)$
 **b** $(3x + 1)(x + 2)$
 **c** $(p + q)^2$

**6** Factorise
 **a** $a^2 - 3a$
 **b** $x^2 + 6x - 7$
 **c** $x^2 - 49$
 **d** $14 + 5x - x^2$

7 a  Given that $p = q + 2r$, find $r$ when $p = 7.8$ and $q = 3$.
  b  Write down the first four terms and the tenth term of the sequence for which the $n$th term is given by $(n + 1)(n + 2)$.
  c  Find, in terms of $n$, an expression for the $n$th term of the sequence 0, 5, 10, 15, ...
  d  $P = x^2 + 3xy$ and $x = y + 2$. Find $P$ in terms of $y$.

8 Solve the equations.
  a  $5x^2 + 4x = 0$
  b  $x^2 - x - 56 = 0$
  c  $x^2 - 5x = 24$
  d  $(x + 5)(x - 4) + 8 = 0$

9 The diagram shows a lean-to workshop.
  Find  a  the area of the cross-section
        b  the amount of space the workshop occupies
        c  the number of litres of rain that falls vertically on the workshop in a storm when the rainfall is 8 mm.

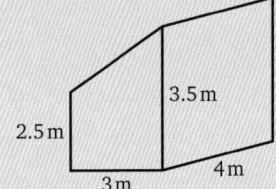

10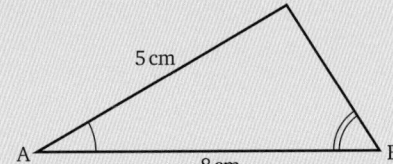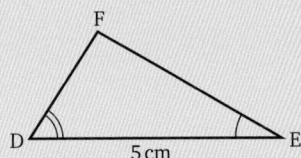

In triangles ABC and DEF, $\hat{A} = \hat{E}$ and $\hat{B} = \hat{D}$. AB = 8 cm, DE = 5 cm and AC = 5 cm. Find EF.

11 In the first round of a golf tournament, the following scores were recorded.

| 70 | 68 | 71 | 67 | 74 | 69 | 69 | 71 | 68 | 70 |
| 71 | 70 | 72 | 69 | 69 | 68 | 71 | 70 | 70 | 72 |
| 72 | 69 | 68 | 70 | 68 | 69 | 67 | 71 | 69 | 70 |
| 68 | 67 | 70 | 70 | 73 | 69 | 71 | 67 | 69 | 68 |

  a  Construct a cumulative frequency table for these scores.
  b  How many rounds of less than 70 were there?
  c  How many rounds of more than 69 were there?
  d  Find the median score.
  e  Explain why you do not need a cumulative frequency curve to estimate the median.

# Mental arithmetic practice 3

1. Write 156 498 correct to the nearest thousand.
2. What is the largest prime number before 40?
3. Express the ratio 8 : 5 in the form $n : 1$
4. Does $x = 2.5$ satisfy $x + 2 \geqslant 4$?
5. Is it true that length is a continuous quantity?
6. Express 0.4 kg in grams.
7. Write 24 as the sum of two prime numbers.
8. Factorise $x^2 - 7x$
9. Find the mean of 7 kg, 9 kg, 10 kg and 10 kg.
10. Find $\frac{3}{4}$ of $\frac{2}{5}$
11. Factorise $x^2 + 2x - 15$
12. Simplify $16a^2 \div 8a$
13. Simplify $9a^2 \times 2a$
14. Express 750 cm in metres.
15. Which is larger, 4% or 0.4?
16. Give 0.36 as a fraction in its lowest terms.
17. Divide $2\frac{1}{2}$ by $\frac{1}{3}$
18. Factorise $x^2 + 7x + 12$
19. What is 5 minus negative 8?
20. Estimate $(3.87)^3$

Questions **21** to **25** refer to the numbers 5, 10, 15, 15, 25.
Is it true or false that

21. the range is 25
22. the mean is 14
23. the mode and median are the same
24. if 5 is added to each number, the median is unchanged
25. if 2 is subtracted from each number, the range decreases by 2?

26. Lauren weighs 63 kg correct to the nearest kilogram. What is the upper bound of her weight?
27. What is the middle value of the group $95 \leqslant h < 100$?
28. Find the lowest common multiple of 18 and 4.
29. Express $42\frac{1}{2}\%$ as a decimal.
30. If you roll an unbiased dice 60 times, about how many times would you expect to score 3?
31. What word is missing from this sentence?
    The running total of frequencies is called the ... frequency.

**32** Express $\frac{3}{8}$ as a decimal.

**33** Express the ratio 5 : 4 in the form $1 : n$

**34** Find $\frac{3}{4} - \frac{2}{5}$

**35** Find 140% of 200 m.

**36** What needs to be added to 3.66 to give 5?

**37** Factorise $x^2 + 9x + 20$

**38** Which is the larger: $3 \times 10^5$ or 3 million?

**39** Find $0.3^3$

**40** Complete this sentence:
The difference between the upper and lower quartiles of a set of data is called the ...

**41** A number $d$ is equal to the difference between the squares of two numbers $b$ and $c$, where $c$ is bigger than $b$. Give a formula connecting $d$, $b$ and $c$.

**42** If $v = u + 8t$, find $v$ when $u = -6$ and $t = 2$.

**43** Make $C$ the subject of the formula $A = B + \frac{1}{2}C$.

**44** If $P = 2Q$ and $Q = 3 - r$, find a formula for $P$ in terms of $r$.

**45** Find, in terms of $n$, an expression for the $n$th term of the sequence 3, 5, 7, 9, 11.

**46** What needs to be multiplied by 6 to give 0.3?

**47** Factorise $x^2 + 7x - 8$

**48** A number $z$ is three times the product of two numbers $x$ and $y$. Give a formula connecting $x$, $y$ and $z$.

**49** Find the value of $p^2 - q^2$ when $p = 2$ and $q = -3$.

**50** Make $P$ the subject of the formula $R = P - Q$

In questions **51** to **55**, it true or false that

**51** the square root of a number is always smaller than the number

**52** discrete values are always whole numbers

**53** any two circles are congruent

**54** a diagonal divides a rectangle into two congruent triangles

**55** a quadrilateral with just one pair of opposite sides parallel is a parallelogram?

**56** Factorise $a^2 - a - 12$

**57** Find the value of $\left(\dfrac{a}{b}\right)^2$ when $a = -3$ and $b = 2$.

**58** Find an expression for the $n$th term of the sequence 4, 9, 14, 19, ...

**59** Make $k$ the subject of the formula $N = \dfrac{m}{k}$

**60** Given that $a = b + c^2$, find $a$ when $b = 7$ and $c = -2$.

# 16 Trigonometry: tangent of an angle

## Consider

Sam wants to know the height of this tree. He can find it by measuring how far he is from the tree and then measuring the angle of elevation of the top of the tree. Showing this information in a diagram gives this right-angled triangle.

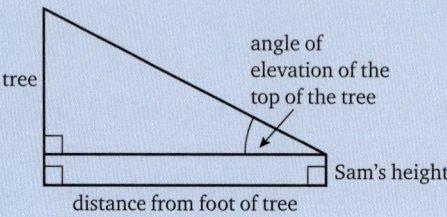

If Sam draws the diagram to scale, he can measure the height of the tree from the scale drawing.

- The disadvantage of this method is that it takes time and precision to get a reasonably accurate result.

Find the height of the tree without making a scale drawing, given that Sam is 1.75 m tall and is standing 40 m from the base of the tree and that he measures the angle of elevation of the tree as 32°.

*You should be able to answer this question after you have worked through this chapter.*

---

In this chapter we find out how to calculate a length such as the height of the tree described in the Consider section by a method which, with the help of a scientific calculator, is fast and accurate. The accuracy depends only on the accuracy of the initial measurements.

We start by investigating the relationship between the size of an angle and the lengths of two sides in a right-angled triangle.

## Exercise 16a

1  **a** Draw the given triangle accurately, using a protractor and a ruler.
   **b** Measure $\hat{A}$.
   **c** Find $\dfrac{BC}{AB}$ as a decimal.

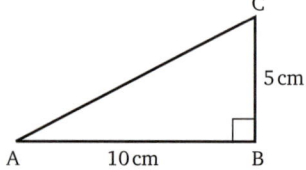

Repeat question **1** for the triangles in questions **2** to **5**.

**2**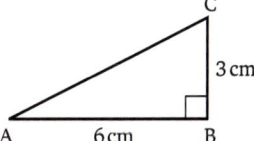

**4**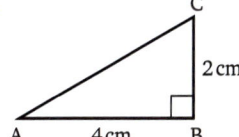

**3**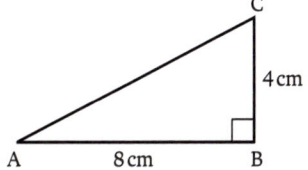

**5**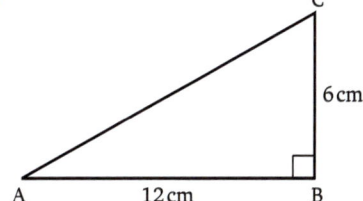

**6** Are the triangles in questions **1** to **5** similar?

Repeat question **1** for the triangles in questions **7** to **10**.

**7**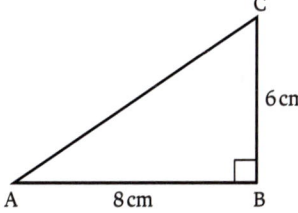

**9**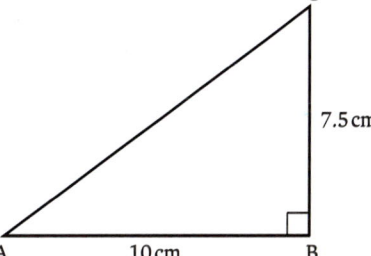

**8**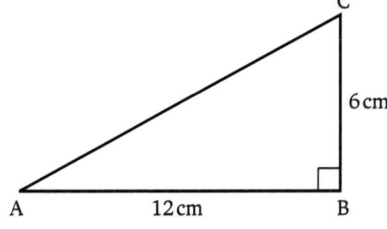

**10**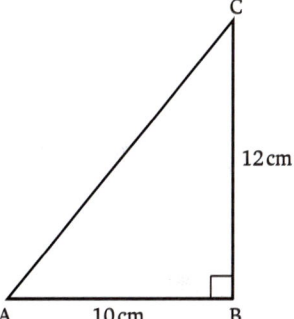

341

**11** Similar triangles can be drawn so that they overlap, as in this diagram. Copy the diagram above onto squared paper. Choose your own measurements, but make sure that the lengths of the horizontal lines are whole numbers of centimetres. Measure $\hat{A}$.

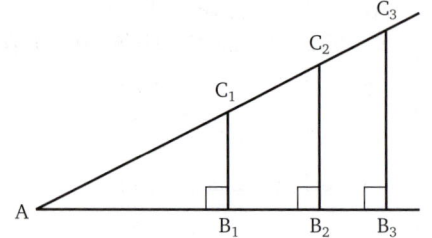

Find $\dfrac{B_1C_1}{AB_1}$, $\dfrac{B_2C_2}{AB_2}$ and $\dfrac{B_3C_3}{AB_3}$ as decimals.

**12** Copy and complete the table using the information from questions **1** to **5**.

| $\hat{A}$ | $\dfrac{BC}{AB}$ |
|---|---|
| 26.6° | 0.5 |

## Tangent of an angle

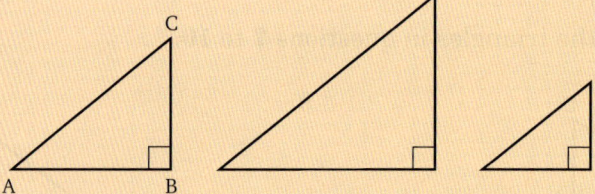

If we consider the set of all triangles that are similar to △ABC then for every triangle in the set,

the angle corresponding to $\hat{A}$ is the same

the ratio corresponding to $\dfrac{BC}{AB}$ is the same

where  BC is the side *opposite* to $\hat{A}$

and  AB is the *adjacent* (or neighbouring) side to $\hat{A}$.

From **Exercise 16a** you can see that in a right-angled triangle the ratio $\dfrac{\text{opposite side}}{\text{adjacent side}}$ is always the same for a given angle, whatever the size of the triangle.

The ratio $\dfrac{\text{opposite side}}{\text{adjacent side}}$ is called the **tangent of the angle**.

$$\textbf{tangent of the angle} = \dfrac{\textbf{opposite side}}{\textbf{adjacent side}}$$

More briefly,

$$\textbf{tan (angle)} = \dfrac{\textbf{opp}}{\textbf{adj}}$$

The information about this ratio is used so often that we need a more complete and more accurate list than the one made in the last exercise. The complete list is stored in scientific calculators.

# 16 Trigonometry: tangent of an angle

## Finding tangents of angles

To find the tangent of 33°, press [tan] [3] [3] [=]. You will obtain a number that fills the display. Write down one more figure than the accuracy required,

e.g.  tan 33° = 0.64940...
            = 0.6494 correct to 4 significant figures.

If you do not get the correct answer, one reason could be that your calculator is not in 'degree mode'. For all trigonometric work at this stage, angles are measured in degrees, so make sure that your calculator is in the correct mode.

## Exercise 16b

### Worked example

→ Find the tangent of 56° correct to 3 significant figures.

tan 56° = 1.482...
       = 1.48 (correct to 3 significant figures)

Find the tangents of the following angles correct to 3 significant figures.

**1** 20°  **2** 28°  **3** 72°  **4** 53°  **5** 59°

**6** Find the tangents of the angles listed in question **12** in **Exercise 16a**. How do the answers you now have compare with the decimals you worked out? If they are different, give a reason for this.

### Worked example

→ Find the tangent of 34.2°.

tan 34.2° = 0.680 correct to 3 s.f.     Press [tan] [3] [4] [.] [2] [=]

Find the tangents of the following angles correct to 3 significant figures.

**7** 15.5°
**8** 29.6°
**9** 11.4°
**10** 60.1°
**11** 3.8°

# STP Maths 9

**The names of the sides of a right-angled triangle**

Before we can use the tangent to find sides and angles, we need to know which is the side opposite to the given angle and which is the adjacent side.

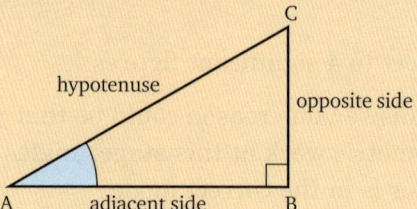

- The longest side is the side opposite to the right angle, and is called the **hypotenuse**.
- The side next to the angle A (not the hypotenuse) is called the *adjacent side*. (Adjacent means 'next to'.)
- The third side is the *opposite side*. It is opposite to the particular angle A that we are concerned with.

Sometimes the triangle is in a different position from the one we have been using, for example:

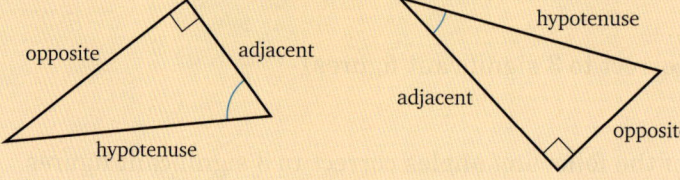

## Exercise 16c

Sketch the following triangles. The angle we are concerned with is marked with a blue arc like this ⌒. Label the sides 'hypotenuse', 'adjacent' and 'opposite'. If necessary, turn the page round so that you can see which side is which.

1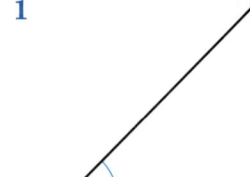

2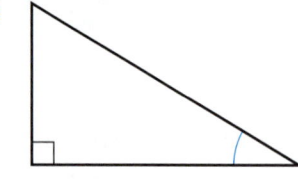

# 16 Trigonometry: tangent of an angle

3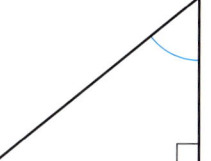

5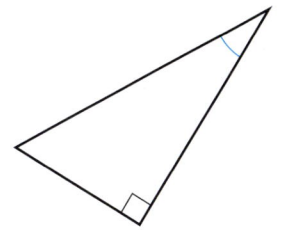

4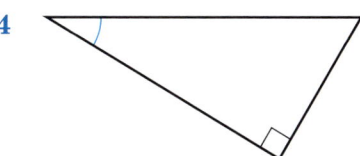

6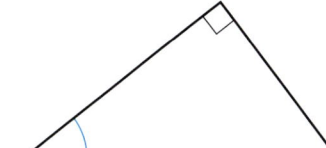

## Finding a side of a triangle

We can now use the tangent of an angle to find the length of the opposite side in a right-angled triangle provided that we know an angle and the length of the adjacent side.

## Exercise 16d

Use a calculator. Give your answers correct to 3 significant figures.

### Worked example

→ In △ABC, $\hat{B} = 90°$, $\hat{A} = 32°$ and AB = 4 cm.
Find the length of BC.

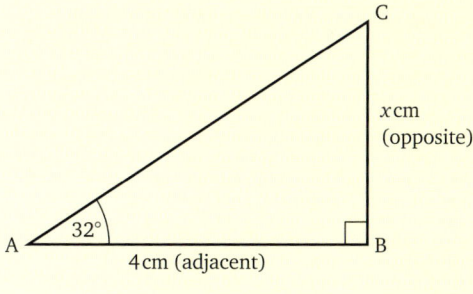

First label the opposite and adjacent sides and use $x$ cm for the length of the side BC.

$$\frac{x}{4} = \frac{\text{opp}}{\text{adj}} = \tan 32°$$

∴ $\quad \frac{x}{4} = 0.6248\ldots$

Do not clear the display.

$4 \times \frac{x}{4} = 0.6248\ldots \times 4$

$x = 2.499\ldots$

Now press $\boxed{\times}$ $\boxed{4}$ $\boxed{=}$

BC = 2.50 cm (correct to 3 s.f.)

In questions **1** to **8**, find the length of BC.

1
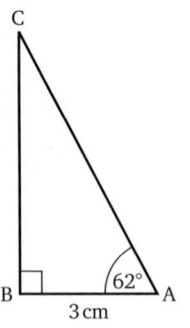

5
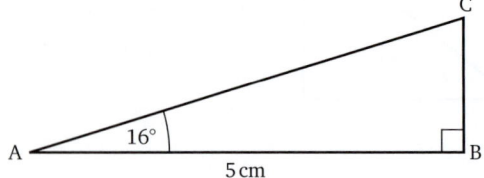

2
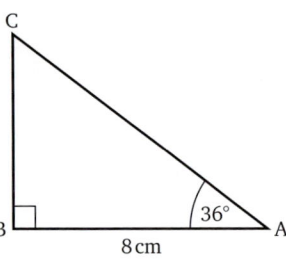

6
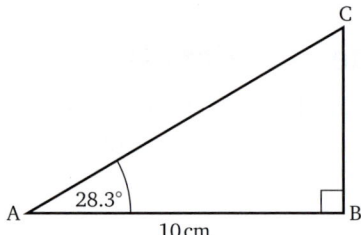

3
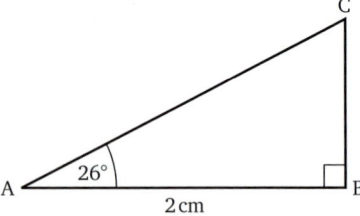

7
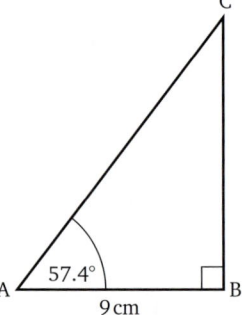

4
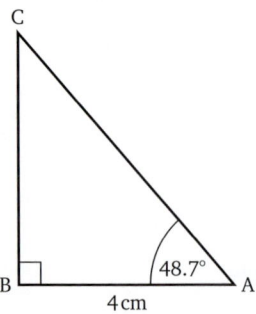

8
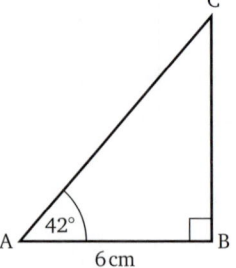

In questions **9** to **12**, different letters are used for the vertices of the triangles. In each case find the side required.

**9** Find PQ.

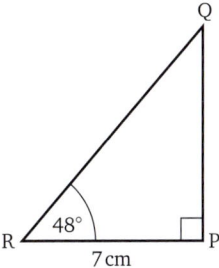

**11** Find AC.

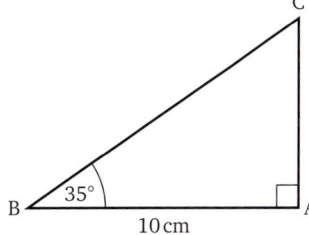

**10** Find YZ.

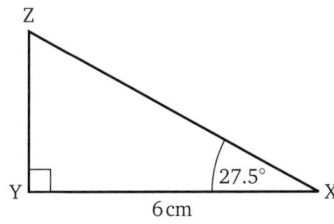

**12** Find AC.

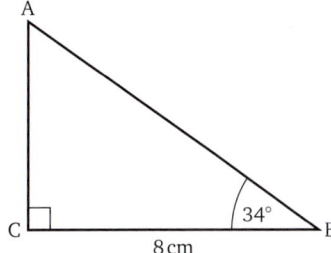

Find BC in questions **13** to **20**. Turn the page round if necessary to identify the opposite and adjacent sides.

**13**

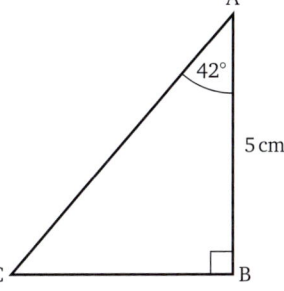

**15**

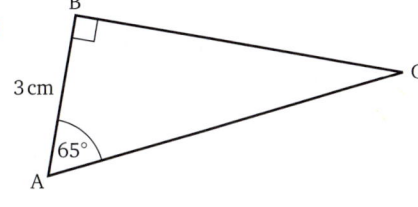

**14**

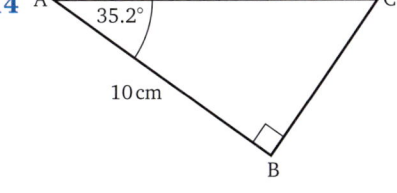

**16**

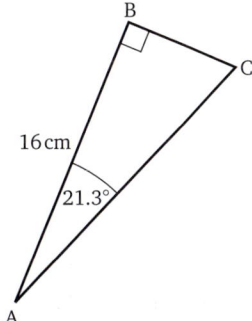

**17**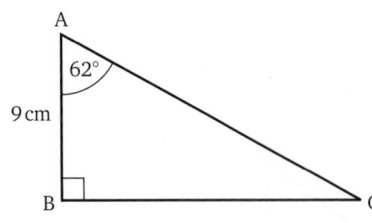

**19**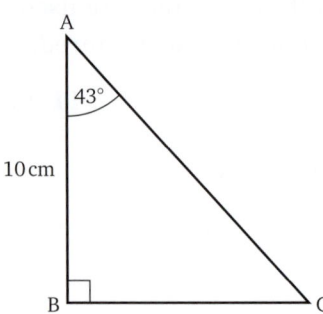

**18**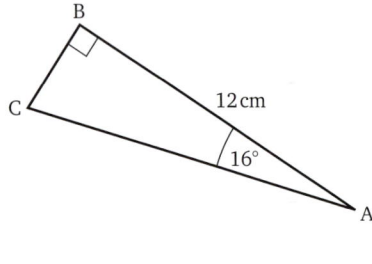

**20**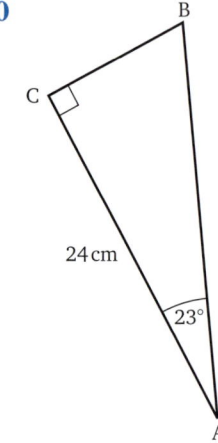

**21** In $\triangle ABC$, $\hat{B} = 90°$, $AB = 6$ cm and $\hat{A} = 41°$. Find BC.

**22** In $\triangle PQR$, $\hat{Q} = 90°$, $PQ = 10$ m and $\hat{P} = 16.7°$. Find QR.

**23** In $\triangle DEF$, $\hat{F} = 90°$, $DF = 12$ cm and $\hat{D} = 56°$. Find EF.

**24** In $\triangle XYZ$, $\hat{Z} = 90°$, $YZ = 11$ cm and $\hat{Y} = 40°$. Find XZ.

### Finding a side adjacent to the given angle

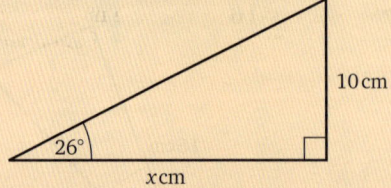

Sometimes the side whose length we are asked to find is adjacent to the given angle instead of opposite to it. Using $\dfrac{10}{x}$ instead of $\dfrac{x}{10}$ can lead to an awkward equation, so we work out the size of the angle opposite $x$ and use this instead. In this case, the other angle is 64° and we label the sides 'opposite' and 'adjacent' to this angle.

**16 Trigonometry: tangent of an angle**

Using 64°,

$$\frac{x}{10} = \frac{\text{opposite}}{\text{adjacent}} = \tan 64°$$

so $\quad \frac{x}{10} = 2.05$, giving $x = 20.5$

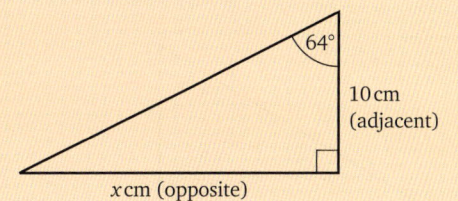

## Exercise 16e

Use a calculator. Give your answers correct to 3 significant figures.

### Worked example

→ In △PQR, $\hat{P} = 90°$, $\hat{Q} = 51°$ and PR = 4 cm.
Find the length of PQ.

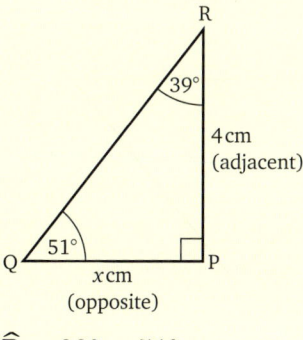

First find the other angle, i.e., $\hat{R}$.

$\hat{R} = 90° − 51°$
$\phantom{\hat{R}} = 39°$

$\frac{x}{4} = \frac{\text{opp}}{\text{adj}} = \tan 39°$

$\frac{x}{4} = 0.8097\ldots \times 4$

Do not clear the display.

$4 \times \frac{x}{4} = 0.8097\ldots \times 4$

Leave the number on the display, then press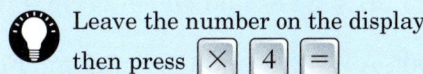

$x = 3.239\ldots$

PQ = 3.24 cm (correct to 3 s.f.)

1  Find ZY.

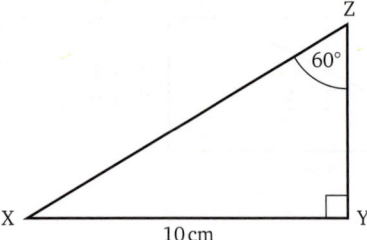

2  Find QP.

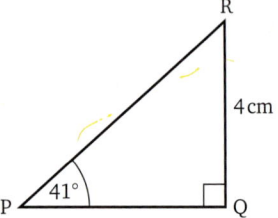

3  Find XZ.

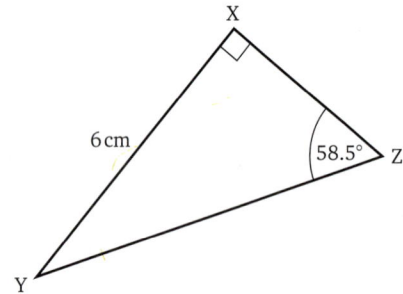

4  Find FD.

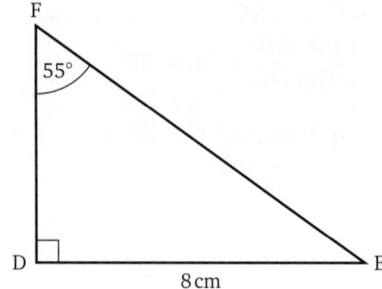

5  Find BC.

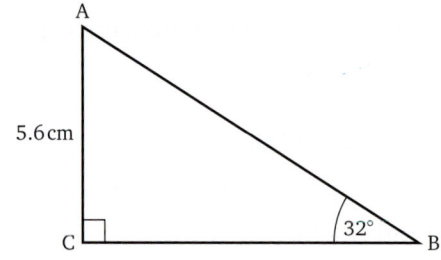

6  Find AB.

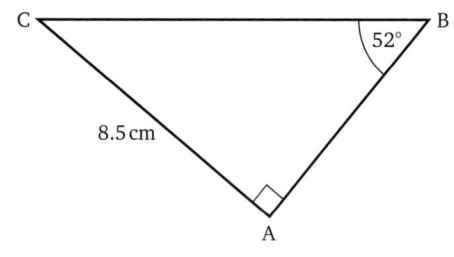

7  In △PQR, $\hat{Q} = 90°$, $\hat{R} = 31°$ and PQ = 6 cm. Find RQ.

8  In △XYZ, $\hat{Z} = 90°$, $\hat{Y} = 38°$ and ZX = 11 cm. Find YZ.

9  In △DEF, $\hat{D} = 90°$, $\hat{E} = 34.8°$ and DF = 24 cm. Find DE.

10  In △ABC, $\hat{C} = 90°$, $\hat{A} = 42.4°$ and CB = 3.2 cm. Find AC.

11  In △LMN, $\hat{L} = 90°$, $\hat{N} = 15°$ and LM = 4.8 cm. Find LN.

12  In △STU, $\hat{U} = 90°$, $\hat{S} = 42.2°$ and TU = 114 cm. Find SU.

## 16 Trigonometry: tangent of an angle

### Worked example

→ Renata is at a point A, 20 metres from the base of a tree which is standing on level ground. From A the angle of elevation of the top, C, of the tree is 23°. What is the height of the tree?

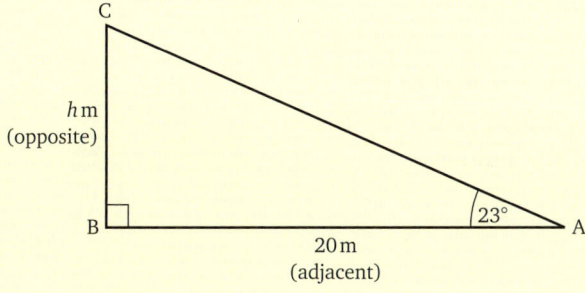

$\dfrac{h}{20} = \dfrac{\text{opp}}{\text{adj}} = \tan 23°$

$\dfrac{h}{20} = 0.4244...$

$20 \times \dfrac{h}{20} = 0.4244... \times 20$

$h = 8.489...$

The height of the tree is 8.49 m correct to 3 significant figures.

---

**13** A pole BC stands on level ground. A is a point on the ground 10 metres from the foot of the pole. The angle of elevation of the top C from A is 27°. What is the height of the pole?

**14** ABCD is a rectangle. AB = 42 m and $B\hat{A}C = 59°$.
Find the length of BC.

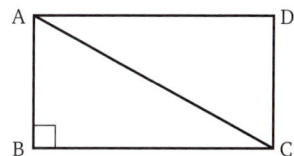

**15** In rectangle ABCD, the angle between the diagonal AC and the side AB is 22°, AB = 8 cm.
Find the length of BC.

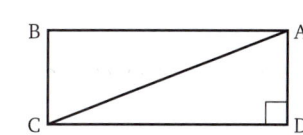

**16** In △PQR, PQ = QR. From symmetry, S is the midpoint of PR. $\hat{P} = 72°$, PR = 20 cm.
Find the height QS of the triangle.

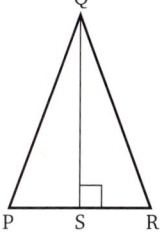

351

**17** A point R is 14 m from the foot of a flagpole PQ. The angle of elevation of the top of the pole from R is 22°.

Find the height of the pole.

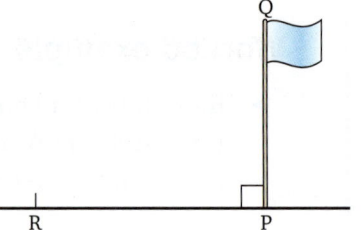

  **18** A ladder leans against a vertical wall so that it makes an angle of 35° with the wall. The top of the ladder is 2 m up the wall. How far from the wall is the foot of the ladder?

  **19** A dinghy B is 60 m out to sea from the foot A of a vertical cliff AC. From C, the angle of depression of B is 16°.

   **a**  Find $\hat{B}$

   **b**  Find the height of the cliff.

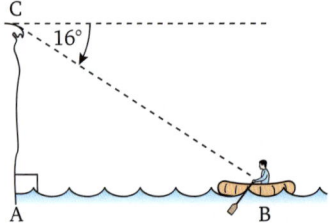

  **20** From a yacht, the angle of elevation of the top of a lighthouse is 26° and the angle of elevation of the base of the lighthouse is 20°. The lighthouse is standing on top of a 40 metre high cliff.
Find

   **a**  how far the yacht is from the base of the cliff

   **b**  the height of the lighthouse.

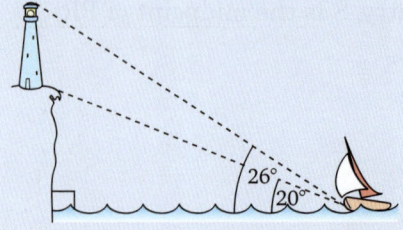

# 16 Trigonometry: tangent of an angle

## Finding an angle given its tangent

If we are given the value of the tangent of an angle, we can use a calculator to find that angle.
For example, if the tangent of angle A is 0.732,

i. e.     $\tan \widehat{A} = 0.732$,

we find $\widehat{A}$ by pressing [shift] [tan] then entering 0.732 = in the calculator. (The 'shift' key or '2nd' key accesses the '$\tan^{-1}$' function, which means 'the angle whose tan is'.)

This gives the size of the angle in degrees.

So when     $\tan \widehat{A} = 0.732$
            $\widehat{A} = 36.20...° = 36.2°$ correct to 1 decimal place.

Check this on your calculator. If it does not give this result, first check that you are using degree mode then, if it is still not correct, consult your manual.

## Exercise 16f

Find, correct to 1 decimal place, the angles whose tangents are given.

1. 2.2
2. 0.36
3. 0.41
4. 4.1
5. 1.4
6. 0.31
7. 0.6752
8. 0.992 93
9. 0.376 24
10. 2.0879
11. 1
12. 0.333 33

### Worked example

→ Find the angle whose tangent is $\frac{3}{4}$

$\tan \widehat{A} = \frac{3}{4} = 0.75$

$\widehat{A} = 36.86...° = 36.9°$ (correct to 1 d.p.)

💡 First express $\frac{3}{4}$ as a decimal.

Find, correct to 1 decimal place, the angles whose tangents are given.

13. $\frac{3}{5}$
14. $\frac{4}{5}$
15. $\frac{1}{2}$
16. $\frac{7}{10}$
17. $\frac{3}{20}$
18. $\frac{5}{4}$
19. $\frac{3}{8}$
20. $2\frac{1}{4}$
21. $\frac{3}{25}$
22. $2\frac{2}{5}$

### Worked example

→ Find the angle whose tangent is $\frac{2}{3}$

$\tan \widehat{A} = \frac{2}{3} = 0.6666...$

$\widehat{A} = 33.69...° = 33.7°$ (correct to 1 d.p.)

💡 Angle A can be found in one step on the calculator. Press

[$\tan^{-1}$] [(] [2] [÷] [3] [)] [=]

# STP Maths 9

Find, correct to 1 decimal place, the angles whose tangents are

**23** $\frac{1}{3}$    **28** $\frac{3}{7}$

**24** $\frac{1}{7}$    **29** $\frac{2}{9}$

**25** $\frac{1}{6}$    **30** $\frac{7}{3}$

**26** $\frac{5}{6}$    **31** $\frac{4}{9}$

**27** $\frac{5}{3}$    **32** $\frac{4}{3}$

## Finding an angle in a right-angled triangle

We can now use tangents of angles to calculate the sizes of the angles in the triangle. The next worked example shows the method.

### Exercise 16g

Give angles correct to 1 decimal place and lengths correct to 3 significant figures.

> **Worked example**
>
> Find angle A in the diagram.
>
>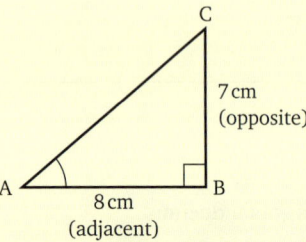
>
> $\tan \widehat{A} = \dfrac{\text{opp}}{\text{adj}} = \dfrac{7}{8}$
>
> $= 0.875$
>
>  On a copy of the diagram, mark the angle to be found. Then label the opposite and adjacent sides to the angle.
>
> $\widehat{A} = 41.18\ldots° = 41.2°$ (correct to 1 d.p.)

Find $\widehat{A}$ in questions **1** to **10**.

**1**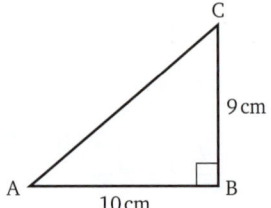

**2**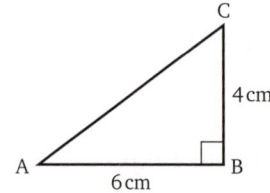

16 Trigonometry: tangent of an angle

**3**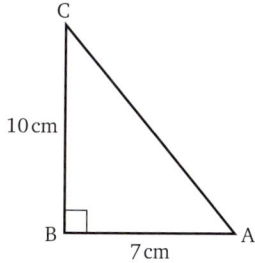

**8**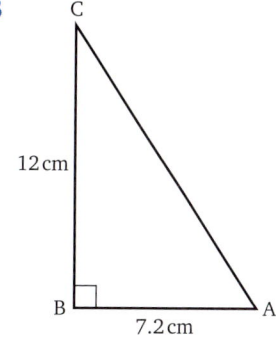

**4**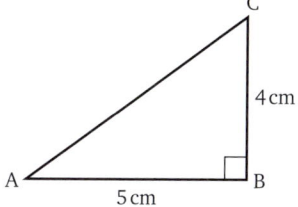

**9**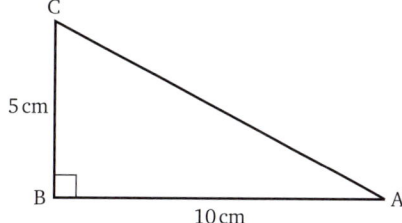

**5**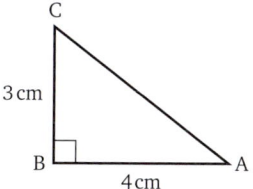

**10**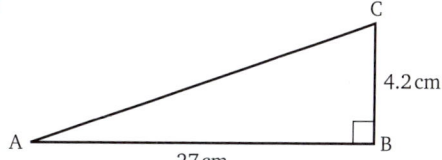

**6**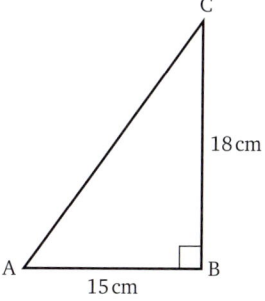

**11** Find $\hat{P}$.
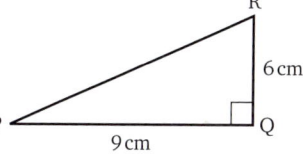

**7**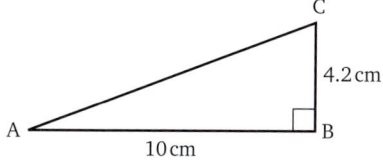

**12** Find $\hat{B}$.
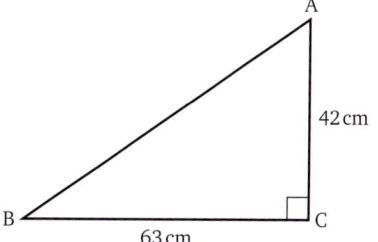

355

**13** Find $\hat{Y}$.

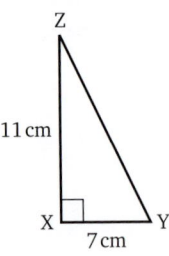

**15** Find $\hat{D}$.

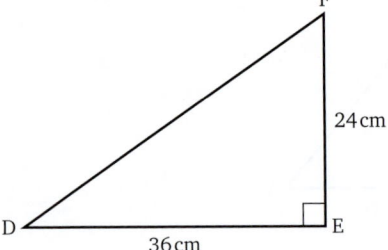

**14** Find $\hat{N}$.

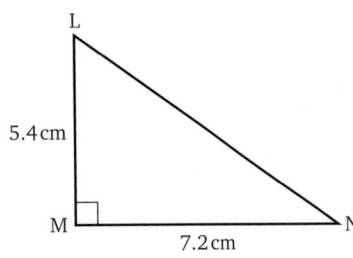

**16** Find $\hat{C}$.

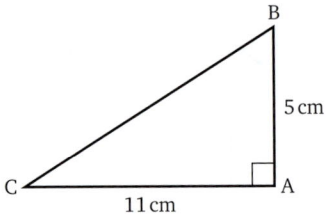

Find $\hat{A}$ in questions **17** to **26**. Turn the page round if necessary before labelling the sides.

**17**

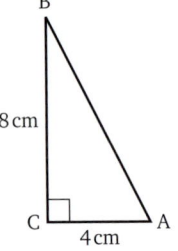

**19**

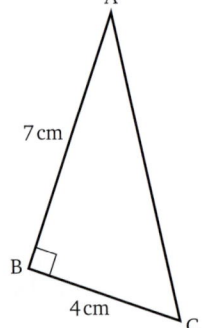

**18**

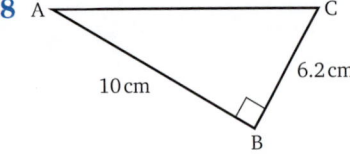

**20**

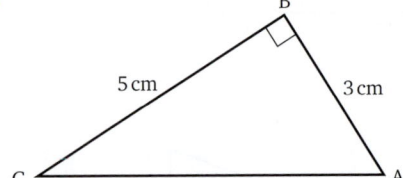

**21**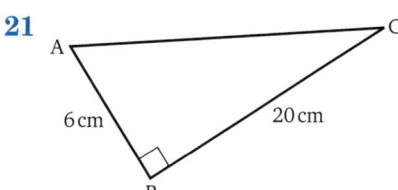

**24**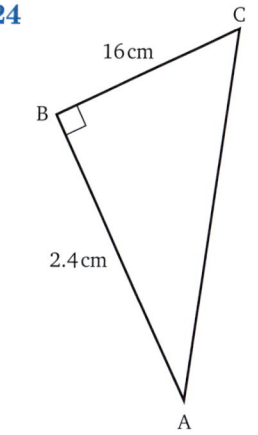

**22**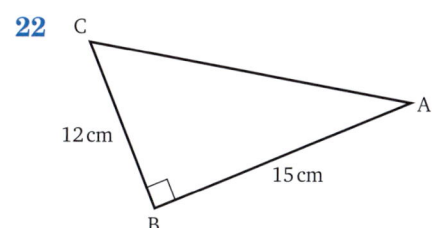

**25**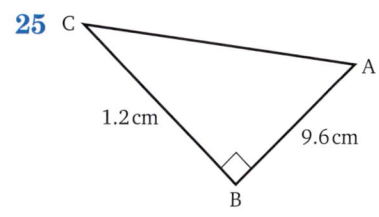

**23**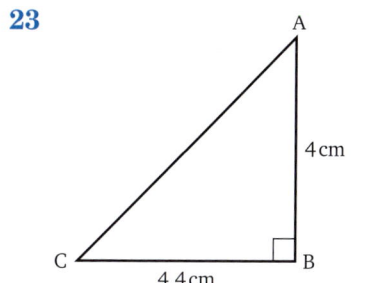

**26**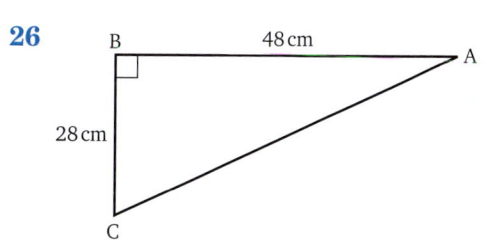

**27** In △ABC, $\hat{B} = 90°$, AB = 12 cm, BC = 11 cm. Find $\hat{A}$.

**28** In △PQR, $\hat{P} = 90°$, PQ = 3.2 m, PR = 2.8 m. Find $\hat{Q}$.

**29** In △DEF, $\hat{D} = 90°$, DE = 108 m, DF = 72 m. Find $\hat{F}$.

If we know two sides in a right-angled triangle, we can use **Pythagoras' theorem** to find the remaining side.

# STP Maths 9

## Worked example

→ A man walks due North for 5 km from A to B, then 4 km due East to C.
  a  What is the bearing of C from A?
  b  How far is he from his starting point?

a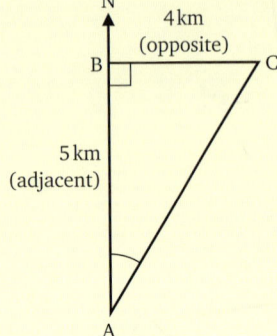

> We start by drawing a diagram showing all the information given. The bearing of C from A is angle A in the triangle, so we mark this angle.

$$\tan \widehat{A} = \frac{\text{opp}}{\text{adj}} = \frac{4}{5}$$
$$= 0.8$$

$\widehat{A} = 38.65...° = 38.7°$ (correct to 1 d.p.)

The bearing of C from A is 038.7°

> Notice that we add '0' to make a three-figure bearing.

b
> The length of AC in the diagram gives the distance required. Tangents will not help to find this, but as we know the other two sides in the right-angled triangle we can use Pythagoras' theorem.

$AC^2 = AB^2 + BC^2$ (Pythagoras)
$\phantom{AC^2} = 25 + 16 = 41$
$AC = \sqrt{41} = 6.403...$

The man is 6.40 km from his starting point (correct to 3 s.f.).

30  ABCD is a rectangle. AB = 60 m and BC = 36 m. Find
  a  the angle between the diagonal and the side AB
  b  the length of the diagonal AC.

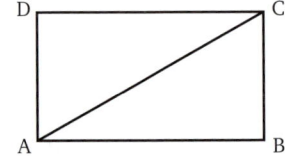

31  A flagpole PQ is 10 m high. R is a point on the ground 20 m from the foot of the pole. Find the angle of elevation of the top of the pole from R (that is, $\widehat{R}$).

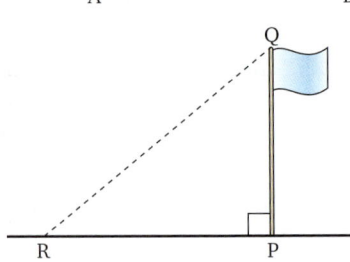

32  In △ABC, AB = BC, AC = 12 cm. D is the midpoint of AC. The height BD of the triangle is 10 cm. Find
  a  $\widehat{C}$ and the other angles of the triangle
  b  the lengths of AB and BC.

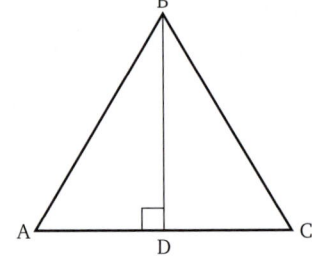

**33** A ladder leans against a vertical wall. Its top, Q, is 3 m above the ground and its foot, P, is 2 m from the foot of the wall. Find the angle of slope of the ladder (that is, $\hat{P}$).

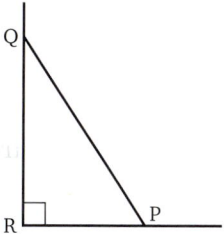

**34** The bearing of town A from town B is 032.4°. A is 16 km North of B.

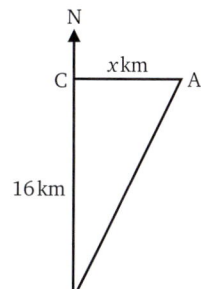

   **a** How far East of B is A?
   **b** How far is A from B?

**35** In a square ABCD of side 8 cm, A is joined to the midpoint E of BC. Find $E\hat{A}B$, $C\hat{A}B$ and $C\hat{A}E$. Notice that AE does *not* bisect $C\hat{A}B$.

**36** A ladder leans against a vertical wall. It makes an angle of 72° with the horizontal ground and its foot is 1 m from the foot of the wall. How high up the wall does the ladder reach?

**37** Sketch axes for $x$ and $y$ from 0 to 5. A is the point (1, 0) and B is (5, 2). What angle does the line AB make with the $x$-axis?

**38** In a rhombus, the two diagonals are of lengths 6.2 cm and 8 cm. Find   **a**  the angles of the rhombus
           **b**  the lengths of the sides of the rhombus.

**39** In △ABC, AB = BC, CA = 10 cm and $\hat{C}$ = 72°. Find the height BD of the triangle.

**40** The diagram shows a section through an underground railway tunnel.

The floor of the tunnel is 2 m wide and the angle of elevation of the highest point, C, of the tunnel from the edge, A, of the floor is 69°.

   **a** Draw a diagram showing just the inside wall and floor of the tunnel. Mark the points A, B, C and O, which is the centre of the circle, part of which forms the inside wall.
   **b** Calculate the height of the tunnel.

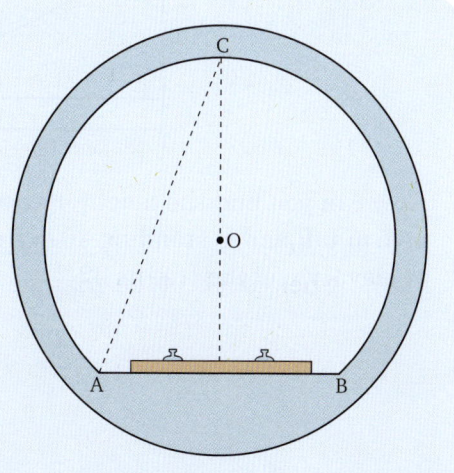

41

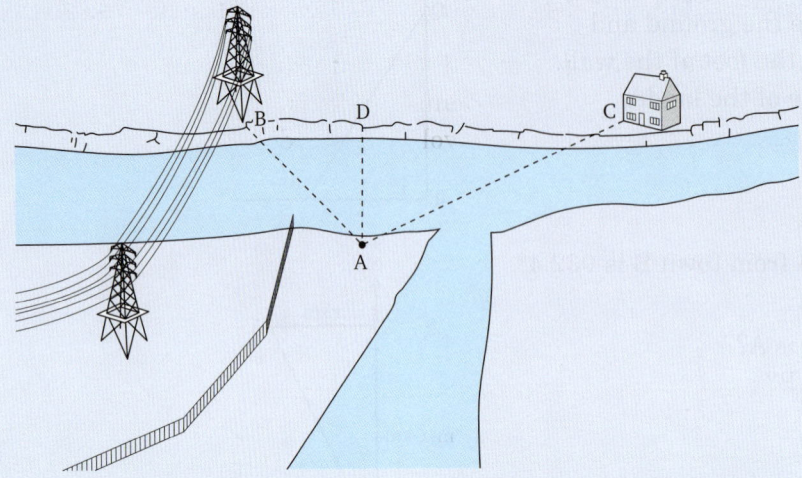

Sula is at point A on the bank of the river and needs to find the distance between the pylon at B and the house at C, both of which are on the opposite bank. She cannot cross the rivers nor can she stand directly opposite the pylon at B.

From a map, she knows that the width, AD, of the river is 60 metres. She measures the angle BAD as 35° and the angle DAC as 58°. Calculate the distance between the pylon and the house. Assume that BDC is a straight line.

## Consider again

Sam wants to know the height of this tree. He can find it by measuring how far he is from the tree and then measuring the angle of elevation of the top of the tree. Showing this information in a diagram gives this right-angled triangle.

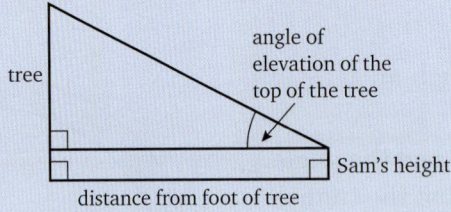

Now can you find the height of the tree without making a scale drawing, given that Sam is 1.75 m tall and is standing 40 m from the base of the tree and that he measures the angle of elevation of the tree as 32°?

# 17 Sine and cosine of an angle

The tangent of an angle is useful when the opposite and adjacent sides of a right-angled triangle are involved, but there are many problems where the hypotenuse and one of the other sides are involved instead.

## Consider

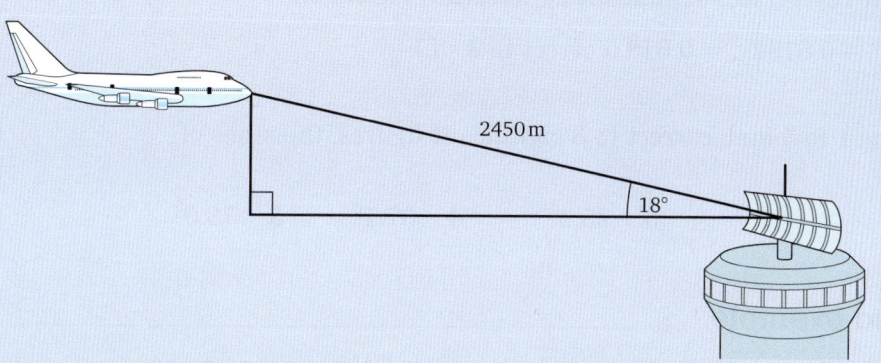

The reading from a radar instrument shows that a plane is 2450 metres from the radar and at an angle of elevation of 18°. Find the height at which the plane is flying above the radar and its horizontal distance from the radar tower.

*You should be able to answer this question after you have worked through this chapter.*

## Sine of an angle

In a right-angled triangle, the ratio of the side opposite an angle to the hypotenuse is called the **sine of the angle**.

In this diagram

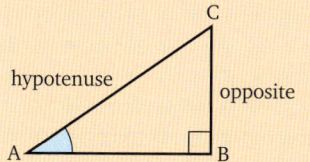

sine of angle A = $\dfrac{\text{opposite}}{\text{hypotenuse}} = \dfrac{CB}{AC}$

or, briefly, $\sin \widehat{A} = \dfrac{\text{opp}}{\text{hyp}} = \dfrac{CB}{AC}$

All right-angled triangles containing, say, an angle of 40° are similar, so the sine ratio, $\dfrac{\text{opp}}{\text{hyp}}$, always has the same value.

The value of the sine ratio of every acute angle is stored in scientific calculators and methods of calculation are similar to those involving tangents.

361

# Exercise 17a

### Worked examples

→ Find, correct to 3 significant figures, the sine of 72°.

sin 72° = 0.951 (correct to 3 s.f.)

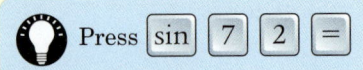 Press sin 7 2 =

→ Find, correct to 3 significant figures, the sine of 38.2°.

sin 38.2° = 0.6184... = 0.618 (correct to 3 s.f.)

In questions **1** to **5** find, correct to 3 significant figures, the sines of the angles.

**1**  26°   **2**  84°   **3**  25.4°   **4**  37.1°   **5**  78.9°

### Worked example

→ Find, correct to 1 decimal place, the angle whose sine is 0.909

sin  = 0.909

 = 65.36...°

 = 65.4° (correct to 1 d.p.)

Press shift sin 0 . 9 0 9 =
(The shift button accesses the sin$^{-1}$ function, which means 'the angle whose sin is'.)

In questions **6** to **10**, find, correct to 1 decimal place, the angles whose sines are given.

**6**  0.834   **7**  0.413   **8**  0.639   **9**  0.704   **10**  0.937

## Using the sine ratio

We can use the sine ratio, in a similar manner to the tangent ratio, to find angles and sides in right-angled triangles.

# Exercise 17b

### Worked example

→ Find the length of .

On a copy of the diagram, label the sides in relation to the angle.

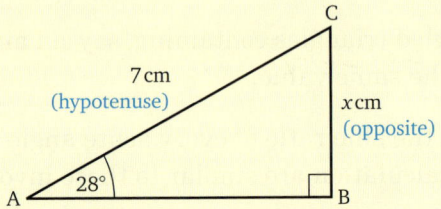

## 17 Sine and cosine of an angle

$\sin 28° = \dfrac{\text{opp}}{\text{hyp}} = \dfrac{x}{7}$

 We write the equation the other way round, as it is easier to handle when the term containing $x$ is on the LHS.

$\dfrac{x}{7} = 0.4694...$

$x = 7 \times 0.4694... = 3.286...$

BC = 3.29 cm (correct to 3 s.f.)

In questions **1** to **10**, give answers correct to 3 significant figures.

**1** Find BC.

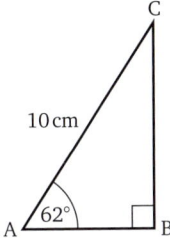

**2** Find BC.

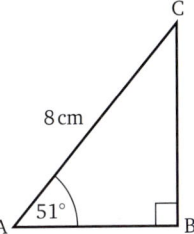

**3** Find AC.

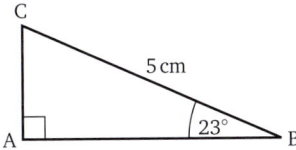

**4** Find BC.

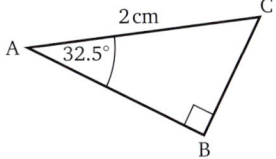

**5** Find QR.

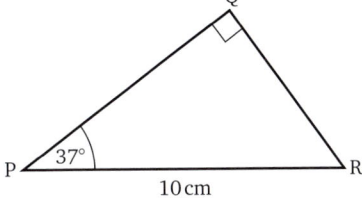

**6** Find RQ.

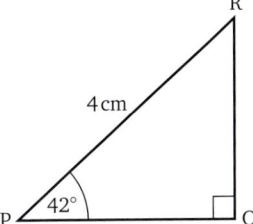

**7** Find PQ.

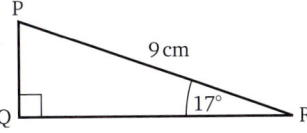

**8** Find XY.

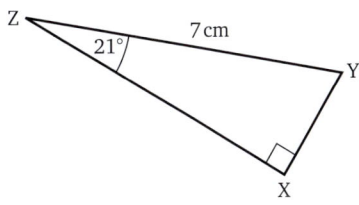

**9** Find LM.

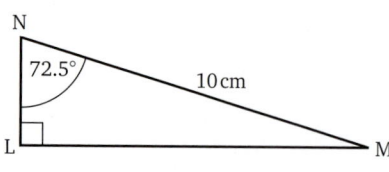

**10** Find PQ.

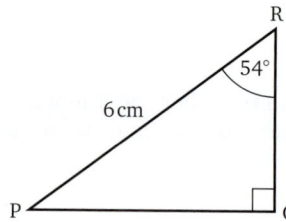

## Worked example

In △ABC, $\hat{B} = 90°$, AC = 4 cm and BC = 3 cm. Find $\hat{A}$.

 Draw and letter the triangle then label the sides in relation to $\hat{A}$.

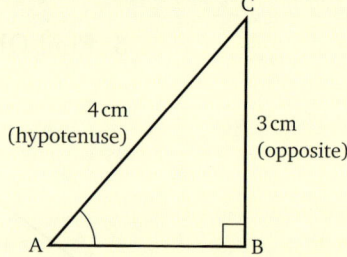

$\sin \hat{A} = \dfrac{\text{opp}}{\text{hyp}} = \dfrac{3}{4}$

$\phantom{\sin \hat{A}} = 0.75$

$\hat{A} = 48.59...°$

$\phantom{\hat{A}} = 48.6°$ (correct to 1 d.p.)

In questions **11** to **20**, give angles correct to 1 decimal place.

**11** Find $\hat{A}$.

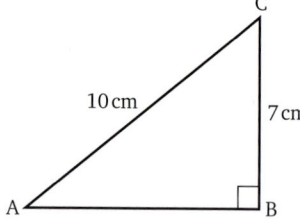

**13** Find $\hat{P}$.

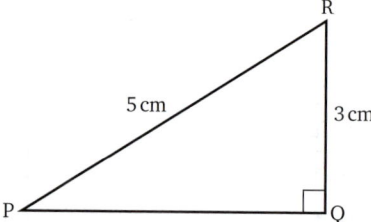

**12** Find $\hat{A}$.

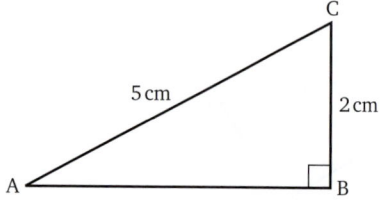

**14** Find $\hat{Q}$.

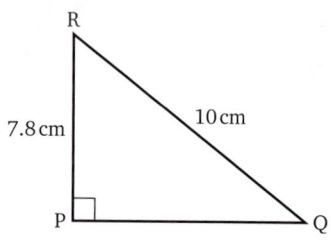

**15** Find $\hat{Y}$.

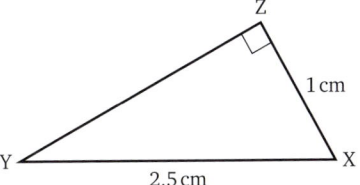

**17** Find $\hat{M}$.

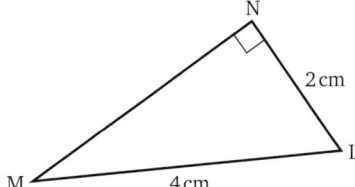

**16** Find $\hat{P}$.

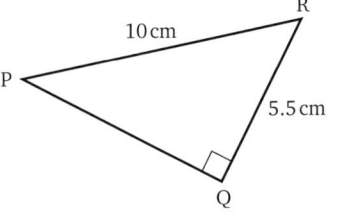

**18** Find $\hat{A}$.

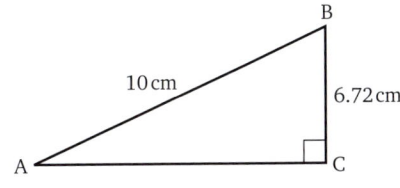

**19** In $\triangle ABC$, $\hat{B} = 90°$, $\hat{C} = 36°$ and $AC = 3.5$ cm. Find $AB$.

**20** In $\triangle PQR$, $\hat{Q} = 90°$, $PQ = 2.6$ cm and $PR = 5.5$ cm. Find $\hat{R}$.

## Cosine of an angle

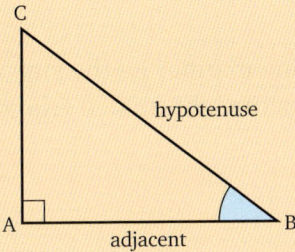

If we are given the adjacent side and the hypotenuse, then we can use a third ratio, $\dfrac{\text{adjacent side}}{\text{hypotenuse}}$.

This is called the **cosine of the angle** (cos for short).

$$\cos \hat{B} = \frac{\text{adj}}{\text{hyp}} = \frac{AB}{BC}$$

Cosines of acute angles are stored in scientific calculators.

### Exercise 17c

**Worked examples**

→ Find, correct to 3 significant figures, the cosine of 41°.

cos 41° = 0.755

→ Find, correct to 3 significant figures, the cosine of 28.7°.

cos 28.7° = 0.877

Find, correct to 3 significant figures, the cosines of the following angles.

**1** 59°   **2** 48°   **3** 4°   **4** 44.9°   **5** 60.1°

**Worked example**

→ Find, correct to 1 decimal place, the angle whose cosine is 0.493

cos $\hat{A}$ = 0.493
$\hat{A}$ = 60.46...° = 60.5° (correct to 1 d.p.)

In questions **6** to **10**, cos $\hat{A}$ is given. Find $\hat{A}$ correct to 1 decimal place.

**6** 0.435   **7** 0.943   **8** 0.012   **9** 0.7   **10** 0.24

## Using the cosine ratio

We can use the cosine ratio in the same manner as the sine and tangent ratios to find angles and sides in right-angled triangles.

### Exercise 17d

**Worked example**

→ In △ABC, $\hat{B}$ = 90°, $\hat{A}$ = 28° and AC = 9 cm. Find AB.

 Draw the triangle and label the sides.

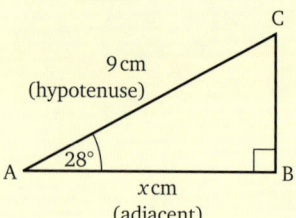

$\dfrac{x}{9} = \dfrac{\text{adj}}{\text{hyp}} = \cos 28°$

$\dfrac{x}{9} = 0.8829...$

$\cancel{9}^1 \times \dfrac{x}{\cancel{9}_1} = 0.8829... \times 9$

$x = 7.946...$

AB = 7.95 cm (correct to 3 s.f.)

## 17 Sine and cosine of an angle

In questions **1** to **8**, find the required lengths, correct to 3 significant figures.

**1** Find AB.

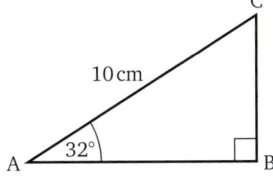

**2** Find PQ.

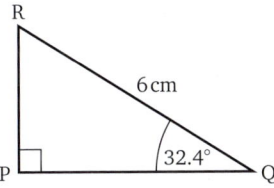

**3** Find PQ.

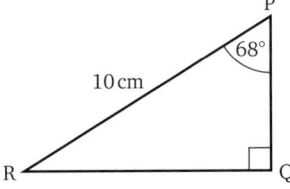

**4** Find AB.

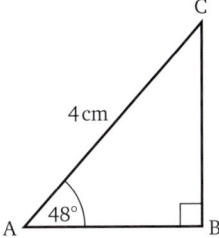

**5** Find AC.

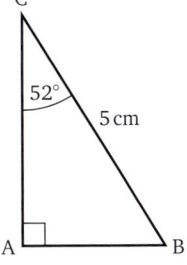

**6** Find XZ.

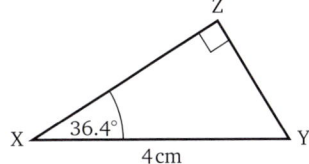

**7** Find PQ.

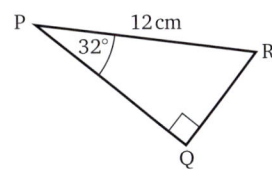

**8** Find PR.

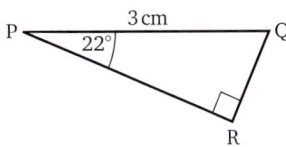

### Worked example

→ In △ABC, $\widehat{B} = 90°$, AB = 4 cm and AC = 6 cm. Find $\widehat{A}$.

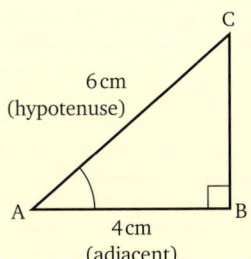

$\cos \widehat{A} = \dfrac{\text{adj}}{\text{hyp}} = \dfrac{4}{6} = 0.6666...$

$\widehat{A} = 48.18...°$
$\phantom{\widehat{A}} = 48.2$ (correct to 1 d.p.)

367

In questions **9** to **16**, give angles correct to 1 decimal place.

**9** Find $\hat{A}$.

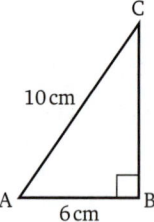

**10** Find $\hat{A}$.

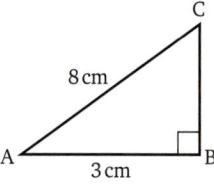

**11** Find $\hat{Y}$.

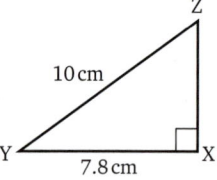

**12** Find $\hat{R}$.

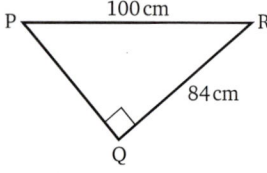

**13** Find $\hat{A}$.

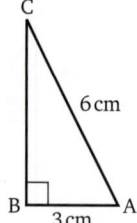

**14** Find $\hat{Q}$.

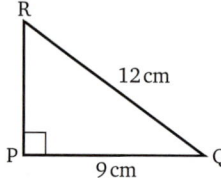

**15** Find $\hat{C}$.

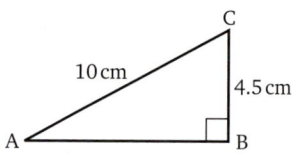

**16** Find $\hat{X}$.

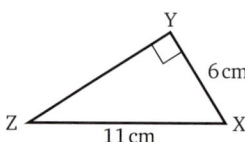

## Use of all three ratios

To remember which ratio is called by which name, some people use the word SOHCAHTOA.

$$\text{Sin } \hat{A} = \frac{\textbf{O}\text{pposite}}{\textbf{H}\text{ypotenuse}} \quad \textbf{SOH}$$

$$\text{Cos } \hat{A} = \frac{\textbf{A}\text{djacent}}{\textbf{H}\text{ypotenuse}} \quad \textbf{CAH}$$

$$\text{Tan } \hat{A} = \frac{\textbf{O}\text{pposite}}{\textbf{A}\text{djacent}} \quad \textbf{TOA}$$

# 17 Sine and cosine of an angle

## Exercise 17e

### Worked example

→ State whether sine, cosine or tangent should be used for the calculation of the marked angle.

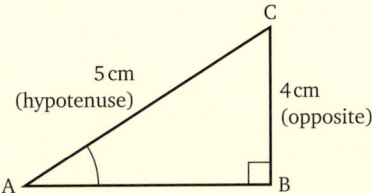

The opposite side and the hypotenuse are given so we should use sin $\hat{A}$.

In questions **1** to **6**, label the sides whose lengths are known, as 'hypotenuse', 'opposite' or 'adjacent'. Then state whether sine, cosine or tangent should be used for the calculation of the marked angle.

**1**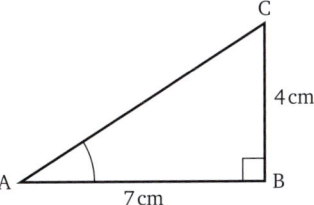

**4**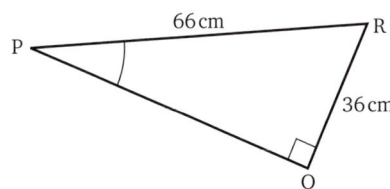

**2**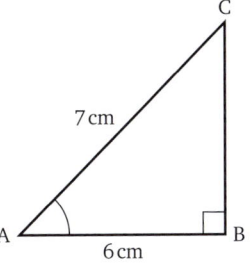

**5**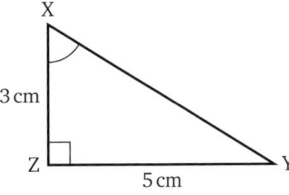

**3**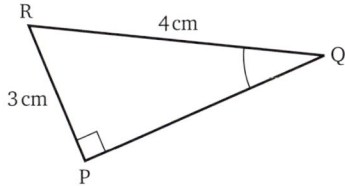

**6**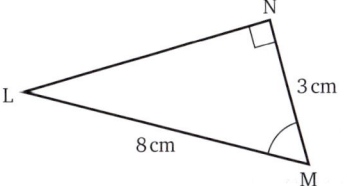

# STP Maths 9

> **Worked example**
>
> → State whether sine, cosine or tangent should be used for the calculation to find $x$.
>
>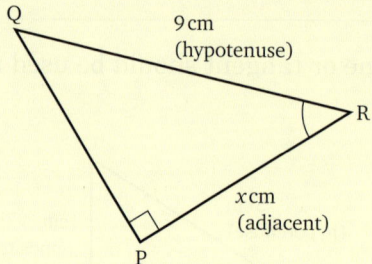
>
> We are given the hypotenuse and wish to find the adjacent side, so we should use $\cos \hat{R}$.

In questions **7** to **12**, use 'opposite', 'adjacent' or 'hypotenuse' to label the side whose length is given and the side whose length is to be found. Then state whether sine, cosine or tangent should be used for the calculation to find $x$.

**7**

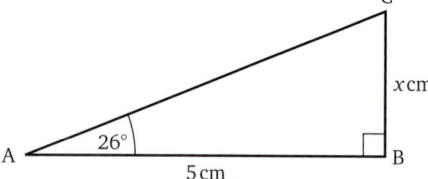

**10**

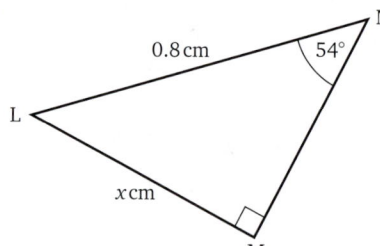

**8**

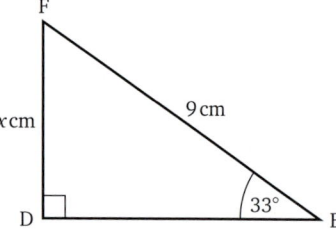

**11**

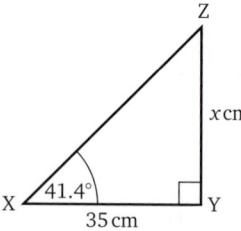

**9**

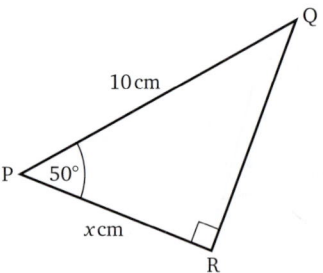

**12**

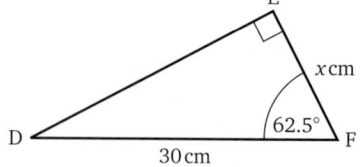

370

## 17 Sine and cosine of an angle

In questions **13** to **18**, find the marked angle correct to 1 decimal place.

**13**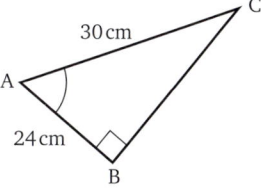

**16**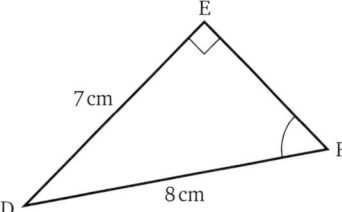

**14**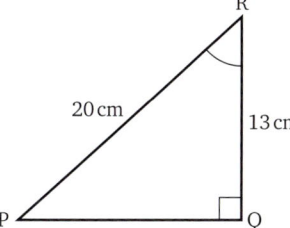

**17**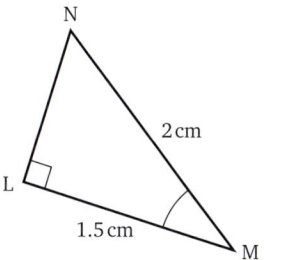

**15**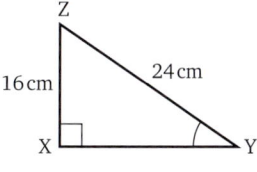

**18**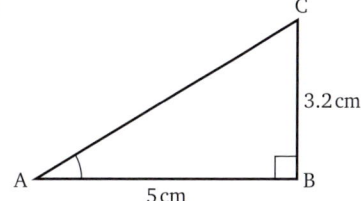

In questions **19** to **24**, find the length of the side marked $x$ cm, giving the answer correct to 3 significant figures.

**19**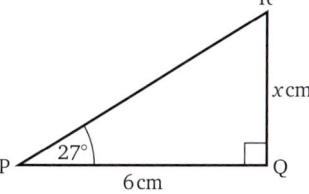

**22**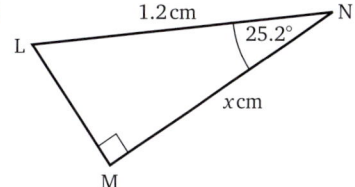

**20**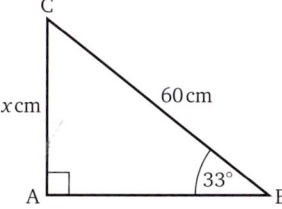

**23**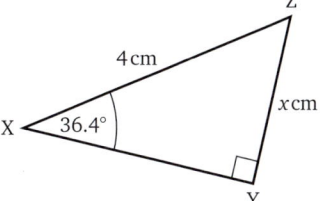

**21**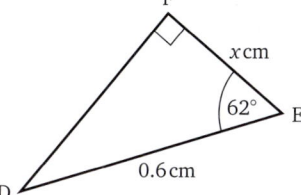

**24**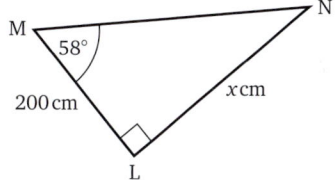

STP Maths 9

## Worked example

→ In △ABC, $\hat{B} = 90°$, AC = 15 cm and BC = 11 cm.
Find $\hat{A}$, then $\hat{C}$.

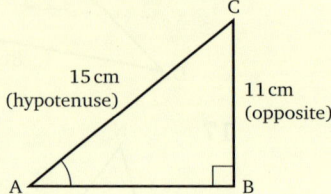

 We are given the hypotenuse and the side opposite to $\hat{A}$ so we use sin $\hat{A}$.

$\sin \hat{A} = \dfrac{\text{opp}}{\text{hyp}} = \dfrac{11}{15}$

$= 0.7333...$

$\hat{A} = 47.16...° = 47.2°$ (correct to 1 d.p.)

$\hat{C} = 90° - 47.2°$
$= 42.8°$

 We know that the three angles of the triangle add up to 180°, so we can use this to find $\hat{C}$.

In questions **25** to **32**, give angles correct to 1 decimal place and lengths correct to 3 significant figures.

**25** Find $\hat{A}$, then $\hat{C}$.

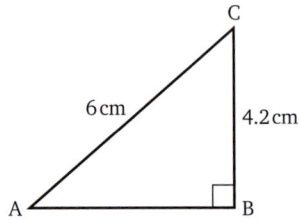

**27** Find $\hat{X}$, then $\hat{Z}$.

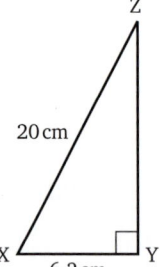

**26** Find AC.

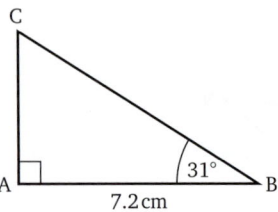

**28** Find PQ.

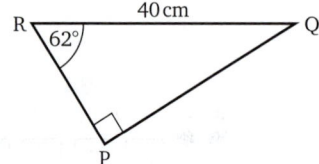

**17** Sine and cosine of an angle

**29** Find $\hat{C}$, then AB.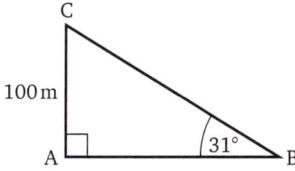

**30** Find AB.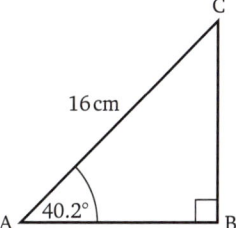

**31** Find XZ.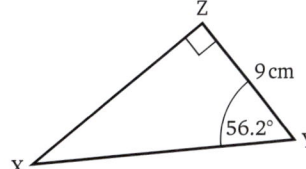

**32** Find $\hat{A}$.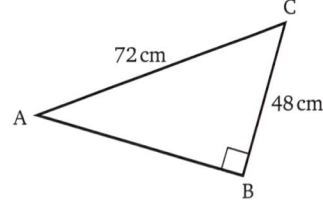

## Finding the hypotenuse

So far when finding the length of a side, we have been able to form an equation in which the unknown length is on the top of a fraction. This is not possible when the hypotenuse is to be found, and the equation we form takes slightly longer to solve.

### Exercise 17f

**Worked example**

→ In triangle ABC, $\hat{B} = 90°$, $\hat{C} = 62°$ and AB = 80 mm.
Find AC.

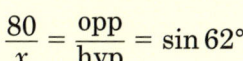

$\dfrac{80}{x} = \dfrac{\text{opp}}{\text{hyp}} = \sin 62°$

$\dfrac{80}{x} = \sin 62°$

$80 = \sin 62° \times x$    💡 Multiply both sides by $x$.

$x = \dfrac{80}{\sin 62°} = 90.60...$    💡 Divide both sides by $sin 62°$. Press

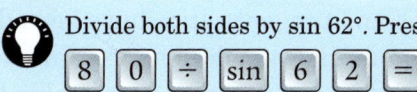

AC = 90.6 mm (correct to 3 s.f.)

In questions **1** to **6**, use the information given in the diagram to find the hypotenuse, correct to 3 significant figures.

**1**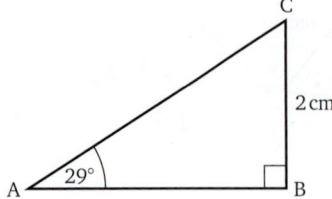

**4**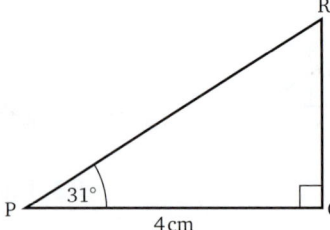

**2**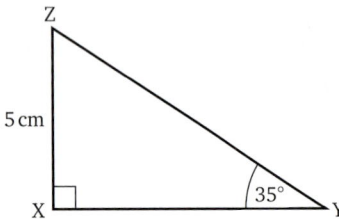

**5**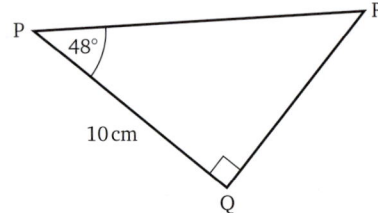

**3**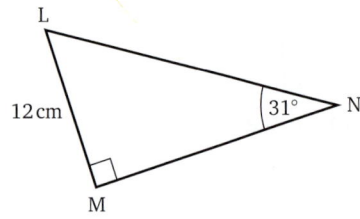

**6**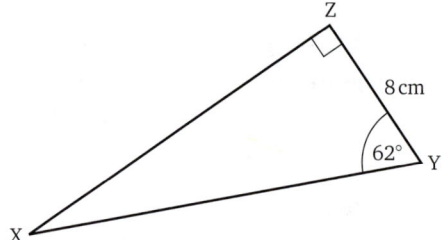

**7** In triangle ABC, $\hat{B} = 90°$, $\hat{A} = 43°$ and BC = 3 cm. Find AC.

**8** In triangle PQR, $\hat{P} = 90°$, $\hat{Q} = 28°$ and PR = 7 cm. Find QR.

**9** In triangle LMN, $\hat{L} = 90°$, $\hat{M} = 14°$ and LN = 8 cm. Find MN.

**10** In triangle XYZ, $\hat{Z} = 90°$, $\hat{Y} = 62°$ and ZY = 20 cm. Find XY.

## Applications

Trigonometry has many applications. These range from finding unknown angles and lengths in geometrical figures to solving problems in surveying and navigation. **Exercise 17g** gives a variety of problems that involve finding angles or sides.

Remember that the sine, cosine and tangent ratios have been defined in right-angled triangles, so if your diagram does not contain a right-angled triangle, you will have to find one by adding a suitable line.

Remember also that we can use Pythagoras' theorem in a right-angled triangle to find one side when we know the other two sides.

# 17 Sine and cosine of an angle

## Exercise 17g

### Worked example

→ In an isosceles triangle PQR, PQ = QR = 5 cm and PR = 6 cm. Find the angles of the triangle.

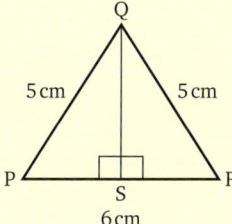

By dividing the triangle down the middle with the line QS, we create two identical right-angled triangles. We can then draw one of these triangles to work with.

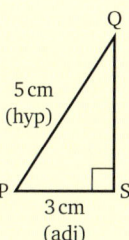

$\cos \widehat{P} = \dfrac{\text{adj}}{\text{hyp}} = \dfrac{PS}{PQ} = \dfrac{3}{5}$

PS = $\tfrac{1}{2}$PR = 3 cm

$\widehat{P} = 53.1°$ (correct to 1 d.p.)

$\widehat{R} = 53.1°$ (isosceles △; base angles equal)

$P\widehat{Q}R = 73.8°$ (angles of a △ add up to 180°)

Unless otherwise stated, give angles correct to 1 decimal place and lengths correct to 3 significant figures.

1  In △ABC, AC = CB = 10 m and $\widehat{A}$ = 64°.
   Find  **a**  the height of the triangle
         **b**  the length of AB
         **c**  the area of the triangle.

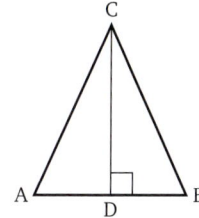

2  An aeroplane climbs at a steady angle. When it has been climbing for 2 minutes, the cockpit instruments show that it has reached an altitude of 1000 m and has travelled a distance of 3 km. At what angle is the plane climbing?

3  An observer at the top of the lighthouse measures the angle of depression of a yacht as 15°. The lighthouse is 50 m high. How far is the yacht from the base of the lighthouse?

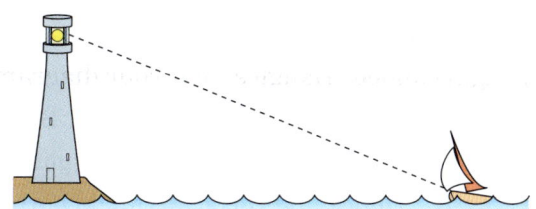

**4** 

The diagram shows an office tower which is 100 m tall (to the top of the roof). The tower is on one bank of a river. Pedro is standing on the opposite bank, 150 m away from the base of the tower. What is the angle of elevation of the top of the tower from Pedro? (Neglect Pedro's height.)

**5** Wei walks 600 m along a road that slopes up uniformly at 5° to the horizontal. How far, to the nearest metre, has she risen?

**6** A ski run is 1500 m long and slopes uniformly at 8° to the horizontal. How far, to the nearest metre, will a skier descend vertically when making this run?

**7** An escalator is to be installed in a shopping mall to raise shoppers through a height of 8 metres. The escalator must be inclined at 20° to the horizontal.
   **a** How long must the escalator be?
   **b** What is the least horizontal distance needed to install it?

**8** A treasure hunter locates two large trees, A and B, which are 250 m apart. He then tries to find a point C such that angle ABC is 90° and angle BAC is 64°.

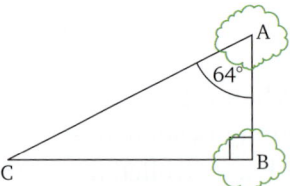

   **a** How far is C from B?
   **b** If he makes an error in measuring $B\hat{A}C$, using 63° instead of 64°, and sets out to walk from A to C, how far will he be from the true position of C when he thinks he has arrived? Give your answer to the nearest metre.

**9** ABCD is a rectangle in which AB = 360 mm and BC = 184 mm.
   **a** Find the angle that the diagonal AC makes with the side BC.
   **b** Hence find the acute angle between the diagonals.

10  The drawing shows one of the series of roof trusses which are to be made to construct a factory roof.

Each length in the truss is to be made from angle iron. What length of angle iron is required to make one truss?

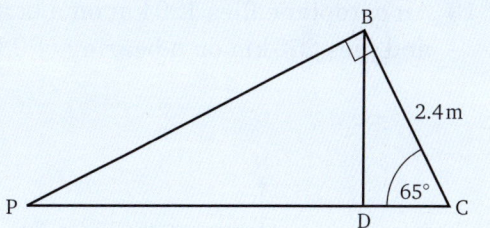

11  The diagram shows a rectangular envelope viewed from the back with the flap opened flat.

How far is B from F when the envelope is sealed?

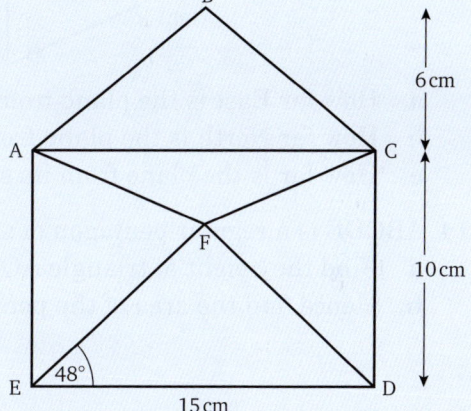

12  The diagram shows a castle tower and a moat. The ground beyond the moat is horizontal.

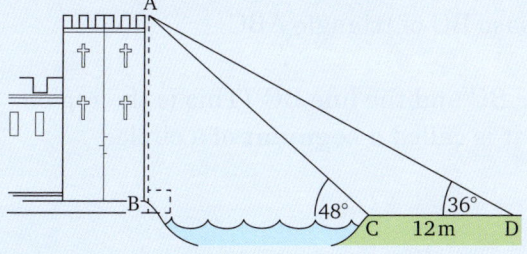

**a**  Draw a diagram showing just triangles ABC and ABD. Mark the given information on your diagram and mark AB as $x$ m and BC as $y$ m.

**b**  Using triangle ABC, find an equation relating $x$ and $y$. Express the equation in the form $x = ...$

**c**  Using triangle ABD, find another equation relating $x$ and $y$. Express this equation also in the form $x = ...$

**d**  Use the equations you found in parts **b** and **c** to form an equation in $y$ only. Solve this equation and hence find the height of the tower and the width of the moat.

**13** An aeroplane flies 120 km on a bearing of 130°. It then alters course and flies 250 km on a bearing of 035°.

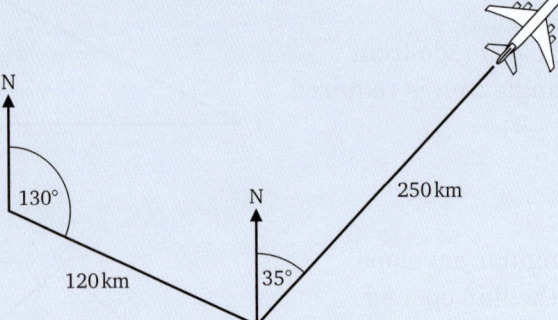

a  How far East is the plane from its starting point?
b  How far North is the plane from its starting point?
c  How far is the plane from its starting point?

**14** ABCDE is a regular pentagon of side 10 cm.
a  Find the height of triangle AOB.
b  Hence find the area of the pentagon.

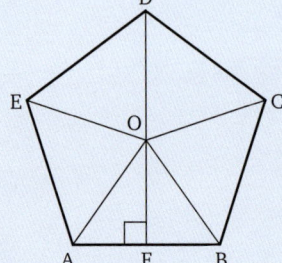

**15** ACB is a sector of a circle of radius 12 cm. The arc CB **subtends** an angle of 28° at the centre A of the circle. Find
a  the area of the sector
b  the distance of A from the base BC of triangle ABC
c  the area of triangle ABC
d  the area enclosed by the arc BC and the line BC. (This is the region shaded in the diagram and it is called a **segment** of a circle.)

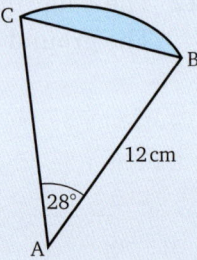

## Consider again

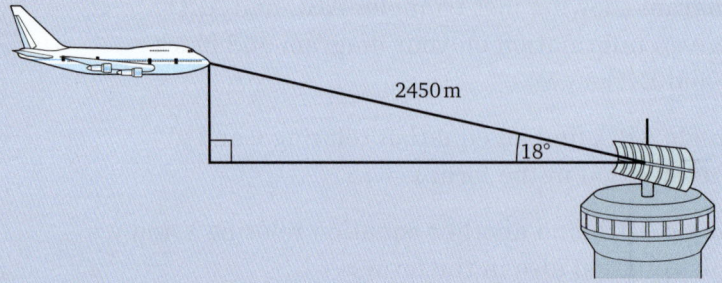

The reading from a radar instrument shows that a plane is 2450 metres from the radar and at an angle of elevation of 18°. Now can you find the height at which the plane is flying above the radar and its horizontal distance from the radar tower?

# 18 Problems in three dimensions

## Consider

The diagram shows a solid that consists of a prism surmounted by a pyramid.

The cross-section of the prism is an equilateral triangle.

The height of the prism is 3 cm and the sloping edges of the pyramid are 6 cm long.

Find    **a**  the surface area of the solid

          **b**  the height of the solid.

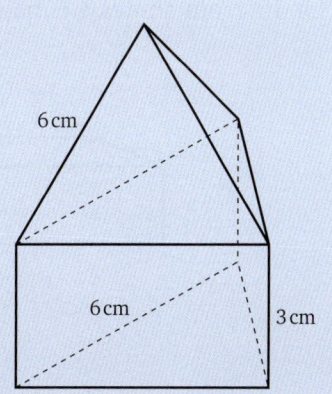

*You should be able to answer these questions after you have worked through this chapter.*

## Solids

A solid is a three-dimensional object.

The faces of a solid may be flat or may be curved.

The edges of a solid are where faces meet.

The vertices of a solid are where edges meet.

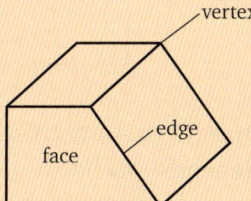

## Cubes, cuboids and prisms

A cube has six faces, each of which is a square.

The diagram shows a cube and a net to make the cube.

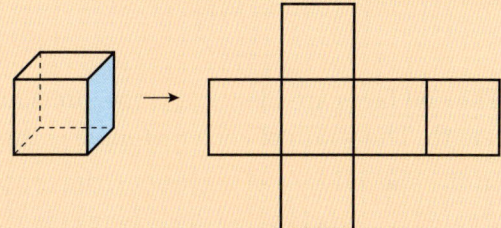

If the length of an edge is $x$ cm, the surface area of the cube is $6x^2$ cm$^2$.

The angle between any two adjacent edges is 90°.

The angle between an edge and the diagonal of a face perpendicular to that edge is also 90°.

So the triangle shown in the diagram has a right angle at D.

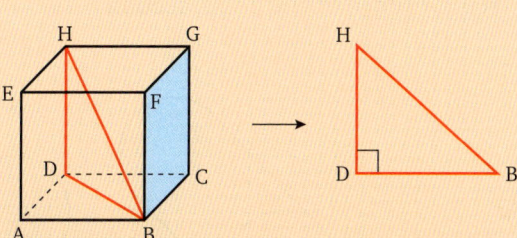

A cuboid also has six faces, each of which is a rectangle. Opposite faces are the same. (Two faces may be squares.)

The diagram shows a cuboid and a net to make it.

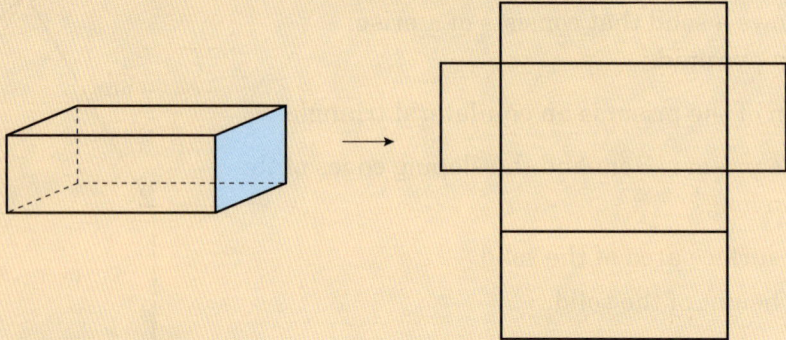

Like a cube, the angle between any two adjacent edges is 90°, so the angle between an edge and the diagonal of a face perpendicular to that edge is also 90°.

Therefore, the triangle shown in the diagram below has a right angle at D.

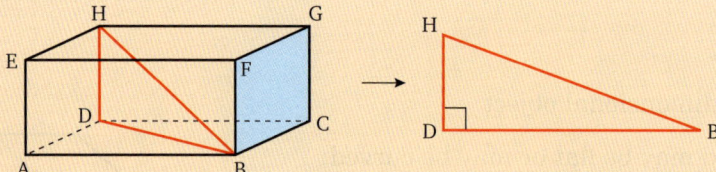

A prism has a constant cross-section. The number of faces depends on the shape of the cross-section.

The diagram shows a prism whose cross-section is a trapezium.

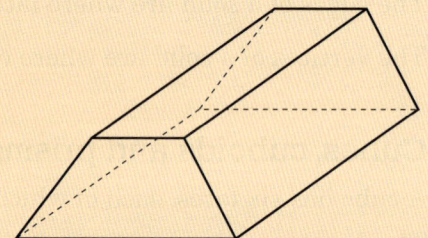

The end faces are identical trapeziums and all the other faces are rectangles of the same length. The width of each rectangle is the same as the edge of the trapezium it is attached to.

In the special case of a cylinder, the cross-section is a circle. The curved surface can be treated as a rectangle whose length is the length of the cylinder and whose width is the circumference of the cross-section.

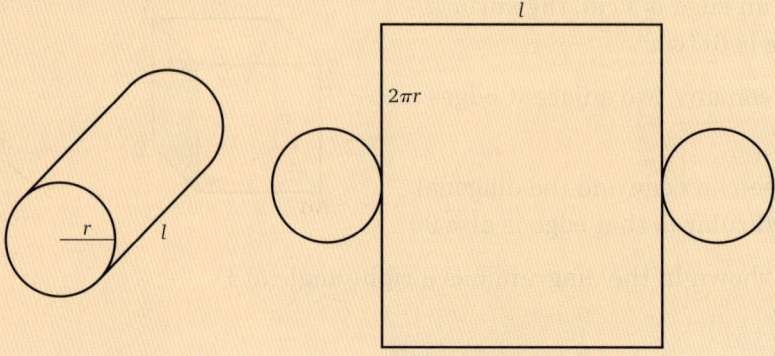

## 18 Problems in three dimensions

### Exercise 18a

1. Find the surface area of each of the following cuboids.

   a 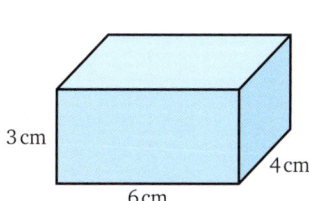   b 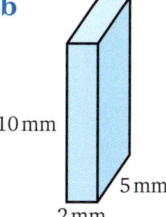   c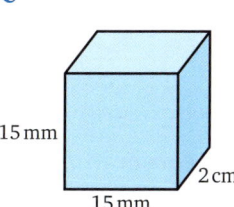

2. Find the area of card used to make this open rectangular (cuboid) box.

   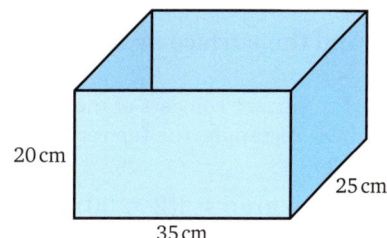

3. Find the surface area of each prism. (The measurements are in centimetres.)

   a Give your answer correct to 3 s.f.

   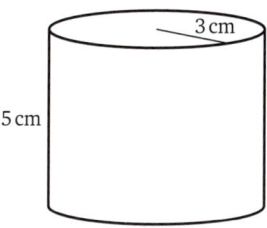

   b

   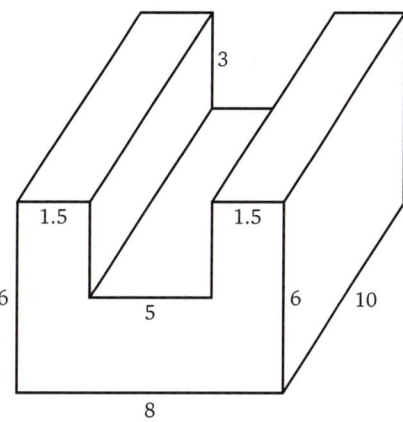

### Worked example

The cross-section of this prism is a trapezium whose non-parallel sides are the same length.

→ Find the area of the cross-section of the prism.

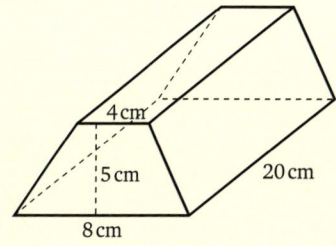

The area of the cross-section is the area of a trapezium
$= \frac{1}{2}(8 + 4) \times 5 \text{ cm}^2 = 30 \text{ cm}^2$

381

→ Find the length of the equal sides of the cross-section of the prism.

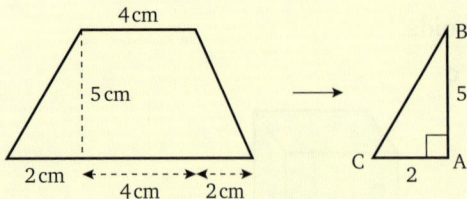

💡 Draw the cross-section separately and use it to identify a triangle that can be used to find the length.

Using Pythagoras' theorem in △ABC gives

$BC^2 = 2^2 + 5^2 = 29$

∴ $BC = \sqrt{29}$ cm = 5.39 cm (correct to 3 s.f.)

→ Find the surface area of the prism.

💡 The surface consists of the two end cross-sections together with the base rectangle, the top rectangle and the two equal side rectangles.

The surface area = $\{(2 \times 30) + (8 \times 20) + (4 \times 20) + (2 \times \sqrt{29} \times 20)\}$ cm²

$= 300 + 40\sqrt{29}$ cm² = 515 cm² (correct to 3 s.f.)

💡 Notice that we used the exact value for the length of BC in the calculation rather than the corrected length. This avoids compounding errors that can creep in when using corrected numbers.

 **4** The cross-section of this prism is a right-angled isosceles triangle.
Find the surface area of the prism.

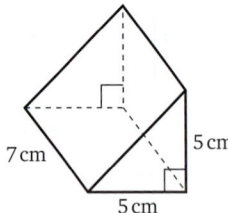

 **5** The cross-section of this prism is an equilateral triangle.
Find the surface area of the prism.

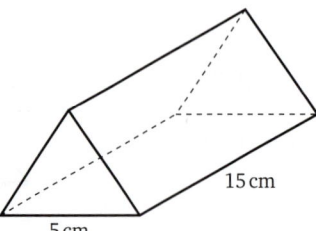

**6** The cross-section of this prism is a square surmounted by an equilateral triangle.
Find the surface area of the prism.

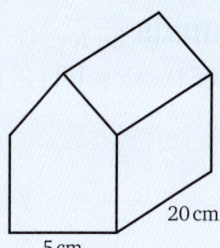

**7** The cross-section of this prism is a semicircle of radius 6 cm.
The prism is 20 cm long.
Find its surface area.

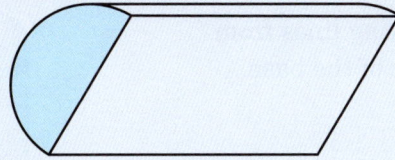

**8** A box in the shape of a prism is to be covered in leather.
The cross-section is a square with a corner cut off.
Find the area of leather required.
(The measurements are in centimetres.)

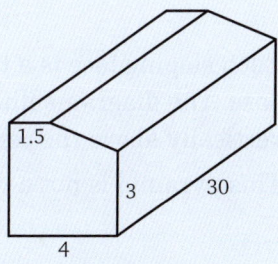

**9** This prism is 20 cm long and its cross-section is a regular hexagon of side 5 cm.
Find the surface area of the prism.

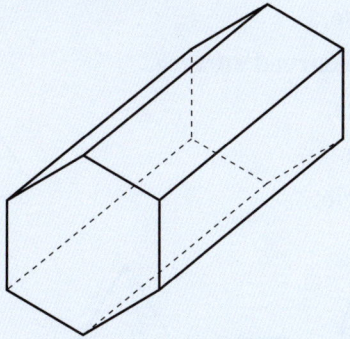

**10** An open box is a cube whose edges are 20 cm long.
Find the length of longest stick that will fit inside the box.

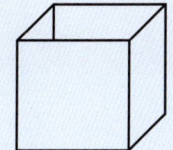

## Pyramids

Each of these solids is a **pyramid**.

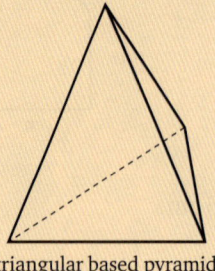
triangular based pyramid
(tetrahedron)

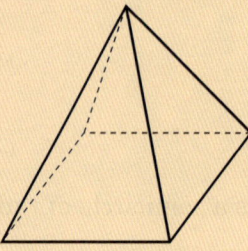
square based pyramid

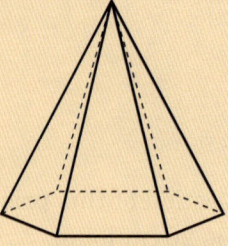
hexagonal based pyramid

The shape of a pyramid is given by drawing lines from a point (called the vertex) to each corner of the base.

Each sloping face is a triangle and the number of these depends on the shape of the base. The diagrams illustrate *right pyramids*. In a right pyramid, the vertex at the top is vertically above the centre of the base.

This pyramid is not a right pyramid.

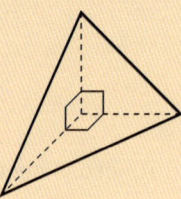

## Cones

A **right circular cone** has a circular base.

A cone has two surfaces, the base and the curved surface.

There is only one vertex and only one edge.

The vertex is above the centre of the base.

A cone can be made from a sector of a circle:

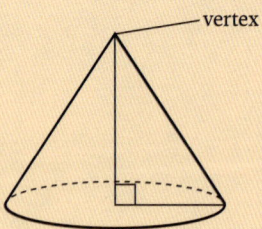
vertex

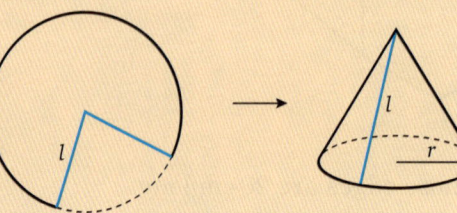

## 18 Problems in three dimensions

If the radius of the sector is $l$, then when the straight edges of the sector are joined to form the cone, $l$ becomes the **slant height of the cone**.

If the radius of the circular base of the cone is $r$, the circumference of this circle is $2\pi r$ and this is equal to the length of the arc of the sector.

The circumference of the circle from which the sector is formed is $2\pi l$.

Therefore,

$$\frac{\text{length of curved edge of the sector}}{\text{circumference of the circle of which it is part}} = \frac{2\pi r}{2\pi l} = \frac{\text{area of sector}}{\text{area of circle of which it is part}}$$

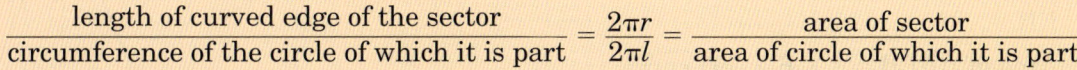

The area of the sector is the curved surface area of the cone.

Therefore,

**the area of curved surface of the cone** $= \dfrac{2\pi r}{2\pi l} \times \pi l^2 = \pi r l$

Using Pythagoras' theorem gives a relationship between the perpendicular height and the slant height of a cone.

If $h$ is the perpendicular height of the cone, then

$l^2 = h^2 + r^2$

### Spheres

A **sphere** has only one surface, no edges and no vertices.
The formula for the surface area of a sphere is given by

$A = 4\pi r^2$

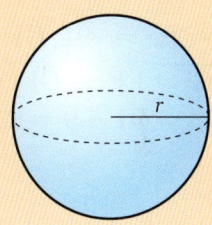

### Exercise 18b

### Worked example

The diagram shows a pyramid ABCDE.
The base is a square of side 6 cm and the sloping edges are 10 cm long.

→ Find the surface area of the pyramid.

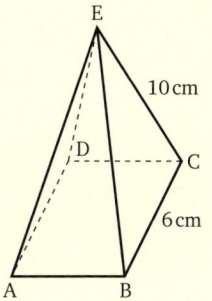

# STP Maths 9

 To find sides and angles in three-dimensional figures, identify a triangle containing the side or angle and draw that triangle separately.

The area of the base is $6 \times 6 \, \text{cm}^2 = 36 \, \text{cm}^2$

 Each sloping face is an isosceles triangle. To find the area of one of these triangles we need to find its height.

Drawing one of these triangles shows that the height of the triangle, EH, can be found using Pythagoras' theorem.

$EH^2 = 10^2 - 3^3 = 91$

$EH = \sqrt{91}$

Area of $\triangle AEB = \frac{1}{2} \times 6 \times \sqrt{91} \, \text{cm}^2$

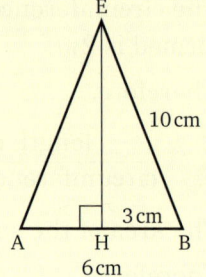

Surface area of the pyramid
$= (36 + 4 \times 3 \times \sqrt{91}) \, \text{cm}^2 = 150 \, \text{cm}^2$ (correct to 3 s.f.)

 We leave the height EH in its exact form until the final calculation.

→ Find the height of the pyramid.

 The height of the pyramid is the perpendicular line from E to the centre of the base, G. EG is the height of the isosceles triangle EHF.

$EH = EF = \sqrt{91}, GF = 3$

$\therefore EG^2 = EF^2 - GF^2$

$\quad = 91 - 9 = 82$

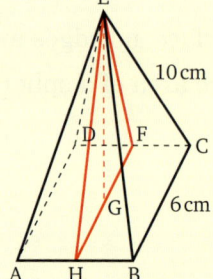
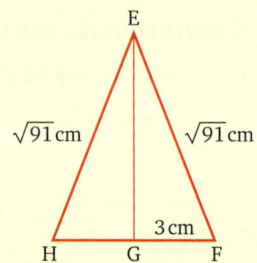

$\therefore$ the height of the pyramid $= \sqrt{82}$ cm

$\quad\quad\quad\quad\quad\quad\quad\quad\quad\quad = 9.06$ cm (correct to 3 s.f.)

→ Find the angle EBD.

Angle EBD is in the triangle EBD, in which

$ED = EB = 10$ and $EG = \sqrt{82}$

$\sin E\widehat{B}G = \dfrac{EG(\text{opp})}{EB(\text{hyp})} = \dfrac{\sqrt{82}}{10} = 0.90553...$

$E\widehat{B}D = E\widehat{B}G$

$\therefore EBD = 64.9°$ correct to 1 d.p.

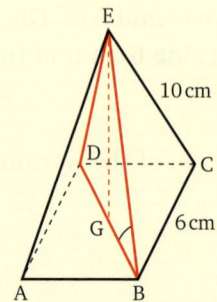
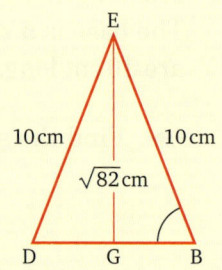

Give lengths and areas that are not exact correct to three significant figures and angles correct to one decimal place.

1  a  Copy and complete the following table.

| Solid | Number of faces (F) | Number of edges (E) | Number of vertices (V) |
|---|---|---|---|
| cube | 6 | 12 | |
| tetrahedron | 4 | | |
| square based pyramid | | | |
| triangular prism | | | |

b  Find a formula connecting $F$, $E$ and $V$.

c  A solid has 8 faces and 6 vertices. How many edges would you expect it to have?

2  Find the surface area of each solid.

a  a sphere of radius 8 cm

b  a cone of base radius 5 cm and height 12 cm

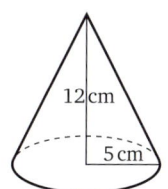

c  a triangular based pyramid, each face of which is an equilateral triangle of side 18 cm. (This is a **regular tetrahedron**.)

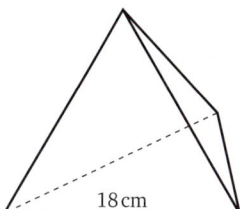

3  An open cone has a base radius of 6 cm and a height of 10 cm.
   a  Find the slant height of the cone.
   b  Find the surface area of the cone.

4  The diagram shows a bollard which is a cylinder surmounted by a **hemisphere** (half a sphere).
   Find the surface area of the bollard.

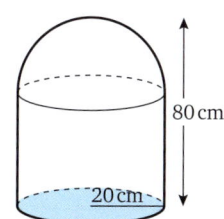

**STP Maths 9**

5. A solid consists of a cube surmounted by a square based pyramid.
   All edges are 5 cm long.
   Find the surface area of the solid.

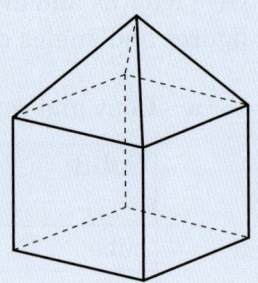

6. The diagram shows a curtain rod which is a cylinder with an identical cone at each end.
   The cylinder is 1.5 metres long with a radius of 1.5 cm. The cones are each 3 cm long.
   Find the surface area of the curtain rod.

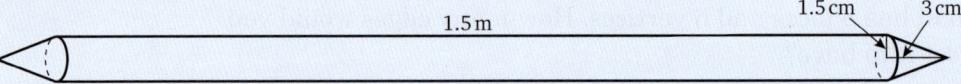

7. The diagram shows a child's toy. It is a hemisphere surmounted by a cone.
   The radius of the hemisphere is 4 cm and the slant height of the cone is 5 cm.
   Find the surface area of the toy.

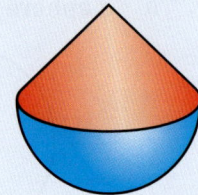

8. ABCDE is a right pyramid on a rectangular base, 8 cm by 6 cm.
   The height of the pyramid is 5 cm.
   H is the midpoint of AB.
   Find
   
   a  the length of EH
   b  the surface area of the pyramid
   c  the angle EHG
   d  the length of DB
   e  the angles in triangle EBD.

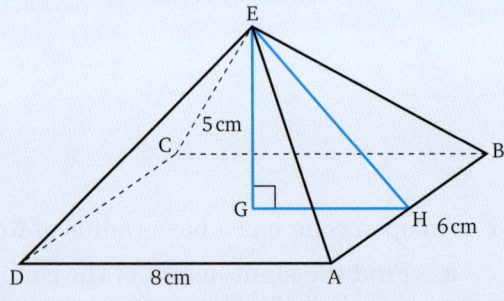

9. The diagram shows an open container for popcorn.
   It is a cuboid on top of a square-based pyramid.
   Find
   
   a  the height of the pyramid
   b  the height of one triangular face of the pyramid
   c  the external surface area of card used to make the container.

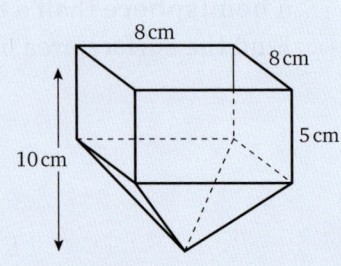

# 18 Problems in three dimensions

10 Each rod in an iron railing is a prism with a triangular cross-section, which is an equilateral triangle, surmounted by a spike which is a tetrahedron.
   The length of the prism is 1 m and the length of each side of the triangular cross-section is 2 cm. The sloping edges of the spike are 6 cm long.
   a  Find the height of one sloping triangular face.
   b  Find the surface area of the rod.
   c  There are 80 of these rods in a length of railing. All the surfaces of the rods are to be painted, except for the bottom ends. What area needs painting?

11 The diagram shows the **frustum** of a pyramid (i.e. a pyramid with its top sliced off).
   ABCD and EFGH are both horizontal squares.
   EA = FB = GC = HD = 6 cm, AB = 10 cm and EF = 6 cm.
   a  Sketch the face FBCG. What shape is it?
   b  P is the point on BC such that $F\hat{P}B = 90°$. Calculate the length of FP.
   c  Find the surface area of the frustum.

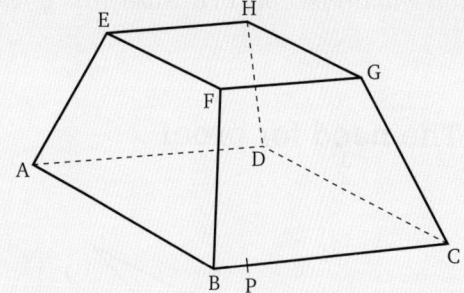

12 The curved surface of a lampshade is in the shape of a frustum of a cone.
   The radius of the circle forming the lower edge is 16 cm and the radius of the circle forming the upper circle is 8 cm. The slant height is 20 cm.
   a  Sketch the net for making the curved surface.
   b  Find the surface area of the curved surface.

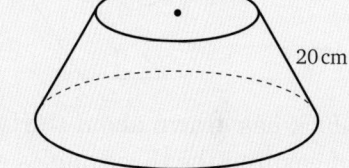

## Consider again

The diagram shows a solid that consists of a prism surmounted by a pyramid.

The cross-section of the prism is an equilateral triangle.

The height of the prism is 3 cm and the sloping edges of the pyramid are 6 cm long.

Find   a  the surface area of the solid
       b  the height of the solid.

Now can you answer these questions?

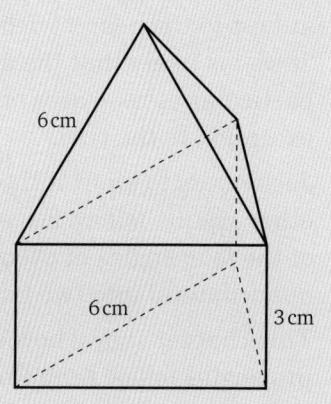

# 19 Geometry and proof

## Consider

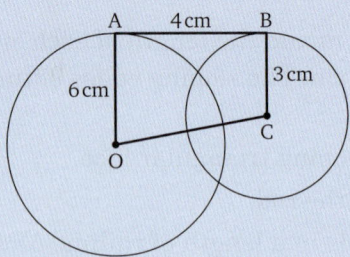

AB is a tangent to the circles with centres O and C, touching them at A and B respectively. Find OC.

*You should be able to answer this question after you have worked through this chapter.*

## The need for proof

Jack has drawn about thirty different triangles of all shapes and sizes. For each one he measured the three angles and found their sum. The results he obtained varied from 178° to 181.5°. It was from demonstrations such as these in Book 7 that we concluded that the sum of the angles of *any* triangle is 180°.

Could it be that this method is 'jumping to conclusions' and is unsatisfactory for many reasons? After all, it is impossible to draw a line because a line has no thickness and if it did not have thickness we could not see it! Furthermore, it is impossible to measure angles with absolute accuracy. The protractor Jack uses is probably capable of measuring angles at best to the nearest degree. The only conclusion that Jack can draw from his results is 'it seems likely that the angle sum of any triangle is 180°'. However, 'proof' by examining particular cases leaves open the possibility that somewhere, as yet unfound, there lurks an exception to the rule.

Jack needs to know if the result can be *proved* to be true for every triangle. If it can, several other results follow. For example, if the angle sum of any triangle is 180°, the angle sum of the four angles in every quadrilateral must be 360° since one diagonal always divides a quadrilateral into two triangles.

This chapter shows how certain geometric properties can be proved and how other properties follow from them.

# 19 Geometry and proof

## Deductive proof

Learning geometrical properties from demonstrations gives the impression that each property is isolated. However, geometry can be given a logical structure where one property can be deduced from other properties. This forms the basis of deductive proof; we quote known and accepted facts and then make logical deductions from them.

For example, if we accept that
- vertically opposite angles are equal
- and corresponding angles are equal,

then, using just these two facts, we can prove that alternate angles are equal.

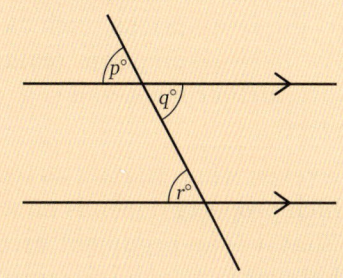

In the diagram, $p° = q°$ (vertically opposite angles)
$p° = r°$ (corresponding angles)
$\Rightarrow q° = r°$

Therefore, the alternate angles are equal.

The symbol $\Rightarrow$ means 'implies that' and indicates the logical deduction made from the two stated facts.

This proof does not involve angles of a particular size; $p°$, $q°$ and $r°$ can be any size. Therefore, this proves that alternate angles are *always* equal whatever their size.

As a further example of deductive proof, we shall prove that in *any* triangle, the sum of the interior angles is 180°.

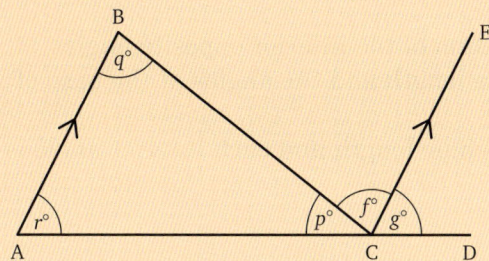

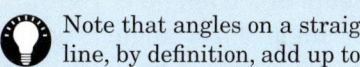

Note that angles on a straight line, by definition, add up to 180°.

If △ABC is any triangle and if AC is extended to D and CE is parallel to AB, then

$p° + f° + g° = 180°$ (angles on a straight line) (1)
$f° = q°$ (alternate angles) (2)
$g° = r°$ (corresponding angles) (3)
$\Rightarrow p° + q° + r° = 180°$

i.e. the sum of the interior angles of *any* triangle is 180°.

Because this proof does not involve measuring angles in a particular triangle, it applies to all possible triangles, thus closing the loophole that there may exist a triangle whose angles do not add up to 180°.

Notice how this proof uses the property proved in the first example, that is, this proof follows from the previous proof. The angle sum property of triangles can now be used to prove further properties.

For example, the statements on the previous page also lead to another useful fact about angles in triangles:

(2) and (3)    $f° + g° = q° + r°$

i.e. an exterior angle of a triangle is equal to the sum of the two interior opposite angles.

Euclid was the first person to give a formal structure to geometry. He started by making certain assumptions, such as 'there is only one straight line between two points'. Using only these assumptions (called *axioms*), he then proved some facts and used those facts to prove further facts, and so on. In this way, the proof of any one fact could be traced back to the axioms.

However, when *you* are asked to give a geometric proof, you do not have to worry about which property depends on which; you can use *any* facts that you know. One aspect of proof is that it is an argument used to convince other people of the truth of any statement, so whatever facts you use must be clearly stated.

It is a good idea to set out your ideas in an orderly way before starting to write out a proof. The easiest way to do this is to mark right angles, equal angles, equal sides, etc. on the diagram.

The exercises in this chapter give practice in writing out a proof.

For **Exercise 19a**, you will need the following facts.
- Vertically opposite angles are equal.
- Corresponding angles are equal.
- Alternate angles are equal.
- Interior angles (of parallel lines) add up to 180°.
- The angle sum of a triangle is 180°.
- An exterior angle of a triangle is equal to the sum of the interior opposite angles.
- An isosceles triangle has two sides of the same length and the angles at the base of those sides are equal.
- An equilateral triangle has three sides of the same length and each interior angle is 60°.

## Exercise 19a

### Worked example

→ In a triangle ABC, the bisectors of angles B and C intersect at I.
Prove that $\widehat{BIC} = 90° - \frac{1}{2}\widehat{A}$

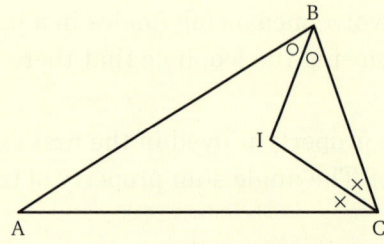

In triangle BIC
$B\hat{I}C + I\hat{B}C + I\hat{C}B = 180°$ (angle sum in a △)
i.e. $B\hat{I}C + \frac{1}{2}\hat{B} + \frac{1}{2}\hat{C} = 180°$ (BI bisects B and CI bisects C) (1)
But $\hat{A} + \hat{B} + \hat{C} = 180°$ (angle sum in a △)
i.e. $\hat{B} + \hat{C} = 180° - \hat{A}$
so $\frac{1}{2}\hat{B} + \frac{1}{2}\hat{C} = 90° - \frac{1}{2}\hat{A}$
Substituting in (1):
$B\hat{I}C + 90° - \frac{1}{2}\hat{A} = 180°$
i.e. $B\hat{I}C = 90° + \frac{1}{2}\hat{A}$

**1** Prove that $A\hat{C}D = A\hat{B}C + D\hat{E}C$

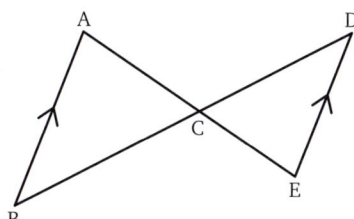

**2** Prove that $A\hat{C}B = 2C\hat{D}B$

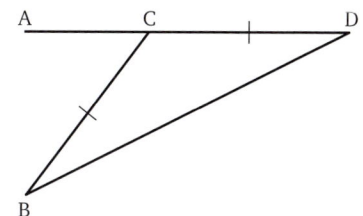

**3** Prove that AD bisects $B\hat{A}C$.

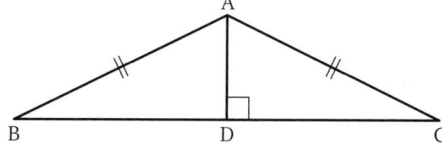

**4** CE bisects $A\hat{C}D$ and CE is parallel to BA. Prove that △ABC is isosceles.

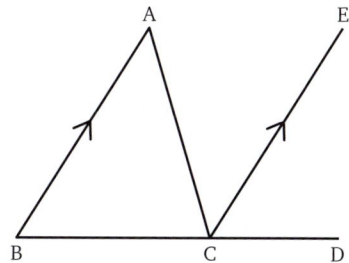

**5** $A\hat{M}L = A\hat{B}C$.

Prove that $A\hat{L}M = A\hat{C}B$.

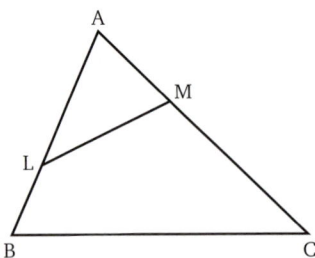

## Showing that a hypothesis is false

We saw in the last section that drawing a few triangles and measuring the angles led us to say that 'it looks as though' the angles of any triangle add up to 180°. At that stage we had a *hypothesis*, which we then *proved* to be true for any triangle.

It is also important to be able to show that certain hypotheses are in fact false.

Suppose that some students are investigating the relationship between the number of lines drawn across a circle and the number of regions that the circle is divided into by those lines.

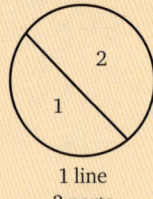        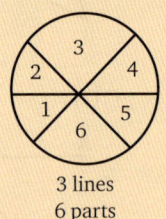

1 line            2 lines            3 lines
2 parts           4 parts            6 parts

These three drawings led Tom to the hypothesis that $n$ lines drawn across a circle give $2n$ regions.

Jess, however, drew the lines in this way, showing that 3 lines can give 7 regions, and therefore that Tom's hypothesis is false.

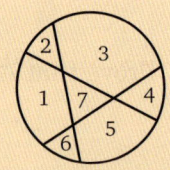

Jess used a **counter example** to disprove the hypothesis.

Now consider the hypothesis 'the square root of an even number is itself even'. This can be shown to be untrue using this counter example: $\sqrt{6} = 2.449...$ which is not an even number.

You may like to see if you can find a counter example to disprove the hypothesis 'all prime numbers are odd'.

Not every hypothesis can be either proved or shown to be false. In mathematics there are several well-known hypotheses in this category, one of which is Goldbach's conjecture. This states that 'every even number greater than or equal to 6 can be written as the sum of two odd prime numbers'. At the time of writing no one has yet proved this to be true; on the other hand no one has yet found a counter example.

## Exercise 19b

In questions **1** to **4**, see if you can find a counter example to disprove each hypothesis.

1. The square root of a positive number is always smaller than the number.

2. If the side of a square is $x$ cm long, the number of units of area of the square is always different from the number of units of length in the perimeter.

3. The diagonals of a parallelogram never cut at right angles.

4. The sum of any two angles in a triangle is always greater than the third angle.

Questions **5** and **6** give 'proofs' that are obviously invalid because they lead to untruths. In each case, find the flaw in the argument.

5. It is a fact that $\qquad 4 - 10 = 9 - 15$

   Adding $\frac{25}{4}$ to each side gives $\qquad 4 - 10 + \frac{25}{4} = 9 - 15 + \frac{25}{4}$

   Factorise $\qquad \left(2 - \frac{5}{2}\right)\left(2 - \frac{5}{2}\right) = \left(3 - \frac{5}{2}\right)\left(3 - \frac{5}{2}\right)$

   i.e. $\qquad \left(2 - \frac{5}{2}\right)^2 = \left(3 - \frac{5}{2}\right)^2$

   Take the square root of each side $\qquad 2 - \frac{5}{2} = 3 - \frac{5}{2}$

   Add $\frac{5}{2}$ to each side $\qquad 2 = 3$ which is nonsense.

6. Let $\qquad x = y$
   and obviously $\qquad x^2 - xy = x^2 - xy$
   Now $\qquad x = y$ so $xy = y^2$
   i.e. line 2 can be rewritten $\qquad x^2 - xy = x^2 - y^2$
   Factorise $\qquad x(x - y) = (x - y)(x + y)$
   Divide both sides by $(x - y)$ $\qquad x = x + y$
   but $x = y$, so $\qquad x = x + x$
   i.e. $\qquad x = 2x$
   i.e. $\qquad 1 = 2$ which is nonsense.

## Circle facts

In Book 8 we defined parts of a circle such as radius, diameter, circumference and arc. In this chapter we summarise the facts we already know and have used in previous chapters, extend a few of them and then investigate some of the basic geometric properties of a circle.

Every point on a circle is the same distance from its centre. This distance is called the *radius* of the circle.

A straight line joining any two points on the circle is called a **chord**.

Any chord passing through the centre of a circle is called a *diameter*.

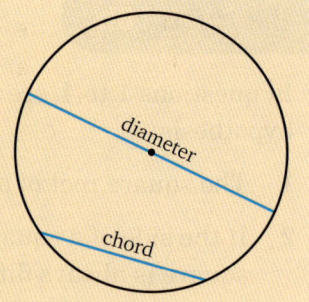

Any part of the circle is called an **arc**. If the arc is less than half the circle, it is called a **minor arc**. If it is greater than half the circle, it is called a **major arc**.

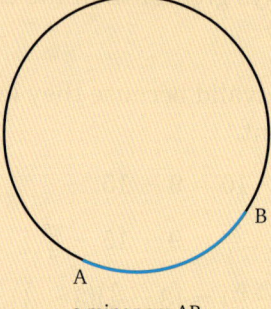
a minor arc AB

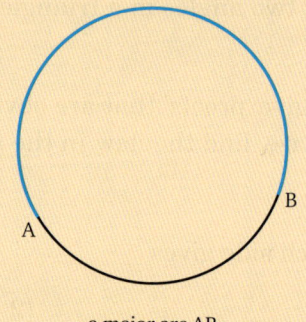
a major arc AB

The shaded area is enclosed by two radii and an arc. It is called a **sector**.

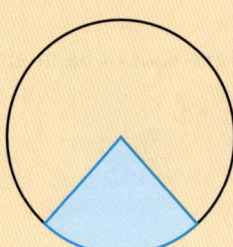

A chord divides the area inside a circle into two regions called **segments**. The larger region is called a **major segment** and the smaller region is called a **minor segment**.

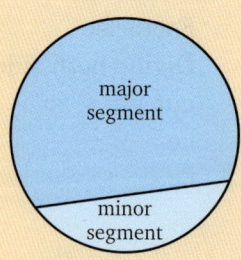

## Tangents

The line PQ cuts the circle at A and B.
AB is a *chord*.

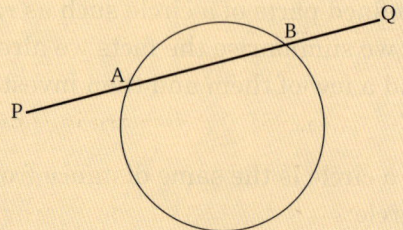

# 19 Geometry and proof

Imagine that this line is pivoted at P. As PQ rotates about P, we get successive positions of the points A and B. As PQ moves towards the edge of the circle, the points A and B move closer together, until eventually they coincide.

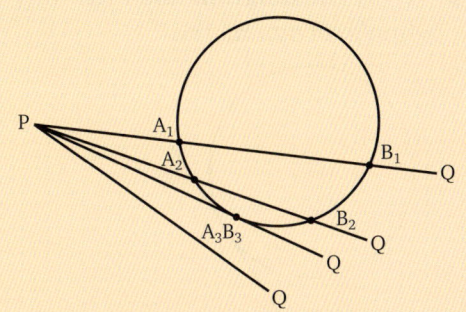

When PQ is in this position, it is called a **tangent to the circle** and we say that PQ touches the circle. (When PQ is rotated beyond this position, it loses contact with the circle and is no longer a tangent.)

We therefore define a tangent to a circle as a straight line which touches the circle.

The point at which the tangent touches the circle is called the *point of contact*.

PT is a tangent to the circle.

T is the point of contact.

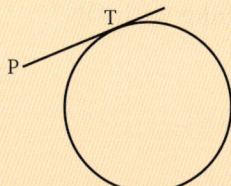

The *length of a tangent* from a point P outside the circle is the distance between P and the point of contact. In the diagram the length of the tangent from P to the circle is the length PT.

## Exercise 19c

1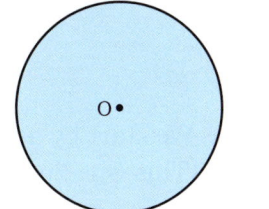

The diagram shows a disc, of radius 20 cm, rolling along horizontal ground.

**a**  Describe the path along which O moves as the disc rolls.

At any one instant,

**b**  how many points on the disc are in contact with the ground

**c**  how far is O from the ground

**d**  how would you describe the line joining O to the ground and what angle does it make with the ground?

397

**2** Copy the diagram and draw any line(s) of symmetry.

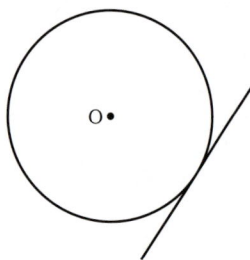

**3** Copy the diagram and draw any line(s) of symmetry.

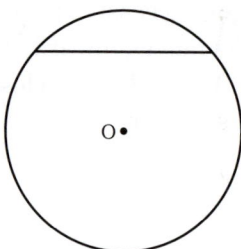

**4 a** Show that the chord AB is perpendicular to the radius ON which bisects AB. (Join OA and OB.)

**b** Now imagine that the chord AB slides down the radius ON.

  **i** When the points A and B coincide with N, what has the line through A and B become?

  **ii** What angle does this line make with ON?

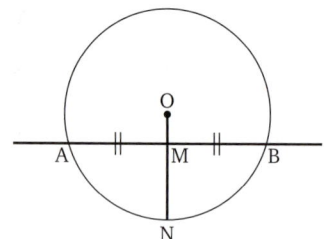

### Tangent property

The investigational work in **Exercise 19c** suggests that

> **a tangent to a circle is perpendicular to the radius drawn from the point of contact.**

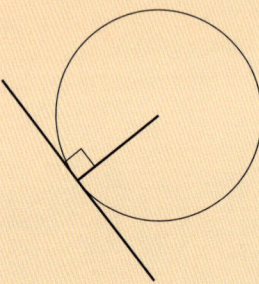

The general proof of this property is an interesting exercise in logic. We start by assuming that the property is *not* true and end up by contradicting ourselves. (This is called 'proof by contradiction'.)

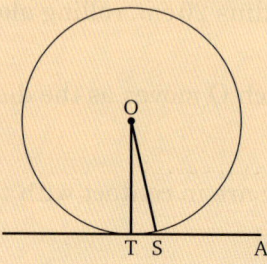

TA is a tangent to the circle and OT is the radius from the point of contact.

If we *assume* that OT̂A is *not* 90°, then it is possible to draw OS so that OS *is* perpendicular to the tangent, i.e. OŜT = 90°.

Therefore, △OST has a right angle at S.

Hence OT is the hypotenuse of △OST

i.e.  OT > OS

⇒  S is inside the circle, as OT is a radius.

⇒  the line through T and S must cut the circle again.

But this is impossible, as the line through T and S is a tangent.

Hence the assumption that OT̂A ≠ 90° is wrong, i.e. OT̂A *is* 90°.


# Exercise 19d

Some of the questions in this exercise require the use of trigonometry.

## Worked example

→ The tangent from a point P to a circle of radius 4.2 cm is 7 cm long. Find the distance of P from the centre of the circle.

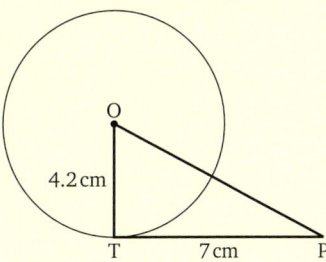

OT̂P = 90°           (tangent perpendicular to radius)

OP$^2$ = OT$^2$ + TP$^2$     (Pythagoras' theorem)

   = (4.2)$^2$ + 7$^2$

   = 17.64 + 49 = 66.64

∴  OP = 8.163...

The distance of P from O is 8.16 cm, correct to 3 s.f.

In questions **1** to **5**, O is the centre of the circle and AB is a tangent to the circle, touching it at A.

**1**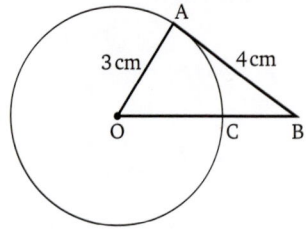

Find OB and CB.

**4**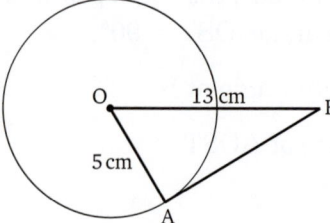

Find AB and OB̂A

**2**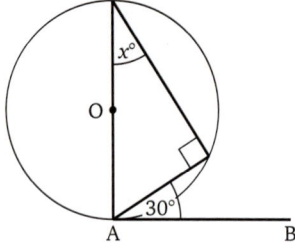

Find the marked angle.

**5**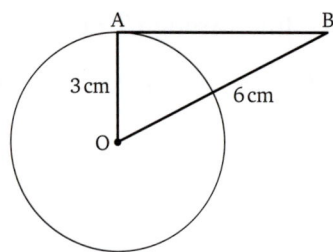

Find AB̂O.

**3**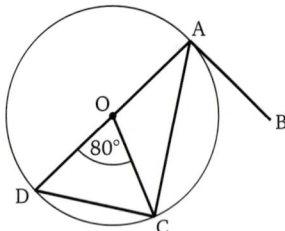

Find    **a** DÂC    **b** BÂC

**6** AB and BC are tangents to the circle centre O, touching it at A and C.

Show that △ABC is isosceles.

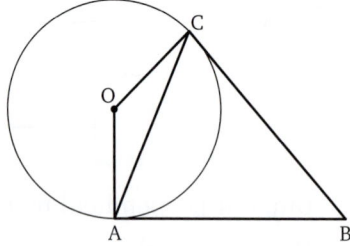

**7** AB is a chord of the larger circle and a tangent to the smaller circle. If O is the centre of both circles, find the length of AB.

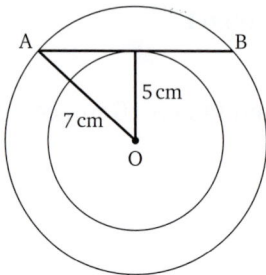

8 Read **Euclid's proof that there must be an infinite number of prime numbers**. Then answer questions **a–g**.
   1  Assume that the largest prime number is $p$.
   2  When the number $2 \times 3 \times 5 \times 7 \times 11 \times 13 \times 17 \times \ldots \times p + 1$ is divided by any prime number, it leaves a remainder of 1.
   3  But every number greater than 1 is exactly divisible by at least one prime number.
   4  This gives a contradiction.
   5  Therefore, the assumption in line 1 is wrong and there cannot be a largest prime number, so there must be an infinite number of prime numbers.

   a  What are the next three numbers after 17 in the second line of the proof?
   b  What is the answer when $2 \times 3 \times 5 \times 7 \times 11 \times 13 \times 17 + 1$ is divided by
      i  13    ii  5?
   c  What do the dots in line 2 represent?
   d  Why is line 2 correct?
   e  Why is line 3 correct?
   f  Prove that 30 031 is a prime number.
   g  Find a prime number that is greater than 1 000 000 000, and state why you are sure that it is prime.

## Consider again

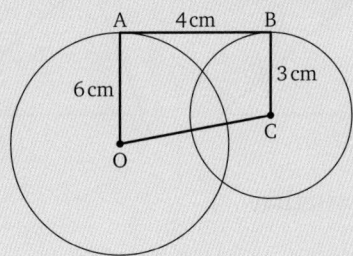

AB is a tangent to the circles with centres O and C, touching them at A and B respectively. Now find OC.

# 20 Congruent triangles

## Consider

ABCD is a trapezium, with BC parallel to AD and AB equal to CD. BE is parallel to CD and angle CDE = 60°.

Prove that △ABE is equilateral.

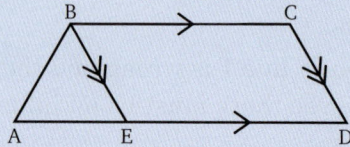

*You should be able to answer this question after you have worked through this chapter.*

## Congruent triangles

Two figures are congruent if one figure is an exact copy of the other. If the figures are drawn on squared paper, it is easy to determine if they are congruent. If the shapes are drawn accurately on plain paper, we can use tracing paper to see whether they appear to be congruent, but we need precise information about the lengths of sides and the sizes of angles to determine whether they really are congruent.

Triangles are simple figures and we do not need very much information to determine whether one triangle is an exact copy of another triangle.

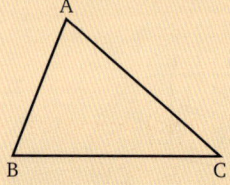 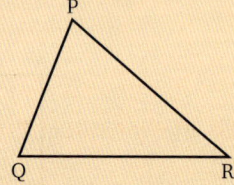

If △s $\frac{ABC}{PQR}$ are congruent, it follows that

$$\left.\begin{array}{l} AB = PQ \\ AC = PR \\ BC = QR \end{array}\right\} \text{ and } \left\{\begin{array}{l} \hat{A} = \hat{P} \\ \hat{B} = \hat{Q} \\ \hat{C} = \hat{R} \end{array}\right.$$

To make an exact copy of these triangles, we do not need to know the lengths of all three sides and the sizes of all three angles. Three measurements are usually enough and we now investigate which three measurements are suitable.

## 20 Congruent triangles

### Exercise 20a

In each of questions **1** to **4**, make a rough sketch of △ABC.
Construct a triangle with the same measurements as those given for △ABC. Is your construction an exact copy of △ABC? (Try to construct a different triangle with the given measurements.)

1. △ABC, in which AB = 8 cm, BC = 5 cm, AC = 6 cm
2. △ABC, in which $\hat{A} = 40°$, $\hat{B} = 60°$, $\hat{C} = 80°$
3. △ABC, in which AB = 7 cm, BC = 12 cm, AC = 8 cm
4. △ABC, in which $\hat{A} = 20°$, $\hat{B} = 40°$, $\hat{C} = 120°$
5. What extra information do you need about △ABC in questions **2** and **4** in order to make an exact copy?

### Three pairs of sides

From **Exercise 20a**, you should be convinced that we can make an exact copy of a triangle if we know the lengths of the three sides.
Therefore,

> **two triangles are congruent if the three sides of one triangle are equal to the three sides of the other triangle.**

However, if the three angles of one triangle are equal to the three angles of another triangle, they may not be congruent (but they are similar).

### Exercise 20b

#### Worked examples

→ Decide whether the triangles in this pair are congruent.
  Give a brief reason for your answer.

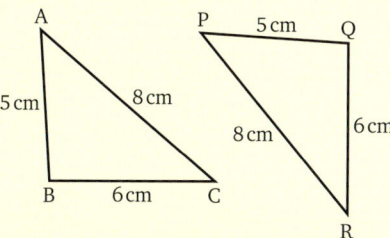

AB = PQ
BC = QR
AC = PR
∴ △s $\begin{matrix} ABC \\ PQR \end{matrix}$ are congruent (three sides).

 Note that we write corresponding vertices one under the other.

STP Maths 9

→ Decide whether the triangles in this pair are congruent.

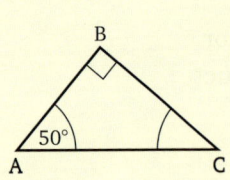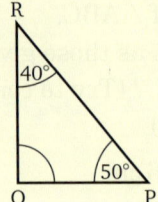

In △ABC, $\hat{C}$ = 40° (angle sum of △)

In △PQR, $\hat{Q}$ = 90° (angle sum of △)

∴ △s $\frac{ABC}{PQR}$ are similar but probably not congruent.

In questions **1** to **6**, state whether or not the two triangles are congruent. Give a brief reason for your answers. All lengths are in centimetres.

**1**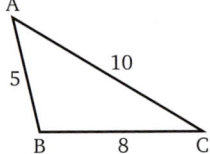

**4**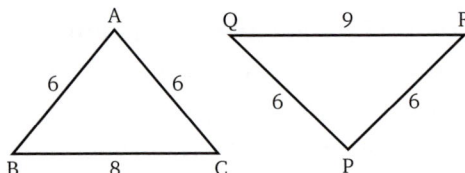

**2**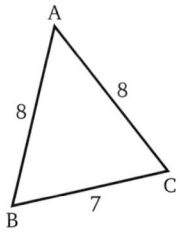

**5**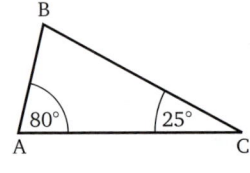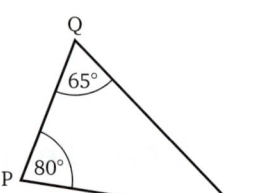

**3**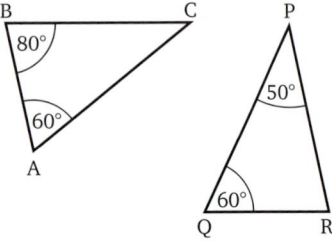

**6**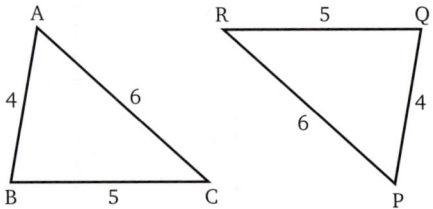

Give brief reasons for your answers to questions **7** to **10**.

**7**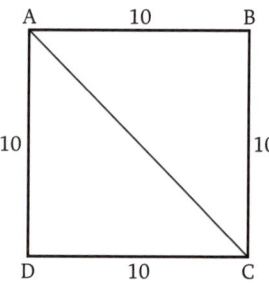

Are △ADC and △ABC congruent?

**9**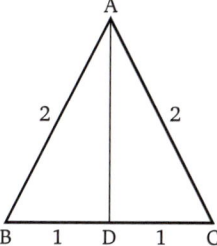

Are △ABD and △ACD congruent?

**8**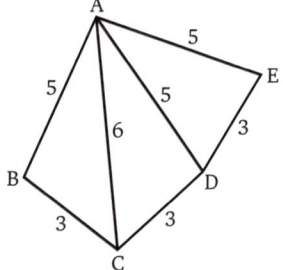

Which triangles are congruent?

**10**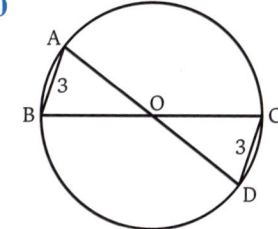

The point O is the centre of the circle and the radius is 5 cm. Are △ABO and △CDO congruent?

## Two angles and a side
To make an exact copy of a triangle, we need to know the length of at least one side.

### Exercise 20c

1  Construct △ABC, in which AB = 6 cm, $\widehat{A}$ = 30°, $\widehat{B}$ = 60°.

2  Construct △PQR, in which PR = 6 cm, $\widehat{P}$ = 30°, $\widehat{Q}$ = 60°.

3  Construct △LMN, in which LM = 6 cm, $\widehat{L}$ = 30°, $\widehat{M}$ = 60°.

4  Construct △XYZ, in which YZ = 6 cm, $\widehat{X}$ = 30°, $\widehat{Y}$ = 60°.

5  How many of the triangles that you have constructed are congruent?

6  How many different triangles can you construct from the following information: one angle is 40°, another angle is 70° and the length of one side is 8 cm?

# STP Maths 9

Now you can see that we are able to make an exact copy of a triangle if we know the sizes of two of its angles and the length of one side, provided that we place the side in the same position relative to the angles in both triangles. In summary,

**two triangles are congruent if two angles and one side of one triangle are equal to two angles and the *corresponding* side of the other triangle.**

## Exercise 20d

### Worked example

→ Decide whether these triangles are congruent. Give a brief reason for your answer.

△s $\genfrac{}{}{0pt}{}{\text{ABC}}{\text{PQR}}$ are similar (angles are equal)

but not congruent (AB and PQ are corresponding sides and are *not* equal).

In questions **1** to **8**, state whether or not the two triangles are congruent. Give brief reasons for your answers. All lengths are in centimetres.

**1**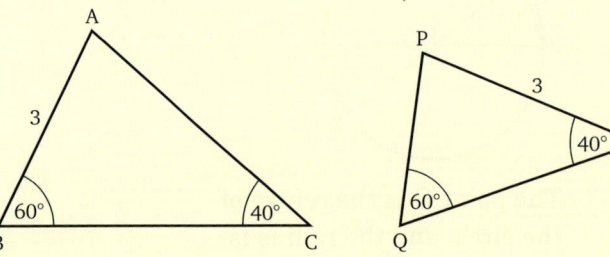

**3**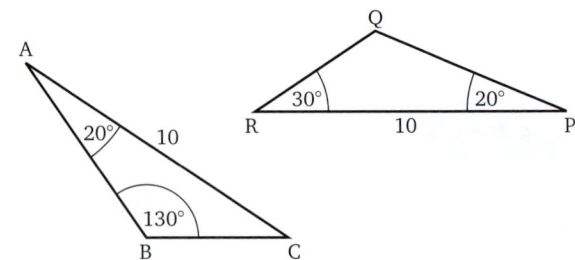

**2**

**4**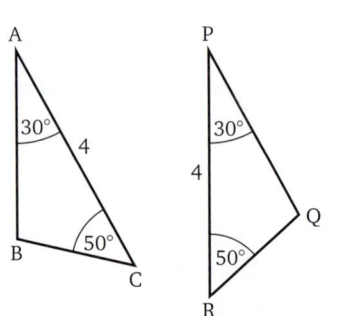

406

**5**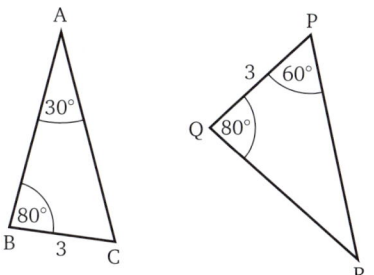

**8**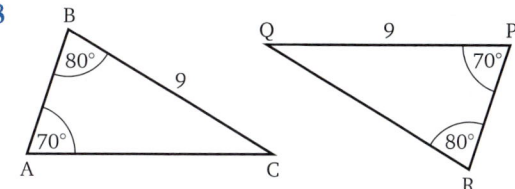

**6**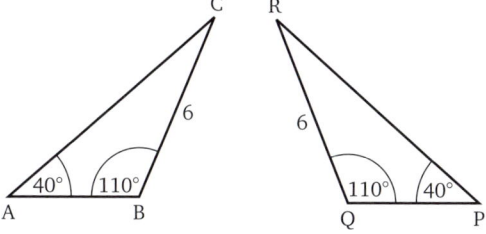

**9** Are △ABC and △ADC congruent?

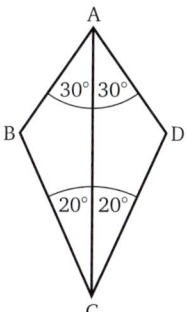

**7**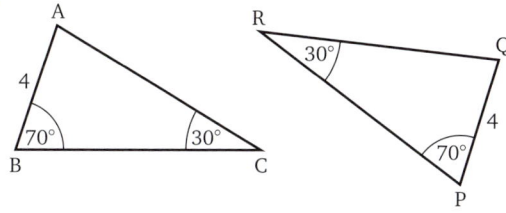

**10** Are △ABC and △ADC congruent?

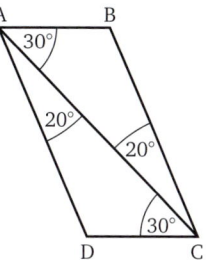

## Two sides and an angle

We are now left with one more possible combination of three measurements: if we know the lengths of two sides and the size of one angle in a triangle, does this fix the size and shape of the triangle?

# STP Maths 9

### Exercise 20e

Can you make an exact copy of the following triangles from the information given about them? (Try to construct each triangle.)

1. $\triangle ABC$, in which $AB = 8$ cm, $BC = 5$ cm, $\hat{B} = 30°$
2. $\triangle XYZ$, in which $XY = 8$ cm, $XZ = 5$ cm, $\hat{Y} = 30°$
3. $\triangle PQR$, in which $\hat{Q} = 60°$, $PQ = 6$ cm, $QR = 8$ cm
4. $\triangle LMN$, in which $LM = 8$ cm, $\hat{M} = 20°$, $LN = 4$ cm
5. $\triangle DEF$, in which $DE = 5$ cm, $\hat{E} = 90°$, $EF = 6$ cm

Now we can see that we can make an exact copy of a triangle if we know the lengths of two sides and the size of one angle, provided that the angle is between those two sides. Therefore,

**two triangles are congruent if two sides and the *included* angle of one triangle are equal to two sides and the *included* angle of the other triangle.**

If the angle is not between the two known sides, then we cannot always be sure that we can make an exact copy of the triangle. We shall now investigate this case further.

## Two sides and a right angle

### Exercise 20f

Can you make an exact copy of each of the following triangles from the information given about them?

1. $\triangle ABC$, in which $AB = 6$ cm, $\hat{B} = 90°$, $AC = 10$ cm
2. $\triangle PQR$, in which $PQ = 8$ cm, $\hat{Q} = 40°$, $PR = 6.5$ cm
3. $\triangle XYZ$, in which $XY = 5$ cm, $\hat{Y} = 90°$, $XZ = 13$ cm
4. $\triangle LMN$, in which $LM = 5$ cm, $\hat{M} = 60°$, $LN = 4.5$ cm
5. $\triangle DEF$, in which $DE = 7$ cm, $\hat{E} = 90°$, $DF = 10$ cm
6. $\triangle RST$, in which $RS = 5$ cm, $\hat{S} = 120°$, $RT = 8$ cm
7. Can you calculate any further information about any of the triangles in questions **1** to **6**?

**20 Congruent triangles**

From question **7**, you can see that, when the given angle is 90°, we can calculate the length of the remaining side of the triangle.

Therefore, if we are told that one angle in a triangle is a right angle and we are also given the length of one side and the hypotenuse, then this information fixes the shape and size of the triangle since it is equivalent to knowing the lengths of the three sides.

**Two triangles are congruent if they both have a right angle, and the hypotenuse and a side of one triangle are equal to the hypotenuse and a side of the other triangle.**

### Exercise 20g

In questions **1** to **8**, state whether or not the two triangles are congruent. Give brief reasons for your answers. All lengths are in centimetres.

**1**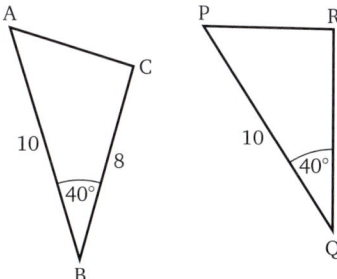

**4**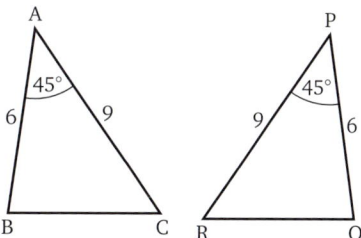

**2**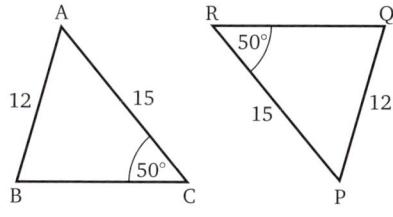

**5**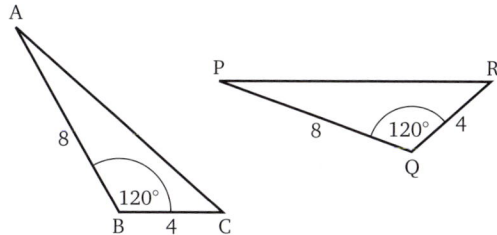

**3**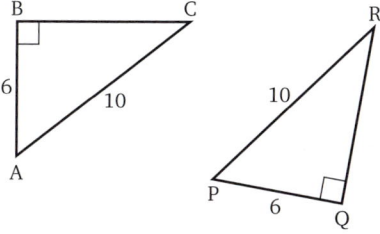

**6**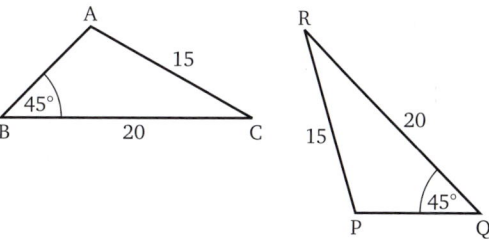

**7**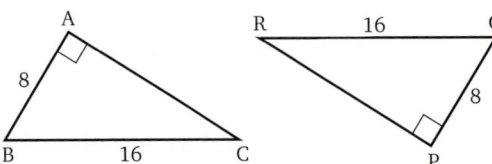

**9** Are △ABD and △ADC congruent?

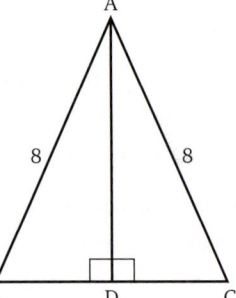

**8**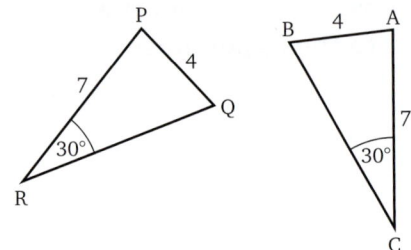

Summing up, two triangles are congruent if:

either   the three sides of one triangle are equal to the three sides of the other triangle (SSS)

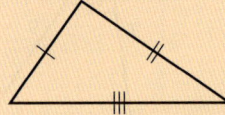 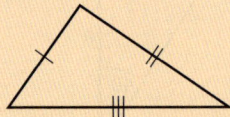

or   two angles and a side of one triangle are equal to two angles and the corresponding side of the other triangle (AAS)

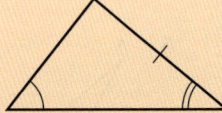 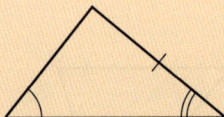

or   two sides and the included angle of one triangle are equal to two sides and the included angle of the other triangle (SAS)

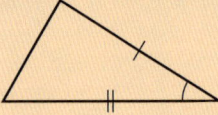 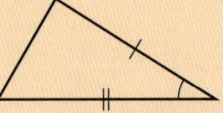

or   the two triangles each have a right angle, and the hypotenuse and a side of one triangle are equal to the hypotenuse and a side of the other triangle (RHS).

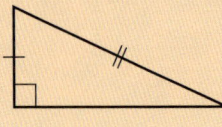 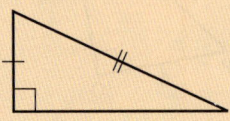

### Exercise 20h

State whether or not each of the following pairs of triangles are congruent. Give brief reasons for your answers. All lengths are in centimetres.

1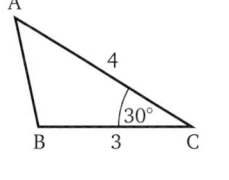

6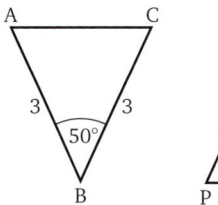

2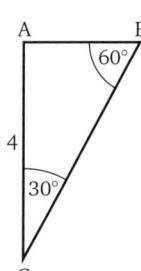

7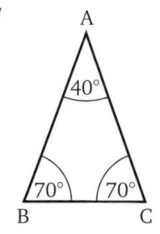

3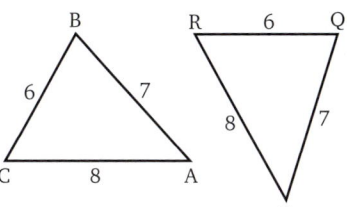

8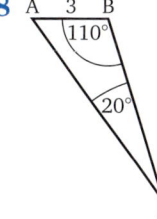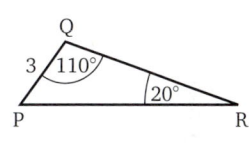

4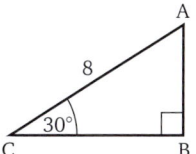

9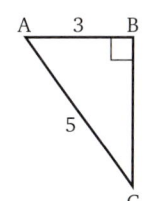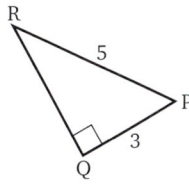

5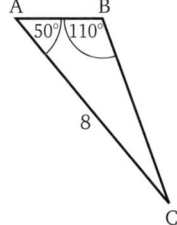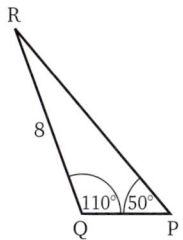

10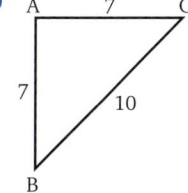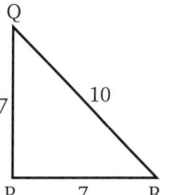

**11**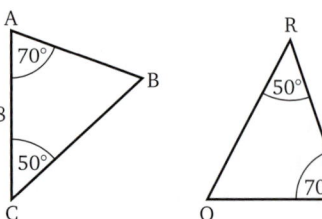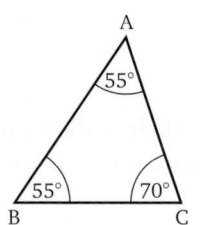

**12**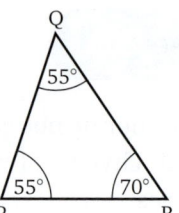

We do not need to know actual measurements to prove that triangles are congruent. If we can show that a correct combination of sides and angles are the same in both triangles, the triangles must be congruent.

## Exercise 20i

### Worked example

→ In quadrilateral ABCD, AB = DC and AD = BC. The diagonal BD is drawn. Prove that △ABD and △CDB are congruent.

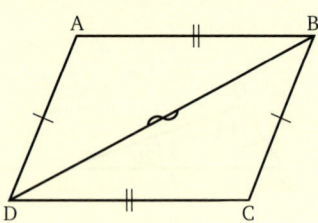

💡 First draw a diagram. Mark on your diagram all the information given and any further facts that you discover. The symbol ~ on the line BD shows that it is common to both triangles.

In △s ABD, CDB,   AB = CD (given)
AD = CB (given)
DB is the same for both triangles.

∴ △s ABD / CDB are congruent (SSS).

**1**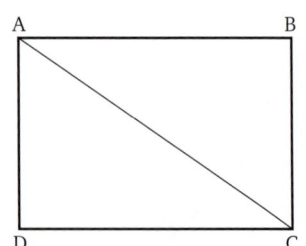

ABCD is a rectangle. Prove that △ABC and △CDA are congruent.

**2** 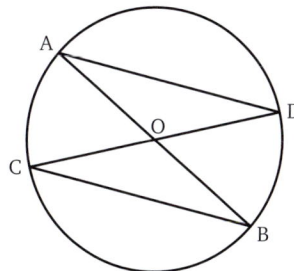 AB and CD are diameters of the circle and O is the centre.

Prove that △AOD and △COB are congruent.

**3** 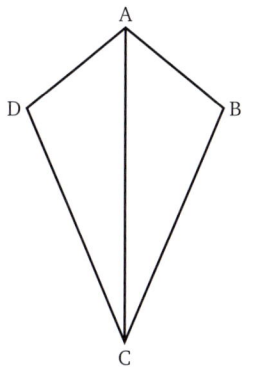 ABCD is a **kite** in which AD = AB and CD = BC. Prove that △ADC and △ABC are congruent.

**4** 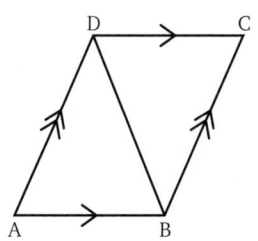 ABCD is a parallelogram. Prove that △ABD and △CDB are congruent.

**5** 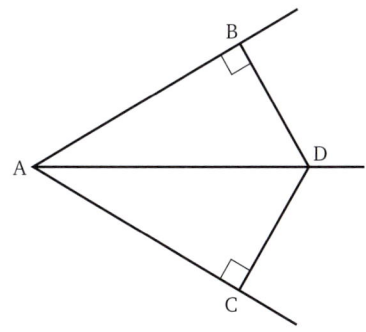 AD bisects angle BAC, DB is perpendicular to AB and DC is perpendicular to AC. Prove that △ABD and △ACD are congruent.

**6** 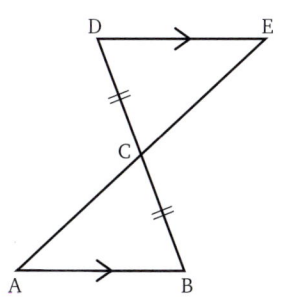 Prove that △ABC and △EDC are congruent.

STP Maths 9

  7  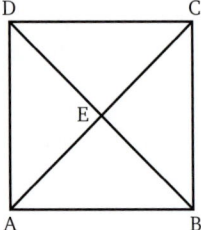  ABCD is a square. Show that △s ABE, BCE, CDE and DAE are all congruent.

  8  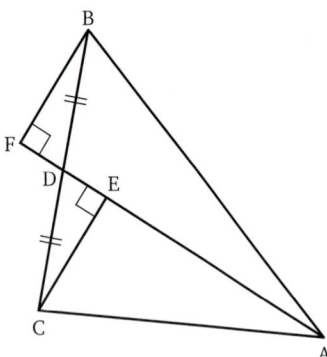  D is the midpoint of BC. CE and BF are perpendicular to AF. Find a pair of congruent triangles.

  9  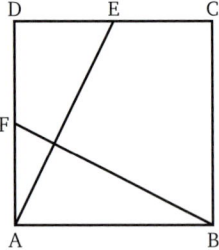  ABCD is a square. E is the midpoint of DC and F is the midpoint of AD. Show that △ADE and △BAF are congruent.

  10  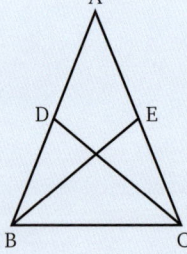  ABC is an isosceles triangle in which AB = AC.
D is the midpoint of AB and E is the midpoint of AC. Prove that △BDC is congruent with △CEB.

  11  ABCD is a rectangle and E is the midpoint of AB. Join DE and CE and show that △ADE and △BCE are congruent.

  12  ABCD is a rectangle. E is the midpoint of AB and F is the midpoint of DC. Join DE and BF and show that △ADE and △CBF are congruent.

414

# 20 Congruent triangles

## Using congruent triangles

Once we have shown two triangles to be congruent, it follows that the other corresponding sides and angles are equal. This gives a good way of proving that certain angles are equal or that certain lines are the same length.

### Exercise 20j

#### Worked example

→ ABCD is a square and AE = DF
Show that DE = CF

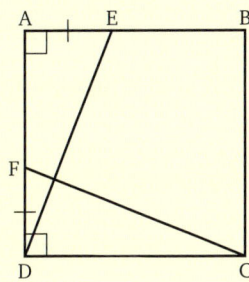

In △s DAE and CDF,
  AE = DF (given)
  DA = CD (sides of a square)
D$\hat{A}$B = C$\hat{D}$F (angles of a square are 90°)
∴ △s $\frac{\text{DAE}}{\text{CDF}}$ are congruent (SAS).
∴ DE = CF

 We have written the triangles so that corresponding vertices are lined up. We can then see the remaining corresponding sides and angles.

**1**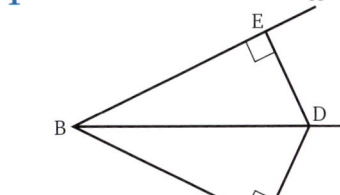

BD bisects A$\hat{B}$C. BE and BF are equal. Show that triangles BED and BFD are congruent and hence prove that ED = FD.

**2**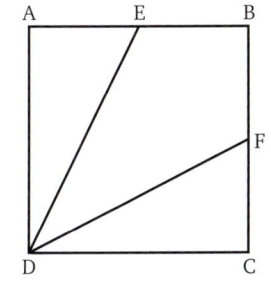

ABCD is a square and E and F are the midpoints of AB and BC respectively. Show that △ADE and △CDF are congruent and hence prove that DE = DF.

415

3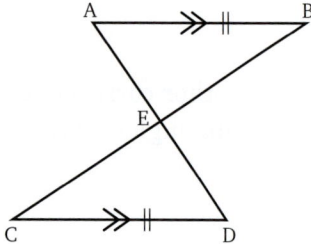

AB and CD are parallel and equal in length. Show that △AEB and △DEC are congruent and hence prove that E is the midpoint of both CB and AD.

## Worked example

→ ABCD is a square. P, Q and R are points on AB, BC and CD respectively such that AP = BQ = CR. Show that $P\hat{Q}R = 90°$.

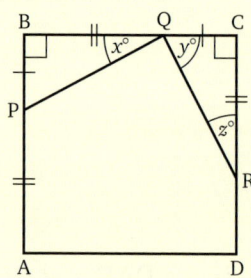

 If you cannot see where to start, work backwards from what you need to prove. In this case, if $P\hat{Q}R = 90°$, then $B\hat{Q}P + C\hat{Q}R = 90°$, so $B\hat{Q}P = C\hat{R}Q$, i.e. △s $\begin{smallmatrix}PBQ\\QCR\end{smallmatrix}$ are congruent.

Therefore, we shall first prove that △PBQ and △QCR are congruent.

In △PBQ and △QCR,

BQ = CR  (given)

$P\hat{B}Q = Q\hat{C}R = 90°$  (angles of a square)

AB = BC (sides of a square) ⎫
AP = BQ (given) ⎬ ⇒ AB − AP = BC − BQ
 ⎭

i.e.  PB = QC

∴ △s $\begin{smallmatrix}PBQ\\QCR\end{smallmatrix}$ are congruent (SAS).

∴       $x = z$

In △QRC,   $y + z = 90$   (angles of triangle)

∴       $x + y = 90$

Also,  $x° + P\hat{Q}R + y° = 180°$ (angles on straight line)

∴       $P\hat{Q}R = 90°$

**4** ABCD is a square and E is the midpoint of the diagonal AC. First show that triangles ADE and CDE are congruent and hence prove that DE is perpendicular to AC.

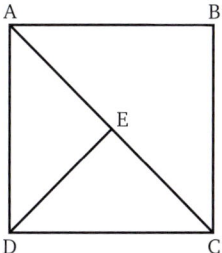

**5** CD is parallel to BA and CD = CB = BA
Show that △CDE and △ABE are congruent and hence show that CA bisects BD.

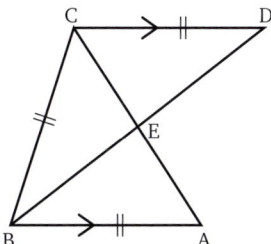

**6** Using the same diagram and the result from question **5**, show that △BEC and △DEC are congruent. Hence prove that CA and BD cut at right angles.

**7** Triangle ABC is isosceles, with AB = AC.
BX is perpendicular to AC and CQ is perpendicular to AB.
Prove that BX = CQ.
(Find a pair of congruent triangles first.)

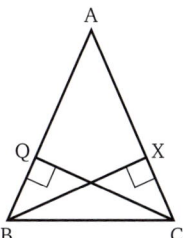

**8**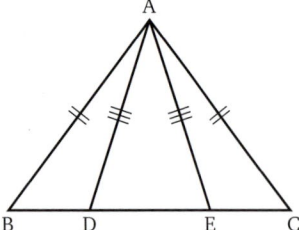

In the diagram, AB = AC and AD = AE.

Prove that BD = EC. (Consider triangles ABD and ACE.)

**9** AB is a straight line. Draw a line AX perpendicular to AB. On the other side of AB, draw a line BY perpendicular to AB so that BY is equal to AX.

Prove that $A\hat{X}Y = B\hat{Y}X$.

## Properties of parallelograms

In Books 7 and 8 we investigated the properties of parallelograms by observation and measurement of a few particular parallelograms.

Now we can use congruent triangles to prove that these properties are true for all parallelograms.

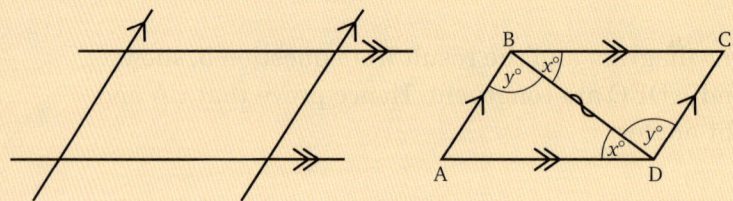

A parallelogram is formed when two pairs of parallel lines cross each other.

In the parallelogram ABCD, joining BD gives two triangles in which:

- the angles marked $x°$ are equal: they are alternate angles with respect to the parallel lines AD and BC
- the angles marked $y°$ are equal: they are alternate angles with respect to the parallel lines AB and DC
- BD is the same for both triangles.

$\therefore \triangle$s $\begin{array}{l}\text{BCD}\\\text{DAB}\end{array}$ are congruent (AAS) $\Rightarrow$ BC = AD and AB = DC

 i.e. **the opposite sides of a parallelogram are the same length.**

Also, from the congruent triangles,

$\hat{A} = \hat{C}$

and $A\hat{B}C = C\hat{D}A$  $(y° + x° = x° + y°)$

 i.e. **the opposite angles of a parallelogram are equal**

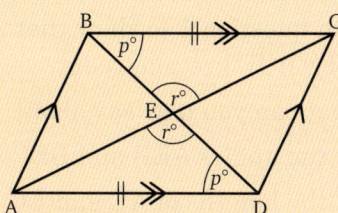

Drawing both diagonals of the parallelogram gives four triangles.
Considering the two triangles BEC and DEA:

BC = AD        (opposite sides of the parallelogram)
EB̂C = ED̂A     (alternate angles)
BÊC = AÊD     (vertically opposite angles).

∴ △s BEC, DEA are congruent (AAS) ⇒ BE = ED and AE = EC

i.e.    **the diagonals of a parallelogram bisect each other**.

The diagrams below summarise these properties.

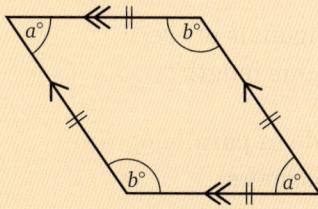  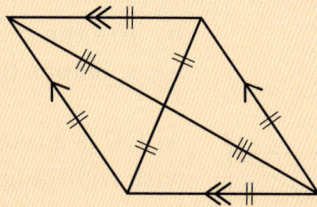

It is equally important to realise that, in general,
- the diagonals are *not* the same length
- the diagonals do *not* bisect the angles of a parallelogram
- the diagonals do *not* cut at right angles.

In **Exercise 20k**, you are asked to investigate the properties of some of the other special quadrilaterals.

## Exercise 20k

1.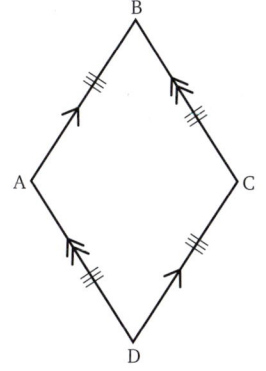

   ABCD is a rhombus (a parallelogram in which all four sides are equal in length).

   Copy the diagram. Join AC and show that △ABC and △ADC are congruent. What does AC do to the angles of the rhombus at A and C?

   Does the diagonal BD do the same to the angles at B and D?

419

**STP Maths 9**

**2** 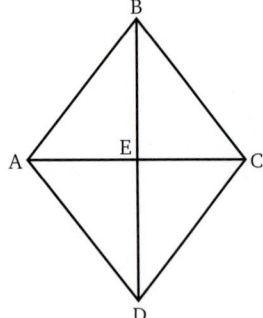 ABCD is a rhombus. Use the results from question 1 to show that △ABE and △BCE are congruent.

What can you now say about

  **a** the angles AEB and BEC
  **b** the lengths of AE and EC
  **c** the lengths of BE and ED?

**3** 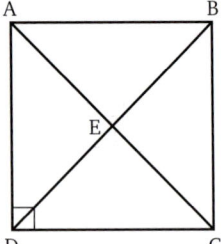 ABCD is a square (a rhombus with right-angled corners).

  **a** Use the properties of the diagonals of a rhombus to show that △AED is isosceles. Hence prove that the diagonals of a square are the same length.
  **b** Are the two diagonals of *every* rhombus the same length?

**4** 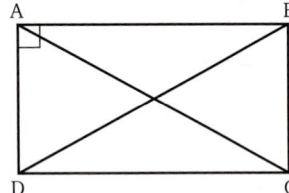 ABCD is a rectangle (a parallelogram with right-angled corners).

  **a** Prove that △s ADB and DAC are congruent.
  **b** What can you deduce about the lengths of AC and DB?

**5** 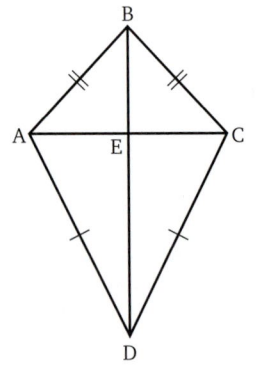 ABCD is a **kite** in which AB = BC and AD = DC.

  **a** Does the diagonal BD bisect the angles at B and D?
  **b** Does the diagonal AC bisect the angles at A and C?
  **c** Is E the midpoint of either diagonal?
  **d** What can you say about the angles at E?

**6** 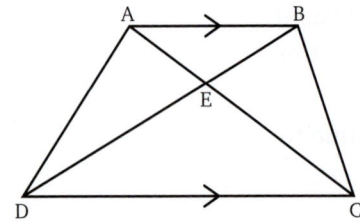 ABCD is a trapezium: it has just one pair of parallel sides. Are there any congruent triangles in this diagram?

420

## Properties of the special quadrilaterals

We can now summarise the results from **Exercise 20k** for special quadrilaterals. The properties of parallelograms are summarised on pages 418 and 419.

A quadrilateral in which each angle is 90° is called a *rectangle*.

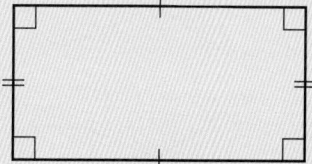

A rectangle has opposite sides that are equal and parallel, and diagonals that are equal and bisect each other.

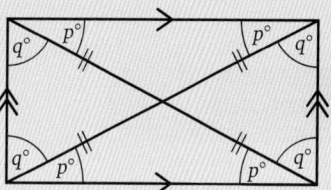

A quadrilateral with four sides of equal length and each angle 90° is called a *square*.

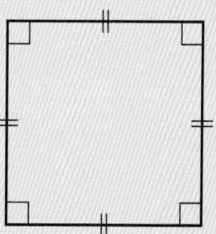

A square also has opposite sides that are parallel and equal diagonals that bisect each other at right angles.

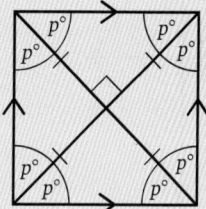

A quadrilateral with four equal sides (but not four equal angles unless it is a square) is called a *rhombus*.

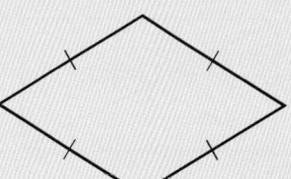

The diagonals of a rhombus bisect each other at right angles (but are not the same length unless it is a square) and bisect the angles of the rhombus.

The opposite sides are parallel and the opposite angles are equal.

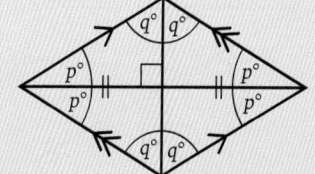

# STP Maths 9

A quadrilateral with two pairs of equal adjacent sides is called a *kite*.

The diagonals cut at right angles, but only one diagonal is bisected. The other diagonal bisects two angles of the kite.

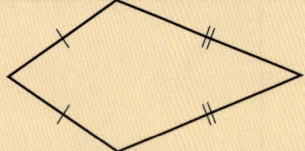 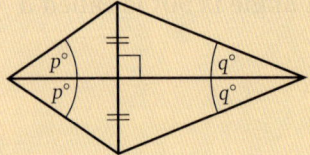

We can quote any of these properties and use them as part of a proof.

## Exercise 20l

**1**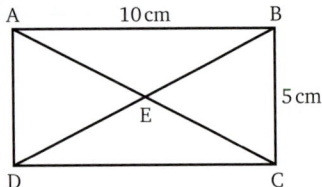

ABCD is a rectangle. The diagonals AC and DB cut at E.

How far is E from BC?

**2**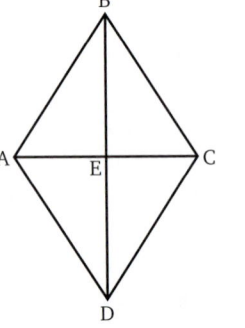

ABCD is a rhombus in which AC = 6 cm and BD = 8 cm.

Find the length of AB.

**3**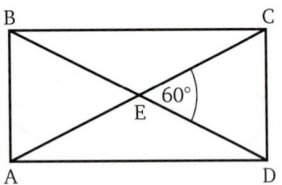

ABCD is a rectangle in which $C\hat{E}D = 60°$. Find $E\hat{C}D$.

**4**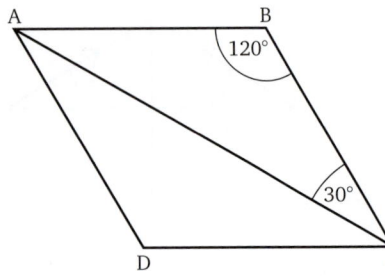

ABCD is a parallelogram in which $A\hat{B}C = 120°$ and $B\hat{C}A = 30°$.

Show that ABCD is also a rhombus.

## 20 Congruent triangles

### Worked example

→ ABCD is a parallelogram. E is the midpoint of BC and F is the midpoint of AD.
Prove that BF = DE.

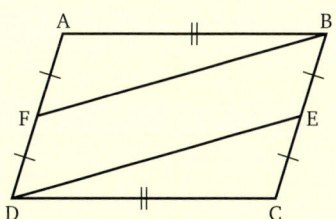

In △s ABF and CDE,
 AF = EC ($\frac{1}{2}$ opposite sides of parallelogram)
 AB = DC (opposite sides of parallelogram)
 $F\hat{A}B = E\hat{C}D$ (opposite angles of parallelogram)

∴ △s $\begin{matrix}ABF\\CDE\end{matrix}$ are congruent (SAS).

∴ BF = DE

---

**5** 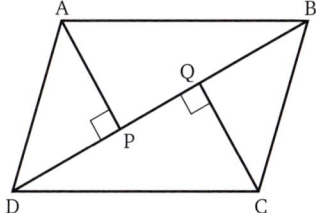 ABCD is a parallelogram. AP is perpendicular to BD and CQ is perpendicular to BD.
Prove that AP = CQ.

**6** 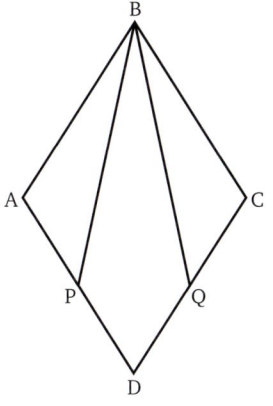 ABCD is a rhombus. P is the midpoint of AD and Q is the midpoint of CD.
Prove that BP = BQ.

**7** 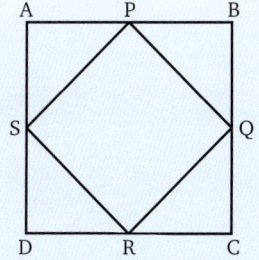 ABCD is a square and P, Q, R and S are the midpoints of AB, BC, CD and DA respectively.
Prove that PQRS is a square.

423

# STP Maths 9

  8    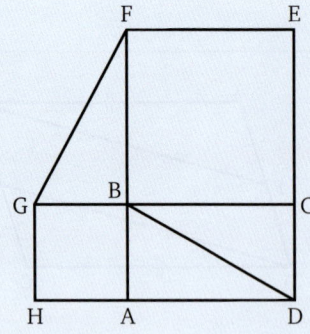   ABCD is a rectangle. ABGH and BCEF are squares.
Show that GF = BD.

## Parallelograms, polygons and congruent triangles

The next exercise uses the following facts in addition to those already used.

In a polygon
- the sum of the exterior angles is 360°
- the sum of the interior angles is $(180n - 360)°$, where $n$ is the number of sides.

If the polygon is regular, all the sides are equal and all the interior angles are equal.

### Exercise 20m

**Worked example**

→ ABC is an isosceles triangle in which AB = AC. Point D is inside the triangle and $D\hat{B}C = D\hat{C}B$. Prove that AD bisects BAC.

$D\hat{B}C = D\hat{C}B$ (given)
⇒ △BCD is isosceles
⇒ BD = CD

In △s ADB and ADC, BD = CD (proved)
AB = AC (given)
AD is common

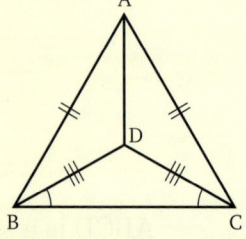

∴ △s $\frac{ADB}{ADC}$ are congruent (SSS).

∴ $B\hat{A}D = C\hat{A}D$,   i.e. AD bisects $B\hat{A}C$.

424

## 20 Congruent triangles

**1**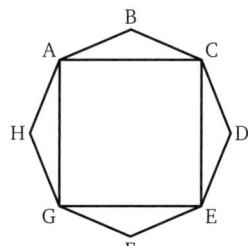

ABCDEFGH is a regular octagon.
Prove that ACEG is a square.

**2**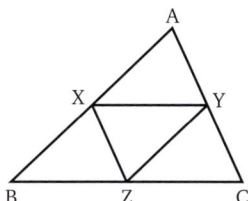

X, Y and Z are the midpoints of sides AB, AC and BC respectively. Prove that △XYZ is congruent with △YZC.
Hence prove that BXYZ is a parallelogram.

**3** ABCDEF is a hexagon in which AB is parallel and equal to ED and BC is parallel and equal to FE.

Join B to E and prove that $A\hat{B}C = F\hat{E}D$.

Hence prove that △ABC and △FED are congruent.
Hence prove that ACDF is a parallelogram.

**4**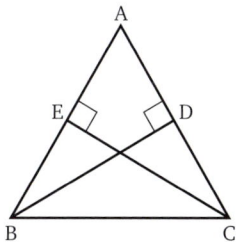

AB = AC, BD is perpendicular to AC and CE is perpendicular to AB.

Prove that △BDC and △CEB are congruent and hence prove that △AED is isosceles.

## Constructions

In this section we use the knowledge gained in this chapter to **construct** some figures using only a straight edge and a pair of compasses.

### Bisecting angles

Remember that **bisect** means 'cut exactly in half'.

The construction for bisecting an angle makes use of the fact that, in a rhombus, the diagonals bisect the interior angles.

To bisect $\hat{A}$, open your compasses to a radius of about 6 cm.

With the point on A, draw an arc to cut both arms of $\hat{A}$ at B and C.

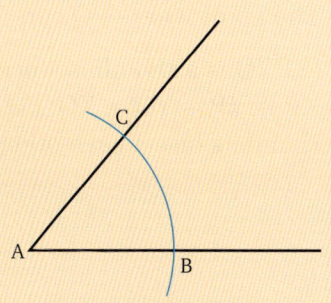

425

With the point on B, draw an arc between the arms of $\hat{A}$.

Move the point to C (being careful not to change the radius) and draw an arc to cut the other arc at D. (If we joined CD and BD, we would have a rhombus ABDC, with AD as a diagonal.)

Join AD.

The line AD then bisects $\hat{A}$.

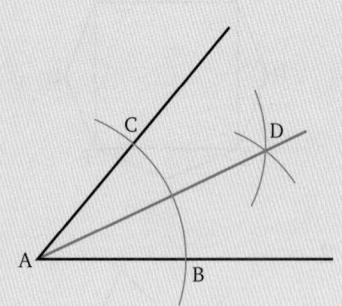

## Constructing the perpendicular bisector of a line

The perpendicular bisector of the line segment XY cuts XY in half at right angles. To find this, we construct a rhombus with the given line as one diagonal, but we do not join the sides of the rhombus.

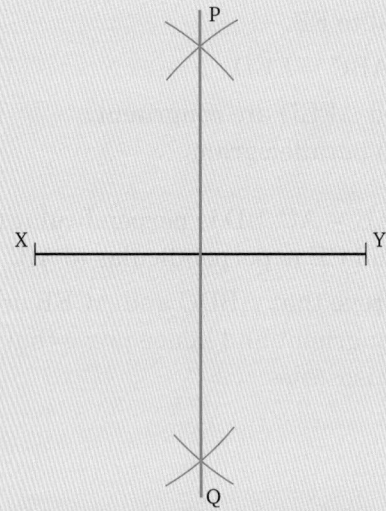

To bisect XY, open your compasses to a radius that is about $\frac{3}{4}$ of the length of XY.

With the point on X, draw arcs above and below XY.

Move the point to Y (being careful not to change the radius) and draw arcs to cut the first pair at P and Q.

Join PQ.

The point where PQ cuts XY is the midpoint of XY, and PQ is perpendicular to XY.

(XPYQ is a rhombus since the same radius is used to draw all the arcs, that is, XP = YP = YQ = XQ. PQ and XY are the diagonals of the rhombus. The diagonals of a rhombus bisect each other at right angles so PQ is the perpendicular bisector of XY.)

## Dropping a perpendicular from a point to a line

If you are told to drop a perpendicular from a point C to a line AB, this means that you have to draw a line through C which is at right angles to the line AB.

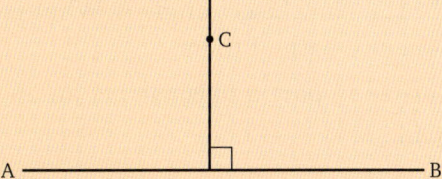

To drop a perpendicular from C to AB, open your compasses to a radius that is about $1\frac{1}{2}$ times the distance of C from AB.

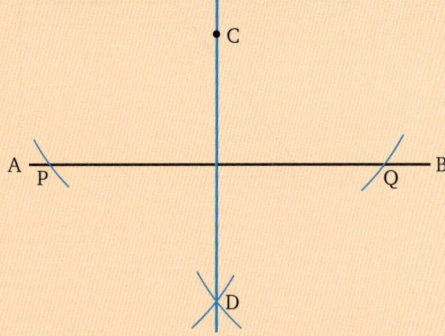

With the point on C, draw arcs to cut the line AB at P and Q.

Move the point to P and draw an arc on the other side of AB. Move the point to Q and draw an arc to cut the last arc at D.

Join CD.

CD is then perpendicular to AB.

Remember to keep the radius unchanged throughout this construction: you then have a rhombus, PCQD, of which CD and PQ are the diagonals.

### Exercise 20n

Remember to make a rough sketch before you start each construction. Use suitable instruments, including a *sharp* pencil.

1. **a** Construct an equilateral triangle of side 8 cm. What is the size of each angle in this triangle?

   **b** Use what you have learned in part **a** to construct an angle of 60°. Now bisect this angle. What size should each new angle be? Measure both of them.

   **c** Use what you have learned in part **b** to construct an angle of 30°.

**2 a** Draw a straight line about 10 cm long and mark a point A near the middle.

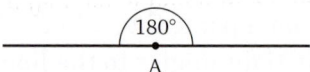

You have an angle of 180° at A. Now bisect angle A. What is the size of each new angle? Measure both of them.

**b** Use what you have learned in part **a** to construct an angle of 90°.

**c** Construct an angle of 45°.

**3** Construct the following figures using only a ruler and a pair of compasses.

**a**

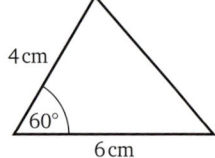

**c**

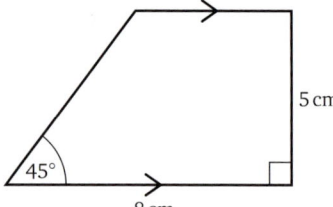

**b**

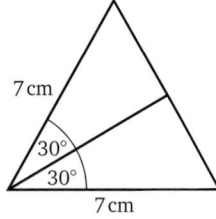

**4** Construct a triangle ABC, in which AB = 6 cm, BC = 8 cm and CA = 10 cm. Using a ruler and compasses only, drop a perpendicular from B to AC.

**5** Construct a triangle PQR, in which PQ = 6 cm, PR = 6 cm and RQ = 10 cm. Using a ruler and compasses only, drop a perpendicular from R to QP (if necessary, extend QP).

**6** Construct the isosceles triangle LMN, in which LM = 6 cm, LN = MN = 8 cm. Construct the perpendicular bisector of the side LM. Explain why this line is a line of symmetry of △LMN.

**7** Draw a circle of radius 6 cm and mark the centre C. Draw a chord AB about 9 cm long. Construct the line of symmetry of this diagram.

**8** Construct a triangle ABC, in which AB = 8 cm, BC = 10 cm and AC = 9 cm. Construct the perpendicular bisector of AB. Construct the perpendicular bisector of BC. Mark G where these two perpendicular bisectors intersect. With the point of your compasses on G and with a radius equal to the length of GA, draw a circle.

This circle should pass through B and C, and it is called the *circumcircle* of △ABC.

## 20 Congruent triangles

The remaining questions require the accurate drawing of a line parallel to a given line and at a given distance from it. A straightforward method uses a ruler and a set-square.

To draw a line parallel to AB and 2 cm from AB, place the set-square on AB with the right-angled corner on the line.

Then place a ruler against the edge of the set-square on the other arm of the right angle. Hold the ruler firmly and slide the set-square along the ruler until the lower edge is 2 cm above the line. Draw the line against this edge, as shown in the diagram.

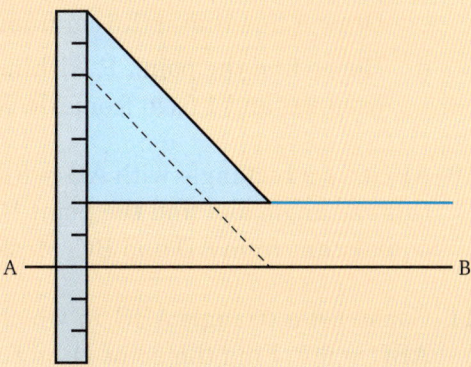

## Worked example

 Construct a triangle ABC, in which AB = 9.5 cm, BC = 7 cm and $A\hat{B}C$ = 60°. Find the point D, within the triangle, that is 2 cm from AB and 5 cm from C. Measure BD.

Points that are 2 cm from AB lie on one or other of the two lines, PQ and RS, which are shown parallel to AB. Points that are 5 cm from C lie on the circle, centre C, radius 5 cm.

This circle cuts PQ at D and E but cannot cut RS. Therefore, we do not need to draw RS in the accurate construction.

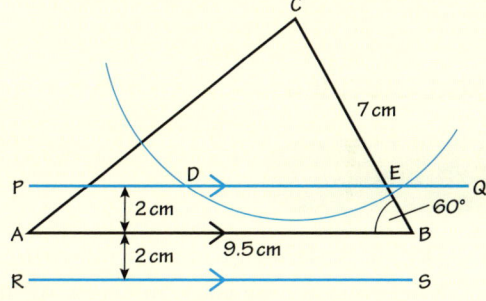

From the sketch, D satisfies the given conditions, but we cannot be certain whether E lies inside or outside the triangle until we do the construction.

BD = 6.8 cm

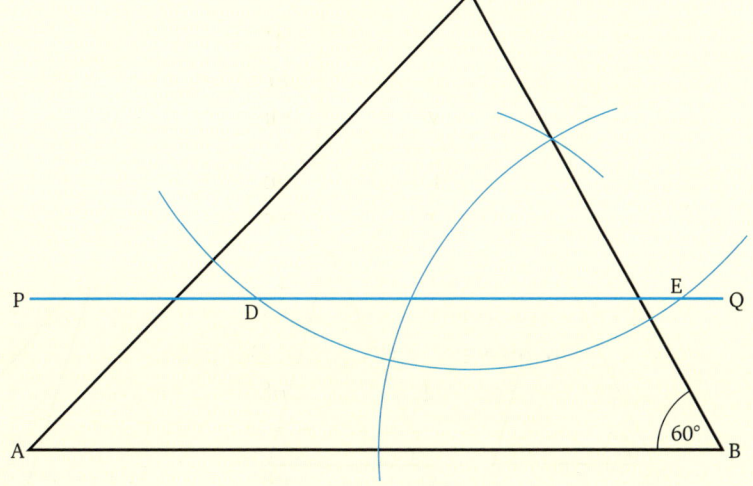

**9 a** Construct a triangle ABC, in which AB = 13 cm, $A\hat{B}C = 45°$ and $B\hat{A}C = 30°$

**b** Draw a line that is 2.5 cm from BC.

**c** Draw a line that is 1.5 cm from AB.

**d** Hence find the point, P, within the triangle, that is 2.5 cm from BC and 1.5 cm from AB. Measure AP.

**10** ABCD is a rectangle with AB = 8 cm and BC = 5 cm. Construct this rectangle and find the point P which is 2 cm from AB and equidistant from AD and BC. Measure PB.

**11** Construct a triangle ABC, in which AB = 9.5 cm, BC = 8 cm and $A\hat{B}C = 60°$. Find the point D, on the opposite side of AB from C, that is 7 cm from BC and 4.5 cm from AC. Measure CD.

**12** Construct a rectangle ABCD, in which AB = 6.5 cm and AD = 8 cm. Find the point X which is 3 cm from AD and equidistant from AD and DC. Measure DX.

**13** ABC is a triangle, in which AB = 12 cm, BC = 9 cm and $\hat{B}C = 90°$. Construct this triangle and find a point D that is 4.5 cm from BC and equidistant from A and C. Measure AD.

## Mixed exercise

### Exercise 20p

**1** Decide whether or not the following pairs of triangles are congruent. Give brief reasons for your answers. All lengths are in centimetres.

**a**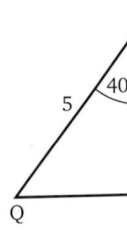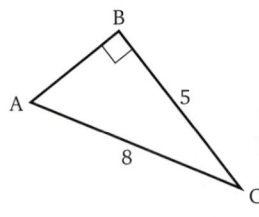

**c**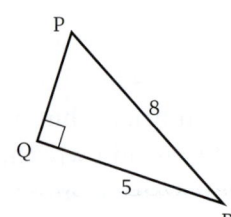

**b**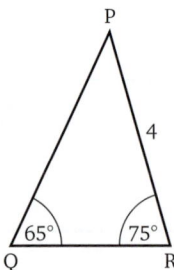

**d**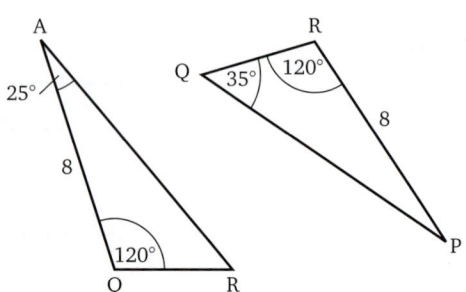

**2** AD = BC.
Prove that △ADE and △BCE are congruent.

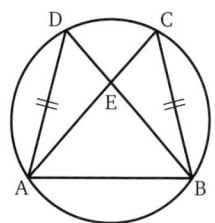

**3** 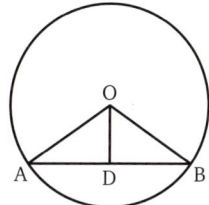 O is the centre of a circle and D is the midpoint of AB. Prove that OD is perpendicular to AB.

**4** ABCDEF is a regular hexagon. Prove that ABDE is a rectangle.

**5** M is the midpoint of the side AB of △ABC.
The line through M parallel to BC meets AC at point H.
The line through M parallel to AC meets BC at point K.
Show that   **a**  MH = BK   **b**  MK = AH

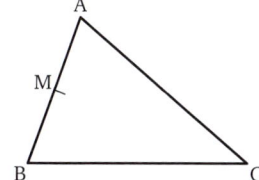

**6** Construct a triangle XYZ, in which XY = 12 cm, XZ = 5 cm and YZ = 9 cm. Using only a ruler and a pair of compasses, drop a perpendicular from Z to XY. Hence give the distance of Z from XY.

**7** Draw a circle of radius 5 cm. Label the centre O and mark a point T on the circumference. Construct the tangent to the circle at T.

## Consider again

ABCD is a trapezium, with BC parallel to AD and AB equal to CD. BE is parallel to CD and angle CDE = 60°.

Now prove that △ABE is equilateral.

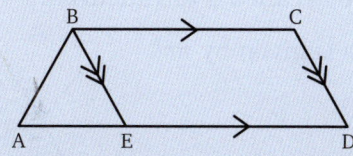

# Summary 4

## Trigonometry

In a right-angled triangle,

the *tangent of an angle*

$$= \frac{\text{side opposite the angle}}{\text{side adjacent to the angle}}$$

the *sine of an angle*

$$= \frac{\text{side opposite the angle}}{\text{hypotenuse}}$$

the *cosine of an angle*

$$= \frac{\text{side adjacent to the angle}}{\text{hypotenuse}}$$

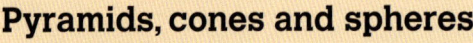

or more briefly,

$$\tan \widehat{A} = \frac{\text{opp}}{\text{adj}} = \frac{BC}{AB}, \quad \sin \widehat{A} = \frac{\text{opp}}{\text{hyp}} = \frac{BC}{AC}, \quad \cos \widehat{A} = \frac{\text{adj}}{\text{hyp}} = \frac{AB}{AC}$$

## Pyramids, cones and spheres

A *pyramid* has a base which is a polygon. The shape of a pyramid is given by drawing lines from a point, called the *vertex* of the pyramid, to each corner of the base. Each sloping face is a triangle and the number of these depends on the shape of the base.

This is a square based pyramid.

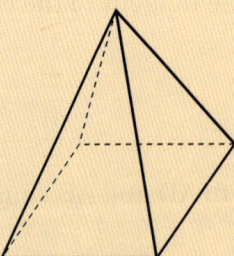

A pyramid with a triangular base is called a tetrahedron.

The curved surface area of a *cone* is given by $\pi r l$.

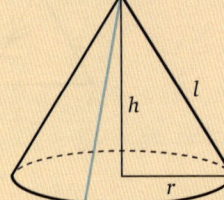

# Summary 4

The surface area of a *sphere* is given by $4\pi r^2$

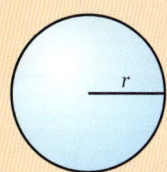

## Congruent triangles

Two triangles are *congruent* if

either the three sides of one triangle are equal to the three sides of the other triangle

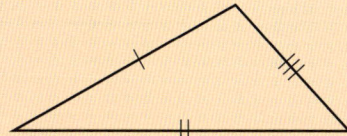 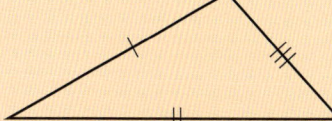

or two angles and a side of one triangle are equal to two angles and the corresponding side of the other triangle

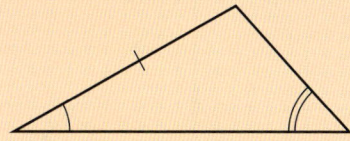 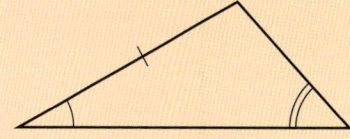

or two sides and the included angle of one triangle are equal to two sides and the included angle of the other triangle

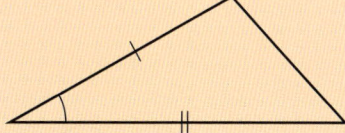 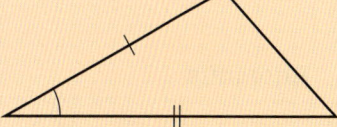

or two triangles each have a right angle, and the hypotenuse and a side of one triangle are equal to the hypotenuse and a side of the other triangle.

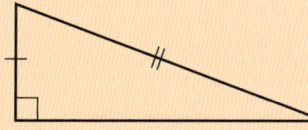 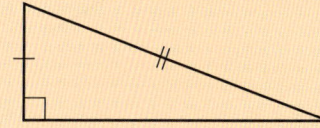

## Special quadrilaterals

The properties of the special quadrilaterals are given in the diagrams below.

### Rectangle

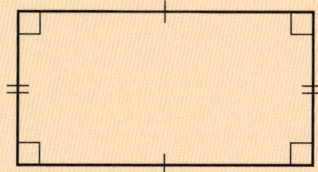 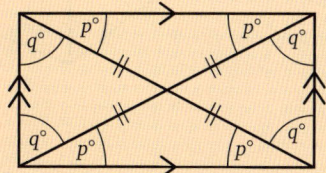

433

## Parallelogram

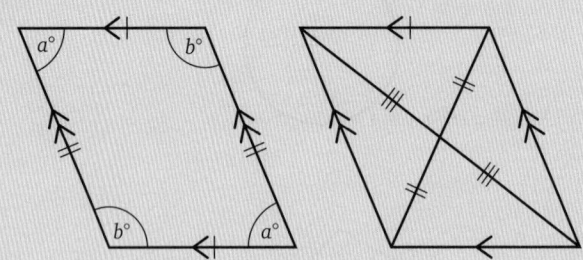

## Square

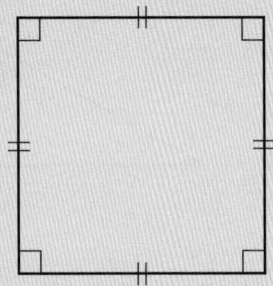 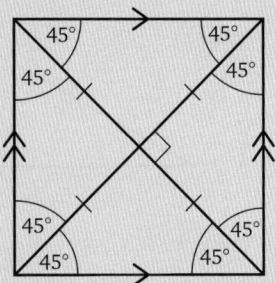

## Rhombus

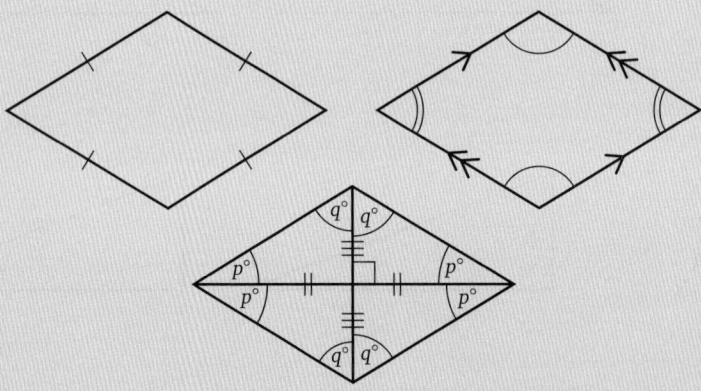

## Kite

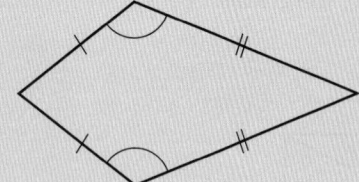 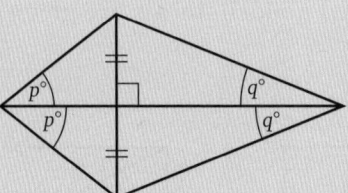

## Constructions

The diagrams show 'ruler and compass only' constructions.

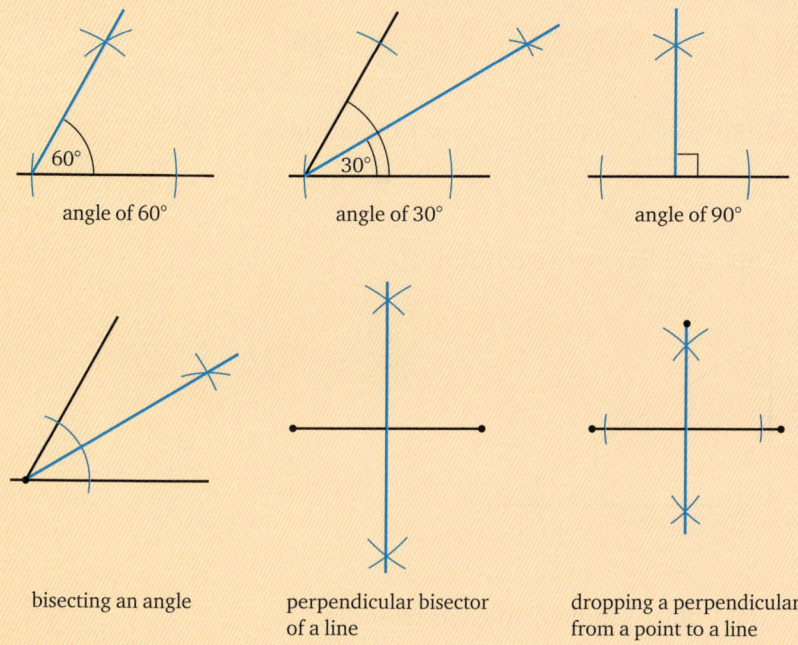

bisecting an angle

perpendicular bisector of a line

dropping a perpendicular from a point to a line

## Circles and tangents

A line joining two points on the circumference of a circle is a *chord*.

A line that touches a circle is a *tangent*.

The point where a tangent touches a circle is called the point of contact.

A *segment* of a circle is the part of the circle cut off by a chord.

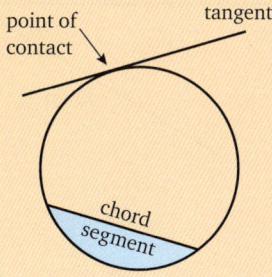

## Tangent property

A tangent to a circle is perpendicular to the radius drawn through the point of contact.

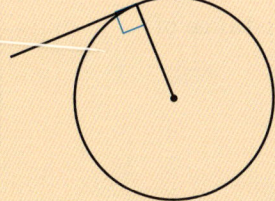

## REVISION EXERCISE 4.1 (Chapters 16 and 17)

**1** Find, correct to 3 significant figures, the tangent of
   **a** 23°    **b** 47°    **c** 36.2°    **d** 72.9°

**2** Find, correct to one decimal place, the angle whose tangent is
   **a** 0.62    **b** 3.4    **c** 1.333    **d** $\frac{1}{5}$

**3** In $\triangle ABC$, $\widehat{B} = 90°$, $\widehat{A} = 59°$ and $BC = 18$ cm. Find $AB$.

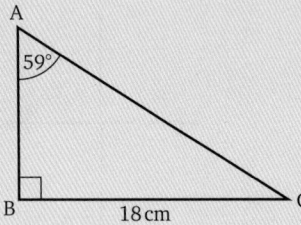

**4** Find, correct to one decimal place, angle A in each triangle.

  **a** 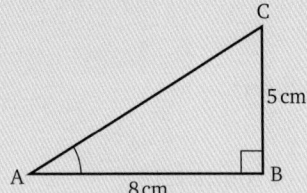  **b**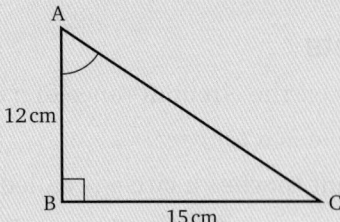

  **c** 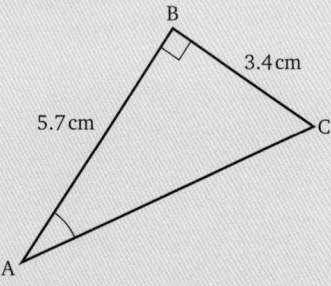  **d**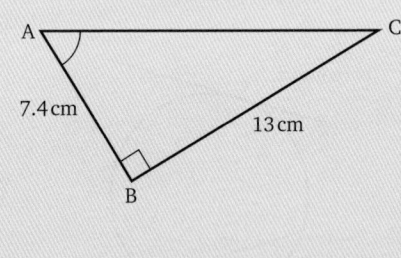

**5** The bearing of a town Q from a town P is 041.7°. R is 14 km due North of P and R is due West of Q.
  **a** How far is Q East of R?
  **b** Use Pythagoras' theorem to find the distance of P from Q.

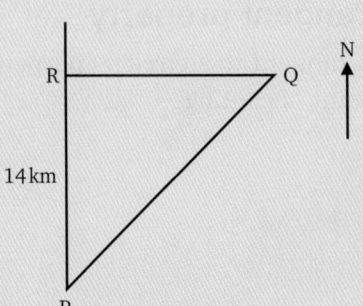

6 Find, correct to 3 significant figures,
   a  sin 67°
   b  cos 24°
   c  sin 42.7°
   d  cos 62.8°

7 Find, correct to 1 decimal place, angle A if
   a  sin Â = 0.474
   b  cos Â = 0.926
   c  sin Â = 0.8682
   d  cos Â = 0.5432

8 a 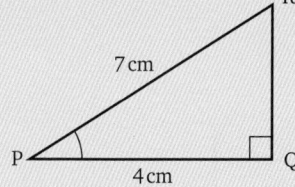   b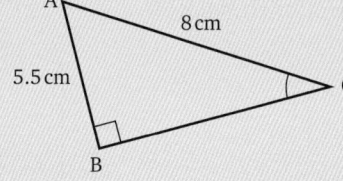

   Find P̂.              Find Ĉ.

9 a 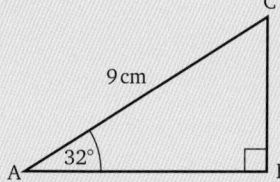   b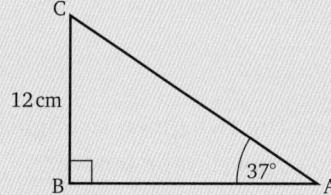

   Find BC and AB.       Find AC.

10 From a point A on the ground, 3 m from the foot of a vertical pole, the angle of elevation of the top is 52°.
   a  How high is the pole?
   b  A supporting wire runs from the top of the pole to A. How long is this wire?

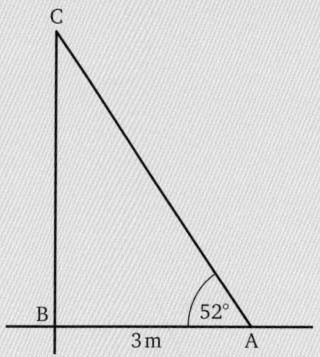

## REVISION EXERCISE 4.2 (Chapters 18 to 20)

1 A solid is formed from two equal cones whose diameters are equal to their heights. The distance between the vertices is 12 cm.
  Find the surface area of the solid.

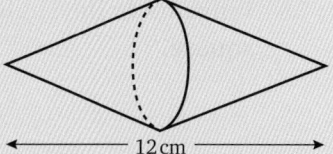

**2** Construct a triangle ABC in which AB = 10.5 cm, BC = 6.4 cm and AC = 8.6 cm. Construct the perpendicular bisector of AB and the bisector of $A\hat{B}C$. Mark the point X where they cross. Measure AX.

**3** The diagram shows a regular tetrahedron whose edges are 4 cm long.

Find the surface area of this solid.

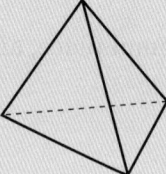

**4** Find the surface area of this solid.

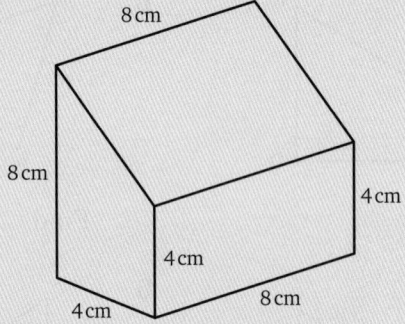

**5** Decide whether or not the two given triangles are congruent. Give brief reasons to support your answer.

**a**

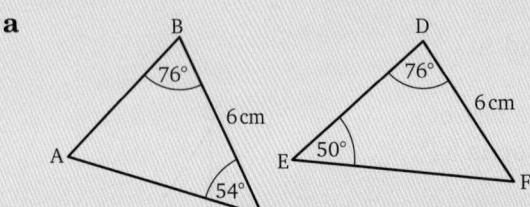

**b**

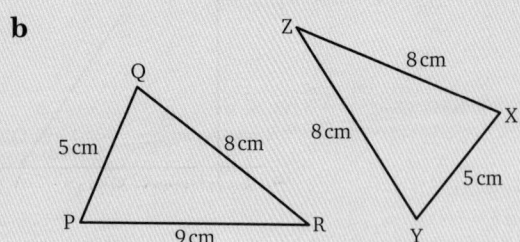

**6** O is the centre of the circle. BT and AT are tangents to the circle. Prove that OATB is a square.

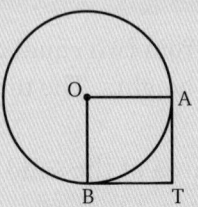

**7 a** O is the centre of the circle. AP and BP are tangents to the circle. Find $A\hat{P}B$.

**b** Prove that triangles OAP and OBP are congruent.

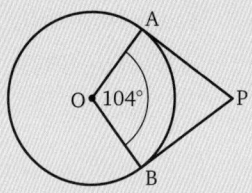

**8** PT and PS are the two tangents from P to a circle centre O. PT = 8 cm and OP = 10 cm. Find the length of

**a** OT

**b** RP.

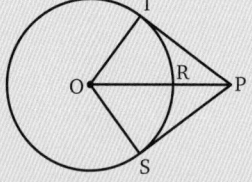

**9** CAB is the tangent at A to a circle centre O.

Find the marked angles.

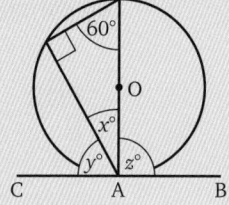

**10** The radius of a cylindrical drum is 20 cm and the drum is 120 cm high. Find as a multiple of $\pi$

**a** the capacity of the drum in    **i** cm³    **ii** litres

**b** the total external surface area of the drum.

## REVISION EXERCISE 4.3 (Chapters 16 to 20)

**1 a** Find, correct to 3 significant figures

    **i** tan 56°    **ii** tan 37.7°

**b** Find, correct to 1 decimal place, the angle whose tangent is

    **i** 0.492    **ii** $\frac{5}{8}$    **iii** $2\frac{1}{7}$

**c** In $\triangle ABC$, $\hat{B} = 90°$, $\hat{A} = 36°$ and AB = 6 cm. Find the length of BC.

**2** The diagram shows the cross-section of a ridge tent. The tent is 1.4 m wide at ground level and $B\hat{A}C = A\hat{B}C = 69°$.

How high is the ridge of the tent above the ground?

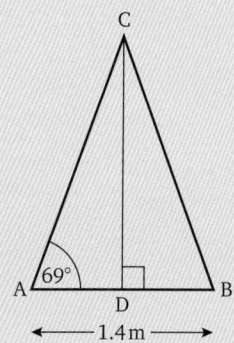

**3 a** In △ABC, $\hat{B} = 90°$, AB = 3.8 cm and AC = 4.5 cm. Find $\hat{A}$ and $\hat{C}$.
**b** In △PQR, $\hat{Q} = 90°$, RQ = 5.6 cm and $\hat{P} = 29°$. Find PR.

**4 a** Find AC.

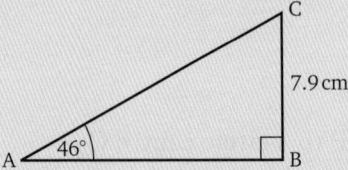

**b** ABCD is a rectangle in which DC = 4 cm and the diagonal BD = 15 cm. Find, correct to the nearest whole number
   **i** $D\hat{B}C$       **ii** BC.

**5** For each triangle, say whether or not you can make an exact copy. If you cannot, give reasons to support your answer.
  **a** △ABC, in which AB = 7 cm, AC = 8 cm, $\hat{C} = 50°$.
  **b** △PQR, in which PQ = 5 cm, PR = 7 cm, $\hat{P} = 90°$.
  **c** △XYZ, in which YZ = 9 cm, $\hat{X} = 40°$, $\hat{Z} = 65°$.
  **d** △RST, in which RS = RT = 5 cm, $\hat{S} = 40°$.

**6** Two straight lines AB and CD intersect at O.
M is the foot of the perpendicular from C to AB and N is the foot of the perpendicular from D to AB.
If CM = DN, prove that triangles OCM and ODN are congruent.

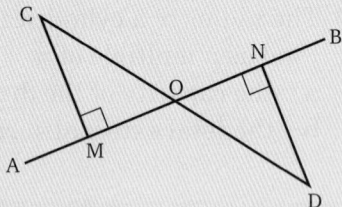

**7** In the diagram, AM = BN and AN = BM.
Prove that
  **a** triangles AMN and BNM are congruent
  **b** $M\hat{A}N = M\hat{B}N$
  **c** triangles AMX and BNX are congruent
  **d** $X\hat{M}N = X\hat{N}M$
  **e** MX = NX.

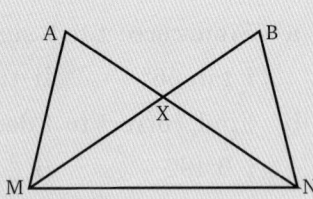

**8** ABCD is a regular tetrahedron whose edges are 8 cm long.
E is the midpoint of BC.
Find the length of DE. Hence find the surface area of the solid.

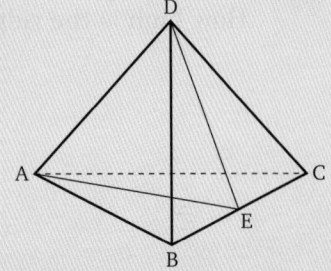

9 PT is a tangent at T to a circle centre O.

Find, giving reasons
- a  QR̂T
- b  OQ̂T
- c  RQ̂T

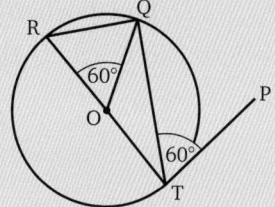

10 AB and EF are tangents to the circle. BE is the diameter.

Prove that AB is parallel to EF.

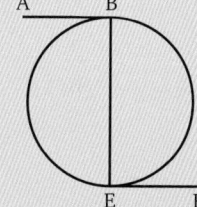

## REVISION EXERCISE 4.4 (Chapters 1 to 20)

1. **a** Use dot notation to write as a decimal
    - **i** $\frac{11}{30}$
    - **ii** $\frac{7}{11}$
   
   **b** Express as a fraction
    - **i** $0.\dot{7}$
    - **ii** $0.0\dot{7}$
   
   **c** Find $\left(6\frac{5}{8} - 4\frac{3}{10}\right) \times 2\frac{2}{3}$

2. **a** A greengrocer buys a box of 150 oranges for £40.50 and sells them at 52 p each. Find his percentage profit.
   
   **b** Find the compound interest on £750 invested for 2 years at 4.55%.

3. Expand
   - **a** $4a(b - 3c)$
   - **b** $(x - 6)(x + 2)$
   - **c** $(4x - 1)(5 + x)$
   - **d** $(5p + 2q)(5p - 2q)$

4. **a** Factorise
    - **i** $a^2 + a^3$
    - **ii** $x^2 - 4x - 21$
    - **iii** $x^2 + 10x + 24$
    - **iv** $x^2 - y^2$
   
   **b** Find the value of $7.54^2 - 2.46^2$ without using a calculator.

**5 a** If $T = x(y - 3z)$, find $T$ when $x = 3$, $y = 4$ and $z = \frac{3}{4}$.

**b** Make the letter in brackets the subject of the formula.
   **i** $a = b - c$     $(b)$
   **ii** $4z = 3x - 2y$     $(x)$
   **iii** $A = b + \frac{1}{5}c$     $(c)$

**c** Find, in terms of $n$, an expression for the $n$th term of the sequence 5, 8, 13, 20, ...

**d** If $A = 2\pi r(r + h)$ and $h = 2r$, find
   **i** $A$ in terms of $r$
   **ii** $A$ in terms of $h$.

**6 a** Solve the equations.
   **i** $x^2 + 5x - 36 = 0$
   **ii** $6x^2 - 18x = 0$
   **iii** $63 = 16x - x^2$

**b** I think of a positive number $x$. If I square it and add this to the number I first thought of, the total is 56. Find the number I first thought of.

**7** The diagram shows the cross-section of a 3 m length of angle iron. Find
   **a** the area of the cross-section in
   **i** cm²
   **ii** m²
   **b** the volume of the angle iron in m³.

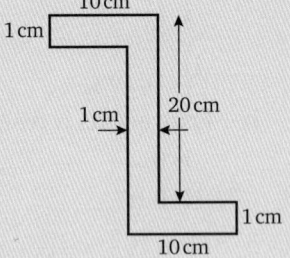

**8 a** Show that triangles ABC and ADE are similar.
**b** Given that AD = 12 cm, BC = 6 cm and DE = 8 cm, find BD.

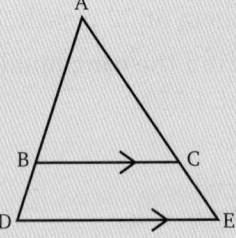

**9 a** In $\triangle ABC$, $\hat{B} = 90°$, $\hat{C} = 32°$ and AC = 12.5 cm. Find
   **i** AB
   **ii** BC.

**b** In $\triangle PQR$, $\hat{Q} = 90°$, $\hat{P} = 63°$ and PQ = 4.3 cm.
   Find
   **i** PR
   **ii** QR
   **iii** the perpendicular distance from Q to PR.

10  Prove that $A\hat{C}D = 2B\hat{A}C$.

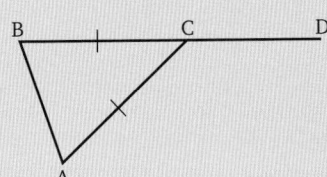

## REVISION EXERCISE 4.5 (Chapters 1 to 20)

1  Daisy is asked to choose a date at random from the year 2010. Calculate the probability that the day is
   a  a day in April
   b  a Sunday in February
   c  a day in November
   d  a day in April or November
   e  either a day in January or a Sunday in February.

   Would any answers have been different if the chosen year was 2012 instead of 2010? Explain your answer.

2  At a steady speed a car uses 8 litres of fuel to travel 96 km. At the same speed
   a  what distance would it travel using 7 litres of fuel
   b  how many litres would the car need to travel 240 km?

3  The length of the arc AB is 5 cm and the arc subtends an angle of 63° at the centre of the circle. Find
   a  the radius of the circle
   b  the area of the sector.

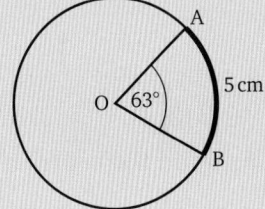

4  The heights of two varieties of broad bean plant were measured five weeks after planting and the following data was obtained.

   **Variety A**

   | Height of plant, $h$ (cm) | $0 \leq h < 5$ | $5 \leq h < 10$ | $10 \leq h < 15$ | $15 \leq h < 20$ |
   |---|---|---|---|---|
   | Frequency | 3 | 5 | 25 | 13 |

   **Variety B**

   | Height of plant, $h$ (cm) | $0 \leq h < 5$ | $5 \leq h < 10$ | $10 \leq h < 15$ | $15 \leq h < 20$ |
   |---|---|---|---|---|
   | Frequency | 4 | 14 | 17 | 6 |

   a  Find the median, the range and the interquartile range for the heights of each variety after five weeks.
   b  Use your results from part **a** to compare the heights of the two varieties after five weeks.

**5** Solve the simultaneous equations

  **a**  $2x + y = 16$
     $3x + 2y = 26$

  **b**  $3x + 5y = 8$
     $5x - 3y = 36$

**6 a** Which of the graphs could represent the equation $y = 2x - x^2 - x^3$?

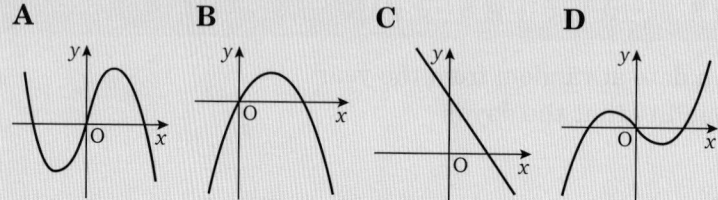

A  B  C  D

  **b** Three cars were observed approaching a set of traffic lights. The graphs show how the speeds of each car changed. Describe the way in which the speed of each car changed.

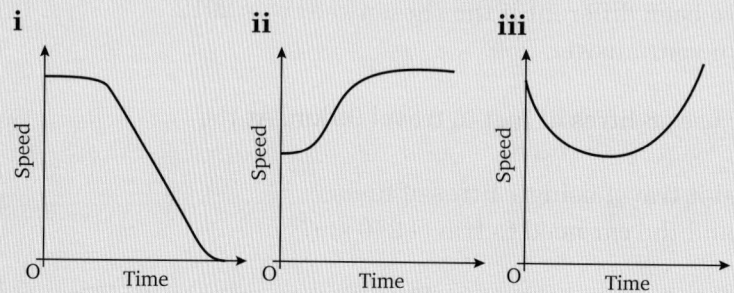

i  ii  iii

**7 a** On $x$- and $y$-axes scaled from $-3$ to $16$, plot the points $A(6, 6)$, $B(5, 4)$, $C(7, 3)$ and $A_1(6, 10)$, $B_1(3, 4)$, $C_1(9, 1)$.

  **b** Find the centre and scale factor of the enlargement that transforms $\triangle ABC$ onto $\triangle A_1B_1C_1$.

  **c** A translation of $\triangle A_1B_1C_1$ is defined by the vector $\begin{pmatrix} -3 \\ 2 \end{pmatrix}$ and the resulting figure is reflected in the line $x = 6$ to give $\triangle A_2B_2C_2$. Write down the coordinates of the vertices of $\triangle A_2B_2C_2$.

**8 a** Find PQ.

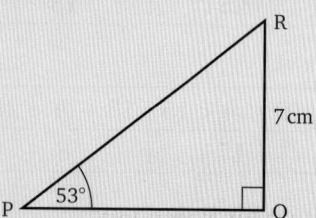

  **b** In $\triangle ABC$, $\hat{B} = 90°$, $AB = 6.4$ cm and $BC = 4.5$ cm. Find $\hat{A}$.

9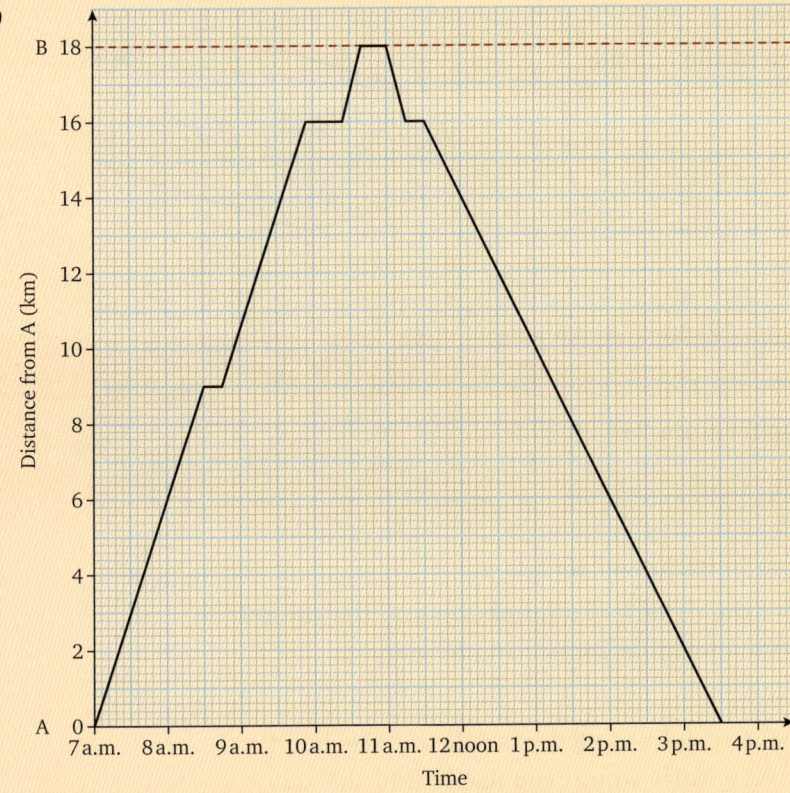

The graph shows the journey of two hikers, Imran and Joe, who set out from A to walk 18 km to a mountain centre B. They walked at a steady pace until they stopped for a rest, then continued towards the mountain centre at the same rate. Before they arrived at B, Imran injured his foot, which forced them to stop. In a short while, Joe decided to press on to B to seek help. He returned to Imran with the rescue party. After attending to Imran, the whole party returned to A, carrying Imran.

Use the graph to answer the following questions:

a  How far did the hikers walk before they stopped to rest?
b  For how long did they rest?
c  How long did Joe stay with Imran before he left to get help?
d  At what time did the rescue party
   i   leave B
   ii  begin their journey to A with Imran
   iii arrive at A?
e  Find Joe's average speed
   i   for the first part of the walk
   ii  for the whole journey from A to B.
f  Find the average speed at which Imran was carried to A.
g  How long was Joe at the mountain centre before the rescue party left?
h  Find Joe's average speed for his whole journey, including stops, from A to the mountain centre and back to A.

10 ABCDEFGH is a cuboid.

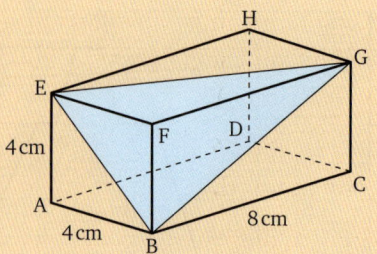

  a  Find the surface area of the cuboid.
  b  Find the length of
     i  EB
     ii BG.
  c  Find the angle GEB.

## Mental arithmetic practice 4

1  Make $c$ the subject of the formula $2a = b + 3c$
2  Increase 44 kg by $\frac{3}{11}$
3  Express 25 cm² in mm².
4  Express 0.48 as a fraction in its lowest terms.
5  A number $R$ is the sum of the square of a number $p$ and the square of a number $q$. Write down a formula connecting $R$, $p$ and $q$.
6  Simplify $2b \times 7b^2$
7  If you are given an angle in a right-angled triangle and the length of the hypotenuse, which trig ratio would you use to find the side next to the angle?
8  Shannon is 156 cm tall to the nearest centimetre. What is her least possible height?
9  Which is the larger: 0.33 or 3.4%?
10 Find $1 \div 0.3^2$ as an improper fraction.
11 Decrease £72 by $\frac{1}{12}$
12 Solve the equation $(x + 5)(x - 11) = 0$
13 Find the value of $(a - 2b)^2$ when $a = -3$ and $b = 2$
14 Find $20 \div 1\frac{2}{3}$
15 Find, in terms of $n$, the $n^{th}$ term of the sequence 6, 9, 12, 15, ...
16 Increase £60 in the ratio 3 : 5.
17 Divide £120 in the ratio 7 : 3.
18 Express 6.05 as a percentage.
19 What is the formula for the surface area of a sphere?
20 Find the coordinates of the point where the line $y = x + 3$ crosses the $x$-axis.
21 Solve the inequality $x - 4 \leqslant 4$
22 What are the prime numbers between 20 and 30?
23 If you are given an angle in a right-angled triangle and the length of the side opposite this angle, which trig ratio would you use to find the side adjacent to the angle?

**24** What is the missing word in this sentence?
Every point on the perpendicular bisector of the join of two points A and B is ... from A and B.

**25** Expand $(2a - 7)(a - 4)$

**26** Solve the inequality $7 < x + 12$

**27** In a right-angled triangle you are given the lengths of two sides, neither of which is the hypotenuse. Which trig ratio would you use to find either acute angle?

**28** Express $0.004\,\text{m}^2$ in $\text{cm}^2$.

**29** What name is given to a polygon with 8 sides?

**30** What is the value of $4^2 + 3^0$?

**31** What is 0.77 in standard form?

**32** One copy of a new book is 25 mm thick. What is the height, in centimetres, of a pile of 14 of these books?

**33** What is the value of $3 - 5x$ when $x = -1.5$?

**34** Express the ratio $\frac{2}{5} : \frac{3}{10}$ in the form $1 : n$.

**35** In a right-angled triangle you are given the length of the side opposite an angle and the hypotenuse. Which trig ratio would you use to find the angle?

**36** Increase 45 cm by $\frac{2}{5}$

**37** Find $445 + 1259$

**38** Expand $(x + 7)(x - 3)$

**39** What is the square root of 1.44?

**40** Express $0.0006\,\text{m}^3$ in $\text{cm}^3$.

**41** Find 30 minutes as a percentage of 3 hours.

**42** Express 0.2 recurring as a fraction.

**43** Find $x$ when $\dfrac{1}{x} = 5$

**44** Pete buys shoes at £50 a pair and sells them at £80 a pair. Find his percentage profit.

**45** Express $\frac{12}{25}$ as a decimal.

**46** Write down the sum of the interior angles of a parallelogram.

**47** If you are given an angle in a right-angled triangle and the length of the side adjacent to this angle, which trig ratio would you use to find the side opposite the angle?

**48** Factorise $x^2 - 3x - 18$

**49** Express 2756 cm in metres.

**50** Simplify $24a \div 6a$

**51** The area of a square is $25\,\text{cm}^2$. Find its perimeter.

**52** How many square metres are there in 1 hectare?

**53** Express 0.055 as a percentage.

**54** If you are given an angle in a right-angled triangle and the length of the side opposite this angle, which trig ratio would you use to find the hypotenuse?

**55** Which is the smaller: 14% or $\frac{1}{5}$?

Are the statements in questions **56** to **60** true or false?

**56** $-3$ is an integer.

**57** Any two rectangles are similar.

**58** The square on the longest side of a triangle equals the sum of the squares on the other two sides.

**59** If the tan of an angle in a right-angled triangle is $\frac{1}{2}$, the tan of the other angle is 2.

**60** The longest side of a triangle is opposite the largest angle in the triangle.

# Key words

| | |
|---|---|
| **acre** | an imperial measure of area used for areas of land such as fields or farms |
| **angle of rotation** | the angle through which an object is rotated |
| **angle subtended by an arc** | an angle whose arms finish at each end of an arc |
| **arc** | part of a circle |
| **arithmetic progression** | a sequence whose successive terms differ by the same amount |
| **bisect** | cut in exact halves |
| **centre of rotation** | the point about which an object is rotated |
| **chord** | a line joining two points on a circle |
| **coefficient** | the number which multiplies a letter or combination of letters, e.g. 4 is the coefficient of $4x$, $4xy$ and $4x^2$ |
| **common fraction** | a fraction in which both the numerator and denominator are whole numbers, e.g. $\frac{2}{5}$ |
| **compound interest** | interest which is added to the total so that the interest for the next period of time is calculated on the new total |
| **compound transformation** | one transformation followed by one or more other transformations |
| **congruent** | objects are congruent when they are exactly the same shape and size |
| **construct** | draw accurately |
| **cosine of an angle** | in a right-angled triangle, the ratio of the side adjacent to the angle to the hypotenuse |
| **counter example** | an example that proves that a hypothesis is wrong |
| **cubic curve** | a curve whose equation has the form $y = ax^3 + bx^2 + cx + d = 0$, where $a \neq 0$ but $b, c, d$ may be 0 |
| **cubic foot** | a volume equivalent to a cube of side 1 foot |
| **cubic inch** | a volume equivalent to a cube of side 1 inch |
| **cubic yard** | a volume equivalent to a cube of side 1 yard |
| **cumulative frequency** | the total number of values up to a particular value |
| **cumulative frequency curve** | a smooth curve drawn through plots of cumulative frequency against each value or, for grouped data, plots of the highest value in each group in a distribution |
| **cumulative frequency polygon** | straight lines joining plots of cumulative frequency against each value or, for grouped data, plots of the highest value in each group in a distribution |
| **cumulative frequency table** | a table listing total frequencies up to each value in a distribution |
| **decimal fraction** | a fraction with a denominator that is a power of 10, with just the numerator written after a decimal point, e.g. 1 and $\frac{4}{100}$ is written as 1.04 |
| **density** | the mass per unit volume |
| **difference between two squares** | expressions of the form $a^2 - b^2$ |

# Key words

| | |
|---|---|
| **dimensions of a formula** | the dimension of a formula is either length, area or volume and depends on whether the formula involves terms with a single letter, or the product of two letters, or the product of three letters, where the letters represent lengths |
| **direct proportion** | two quantities are in direct proportion when one quantity is always the same multiple of the other quantity, e.g. when pens cost 50 p each, the cost of $n$ pens is always $50n$ p |
| **displacement** | the distance and direction of an object from a fixed position |
| $\Sigma$ | a symbol meaning 'the sum of', e.g. $\Sigma n$ means the sum of values of $n$ from 1 to a given value |
| **enlargement** | making a shape larger or smaller |
| **equation** | an equality between two expressions, e.g. $2x - 5 = \frac{1}{x}$ |
| **exhaustive** | describes the set of all possible outcomes in an experiment |
| **exponential graph** | a graph representing an equation of the form $y = a^x$, where $a$ is a constant |
| **expression** | a collection of algebraic terms that does not include an equals sign, e.g. $2x - 3y + xy$ |
| **factorising** | expressing as the product of two or more expressions |
| **formula** | an instruction for finding one quantity in terms of other quantities |
| **frequency** | the number of times an event or group of events occurs |
| **frustum** | the part of a solid between its base and a plane parallel to its base, usually of a cone or pyramid |
| **geometric progression** | a sequence where each term is the same multiple of the previous term, e.g. 3, 6, 12, 24, 48, … |
| **gradient** | the gradient measures the rate at which the quantity on the vertical axis is changing as the quantity on the horizontal axis increases. The gradient of a straight line is constant.<br><br>The gradient of a curve at a point on the curve is measured by the gradient of the tangent to the curve at that point. |
| **hectare** | a metric measurement of area, usually of fields or farms |
| **hemisphere** | half a sphere |
| **hyperbola** | the shape of the curve whose equation is of the form $y = \frac{a}{x}$, where $a$ is a constant |
| **hypotenuse** | the side opposite the right angle in a right-angled triangle |
| **identity** | the equality between two different forms of the same expression, e.g. $a + 3a = 4a$ |
| **independent events** | events where the occurrence of any one event has no effect on the occurrence or otherwise of the other events |
| **interest** | money added to a deposit or loan, usually given as a percentage of the amount deposited or loaned |
| **interquartile range** | the range of the middle half of a set of figures when they are arranged in order of size |
| **invariant line** | a line that stays the same in size and position when an object is transformed |
| **invariant point** | a point that stays in the same position when an object is transformed |
| **inverse proportion** | two quantities are in inverse proportion when their product is constant, e.g. the product of the time taken, $t$ hours, and the average speed, $s$ km/h, of a car to travel a fixed distance is constant, since distance = speed × time |

# Key words

| | |
|---|---|
| **inverse square law** | two quantities are related by the inverse square law when the relationship between them has the form $yx^2 = a$, where $a$ is a constant. |
| **kite** | a quadrilateral with two pairs of adjacent sides the same length |
| **linear equation** | a linear equation only involves letters to the power one, e.g. $y = 3x + 5$ |
| **lower bound** | the lowest value given in a range of values, which may or may not be included in the range, e.g. 1 is the lower bound of $1 < x < 2$ |
| **lower quartile** | the value one quarter through a set of values when they are arranged in ascending order of size |
| **major arc** | an arc that is greater than half the circle |
| **major segment** | a segment that is greater than half the area of the circle |
| **midclass value** | the value half way through a group of values |
| **minor arc** | an arc that is less than half the circle |
| **minor segment** | a segment that is less than half the area of the circle |
| **mirror line** | the line in which an object is reflected to give an image |
| **mutually exclusive** | two events are mutually exclusive when they cannot both happen at the same time, e.g. when a coin is tossed once, getting a head and a tail are mutually exclusive |
| **natural number** | a counting number, i.e. 1, 2, 3, 4, ... |
| **outlier** | a value which is much larger or smaller than other values in a set of values |
| **parabola** | the shape of a curve whose equation has the form $y = ax^2 + bx + c = 0$, where $a \neq 0$ |
| **perfect square** | an expression that can be expressed as the square of another expression, e.g. $x^2 - 4x + 4 \,(= (x - 2)^2)$ |
| **piece-wise linear graph**  | a straight line graph made up of two or more lines with different gradients |
| **possibility space** | a list or table giving all possible outcomes in an experiment |
| **principal (money)** | the initial amount borrowed or invested |
| **prism** | a solid whose cross-section is constant |
| **probability tree** | a diagram illustrating the probabilities involved in two or more events |
| **pyramid** | a solid whose base is a polygon with triangular faces meeting at a point |
| **Pythagoras' theorem** | in a right-angled triangle, the square on the hypotenuse is equal to the sum of the squares on the other two sides |
| **quadratic equation** | equation that can be written in the form $ax^2 + bx + c = 0$, where $a \neq 0$ |
| **quadratic expression** | an expression of the form $ax^2 + bx + c$, where $a \neq 0$ |
| **raw data** | a set of information that has not been sorted |
| **reciprocal** | 1 divided by a number, e.g. the reciprocal of 2 is $\frac{1}{2}$ |
| **reciprocal curve** | a curve whose equation has the form $y = \frac{a}{x}$, where $a$ is a constant |
| **reciprocal equation** | an equation of the form $y = \frac{a}{x}$, where $a$ is a constant |

## Key words

| | |
|---|---|
| **recurring decimals** | a decimal that does not terminate and consists of a repeating digit or a repeating pattern of digits |
| **reflection** | a transformation giving the mirror image of an object in a line (called the mirror line) |
| **regular terahedron** | a pyramid, each of whose faces is an equilateral triangle |
| **right circular cone** | a cone with a circular base whose vertex is above the centre of the base |
| **roots** | the roots of an equation are the values of the unknown that satisfy the equation |
| **scalar** | a quantity that is fully defined by a number, e.g. length is a scalar |
| **scale factor** | the number that multiplies the lengths of an object to give the enlarged image |
| **sector** | the shape bounded by an arc of a circle and two radii |
| **segment** | the shape bounded by an arc of a circle and a chord |
| **similar** | two objects are similar when they are the same shape but different sizes, e.g. an object and its image under an enlargement are similar |
| **simultaneous equations** | a set of different equations with a common solution |
| **sine of an angle** | in a right-angled triangle, the ratio of the side opposite the angle to the hypotenuse |
| **slant height of a cone** | the distance between the vertex and a point on the circle forming the base |
| **sphere** | a solid in which every point on its surface is the same distance from its centre |
| **standard form** | a number written as the product of a number between 1 and 10 and a power of 10, e.g $2.4 \times 10^3$ |
| **stem-and-leaf diagram** | a method of grouping data without losing the detail of individual items (see Chapter 8) |
| **subject of a formula** | the single letter that is equal to everything else, e.g. $y$ is the subject of the formula $y = 2x + z$ |
| **subtend** | to form an angle that is enclosed by straight lines drawn from a line or arc to a point |
| **tangent of an angle** | in a right-angled triangle, the ratio of the side opposite the angle to the side adjacent to the angle |
| **tangent to a circle** | a straight line that touches a circle but does not cross the circle |
| **trial and improvement** | a method of finding a solution of an equation by guessing a solution and then trying an improved guess |
| **translation** | changing the position of an object without reflecting it or rotating it |
| **trapezium** | a quadrilateral with just one pair of parallel sides |
| **tree diagram** | an illustration giving probabilities for two or more mutually exclusive events |
| **upper bound** | the highest value given in a range of values, which may or may not be included in the range, e.g. 2 is the upper bound of the ranges $1 \leq x < 2$ and $1 < x \leq 2$ |
| **upper quartile** | the value three quarters of the way through a set of values when they are arranged in ascending order of size |
| **vector** | a quantity that requires both a size and a direction, e.g. displacement |
| **Venn diagram** | a diagram illustrating sets with circles inside a rectangle which represents the universal set |
| **vulgar fraction** | a fraction of the form $\frac{a}{b}$ where $a$ and $b$ are integers |

# Index

Note: Key terms are in **bold** type

## A

**acre** **268**, 324
adding 1, 2, 4, 56
adjacent sides 342, 344, 432
algebraic expressions 12–13, 140, 207
algebraic factors 141
algebraic fractions 13
angles 7
　acute 7
　alternate 8, 314, 391
　bisecting 425–6, 435
　of depression 8
　of elevation 8, 340, 351
　exterior 9, 392, 424
　interior 8, 9, 391, 392, 424
　obtuse 7
　at a point 7
　of polygons 9, 424
　in quadrilaterals 8
　**of rotation** 10, **292**, 293
　on a straight line 7, 391
　**subtended by an arc** **269**
　in triangles 8, 391, 392, 424
appreciation 97
**arc** **269**–70, 324, **396**
area 6, 190, 282
　of a cone 385, 432
　of a rectangle 6, 12, 231
　of a sector 270
　of a trapezium 264–5, 324, 380, 381
　units of 268
**arithmetic progression** 192–**3**, 208
　formula for the $n$th term 193
average speed 36, 37, 38–9, 40–1, 116
axioms 392
axis of symmetry 10

## B

bacterial growth 254
bar charts 178, 179
base number 5
bearings 8, 358
**bisect** **425**
bisecting an angle 425–6, 435
bisecting a line 426
body mass index 189
bollards 387
bonds 100
**bounds** **52**, 53–4, 116
brackets
　collecting numbers in 131, 138–9, 145, 146
　expanding 130, 131, 132, 134, 138, 139
　product of two 128–9, 130, 131
　squared expressions 134, 135, 136, 138
　summary 134, 135, 136, 137, 138

## C

calculator use 61, 343, 345, 353, 362, 373
capacity 6, 7
centre of enlargement 11, 305
**centre of rotation** 10, **292**
changing the subject of formula 185–6, 187, 188, 208
**chord** **396**, 435
circle 6, 269–70, 395–7, 435
circumcircle 428
circumference 6, 269–70, 435
**coefficient** **197**, 207
coin tossing 67–8, 70, 73
collecting numbers in brackets 131, 138–9, 145, 146
common denominator 56
common factor 141, 143, 207
**common fraction** **59**
**compound interest** **96**–7
compound percentage 96, 117
compound transformation 294, 326
conduit 264
cones 282, 384–5
　surface area 385, 432
　frustum 389
　**slant height** **385**
**congruent** 10, **272**
congruent triangles 402, 403, 404, 410, 412, 415, 433
　three sides 403, 410, 433
　two angles and a side 405, 406, 410, 433
　two sides and an angle 407, 408, 409, 410, 416, 433
constant $k$ 251, 252
constant product method 112
**constructions** **425**–7, 429, 435
continuous data 156
continuous values 11
corrected numbers 52–4, 56, 382
correlation 12
corresponding angles 8, 308
corresponding sides 317
corresponding vertices 310
**cosine of an angle** **365**–6, 368, 432
　to find angles 367
　to find sides 366
**counter examples** **394**
cubes (solids) 379
**cubic curves** **243**, 244–5, 246, 323
**cubic foot** **273**
**cubic inch** **273**
**cubic yard** **273**
cuboids 6, 380
**cumulative frequency** **163**, 209–10
　**curve** **166**–8, 174, 209, 210
　**polygon** **166**–7, 209
　**table** **163**–4, 166
current flow 236, 251–2
curve, equation of 14, 15
curve, gradient of 257, 258–9, 324
cylinders 7, 277, 380

## D

data, summarising 154, 155
**decimal fractions** **59**
decimals 2–3, 56
　to fractions 59, 60
　to percentages 3
　**recurring** **60**, 116
denominator 1, 57
denominator, common 56
**density** 7, **264**
depreciation 96, 98
diagonals 419, 421, 425, 427
diameter 6, 396
dice rolling 64, 65–6, 67–8
**difference between two squares** **137**, 149–50, **207**
**dimensions of a formula** **279**–80, 282
**direct proportion** 5, **102**, 106–7, 108, 110, 117
　ratio method 108
　unitary method 107, 108, 109, 110
directed numbers 4
discount 83, 88–9
discrete values 11
**displacement** **300**–1
distance table 38
distance/speed/time formula 33, 35, 36, 116
division 2, 4, 13, 57
　in a given ratio 105, 117

## E

$\Sigma$ **158**
$E = mc^2$ 62
edges of a solid 379
electricity bills 181
electronic chip 219
elements of sets 7
**enlargement** 10, **286**, 305, 308, 319, 325
　centre of 11, 305

453

# Index

**equations** 140
  of a curve 14, 15
  forming 229–30
  **linear** 14–15, **220**
  polynomial 14
  quadratic 220, 221–2, 223–9, 233, 239–41, 323–4
  **simultaneous** 13–14, **198**–9, 200–6, 208–9
equilateral triangles 8, 382
equivalent fractions 1
Euclid's proof 392, 401
events 12, 65, 117–18
  **independent** **64**, 68, 70, 118
  **mutually exclusive** **64**, 65–6, 70, 117–18
  two or more 70
**exhaustive** outcome 66
**exponential graphs** 255
expressions 12–13, **140**, 207
  for $n$th term of a sequence 192–3, 194
exterior angles
  polygons 9, 424
  triangles 392

## F

faces of a solid 379
**factorising** 141, 142, 143, 151, 152, 207–8
  **difference between two squares** 137, 149–50, **207**
  **quadratic equations** 223, 224, 225–6, 227, 228, 229, 323
  **quadratic expressions** 144, 145–6, 147, 148, 149, 150, 151
factors 1
**formula** 12, **180**, 181, 208
  changing the subject 185–6, 187, 188, 208
  **dimensions of** **279**–80, 282
  distance/speed/time 33, 35, 36, 116
  for $n$th term 193
  **subject of** **186**, 208
  substituting numbers into 182, 184
  substituting one into another 190
fractions 1–2, 4, 13, 56, 57
  algebraic 13
  **common** 59
  **decimal** 59
  to decimals 2, 59, 116
  dividing by 2, 13, 57
  to percentages 2
  **vulgar** 59
**frequency** 12, **158**, 159
frequency distribution 11, 209
  finding mean 158–9, 209
  finding median 169–70
frequency tables 156, 158, 159
fruit machines 63
**frustum** 389

## G

**geometric progression** **194**, 208
Goldbach's conjecture 394
**gradient** **236**, 237, **258**–9
  of a curve 257, 258–9, 324
  of a line 14–15
graphical solutions to equations 205–6, 239–41
graphs 14–15, 323
  cubic 243, 244–5, 246
  **exponential** 255
  **linear** 236–7, 239, 323
  reciprocal 246–7, 251
  travel 30, 32, 40, 46
  using 251–2
grouped frequency distribution 11, 209, 210
  finding median 209

## H

**hectare** **268**, 324
height 282
  of 5-year-olds 156
  of pyramids 386
  **slant (cone)** **385**
  of a tree 340, 351
**hemisphere** 387
hexagon 9, 383, 384
**hyperbola** **248**, 252, 324
**hypotenuse** 344, **362**, 373, 432
hypothesis 11, 394, 395

## I

**identity** 140
image 10, 289, 325
Imperial units 5, 6, 268, 273
Imperial/metric conversion 5–6
improper fractions 2, 57
income tax 83, 86, 87
**independent events** **64**, 68, 70, 118
indices (powers) 5
inequalities 14
**interest** **95**, 100, 117
  **compound** **96**–7
interest rate 95
interior angles in
  parallel lines 8
  polygons 9, 424
  triangles 391, 392
**interquartile range** **173**–4, 210
intersection of sets 7
**invariant line** 289
**invariant point** 289
**inverse proportion** **102**, 111, 117, 251–2, 324
  constant product method 112
  unitary method 112
**inverse square law** 257
isosceles triangle 8, 375

## K

key words 449–52
**kite** 23, **413**, **420**
  properties 422, 434
knots 40

## L

**linear equation** 220
**linear graphs** 46, 236–7, 239, 323
lines
  equation of 14–15
  of best fit 11–12
  gradient of 14–15
  parallel 8, 15, 429
  perpendicular 15, 398
  segment of 52, 53–4
loss 83, 84, 90
**lower bound** **52**, 53–4, 116
**lower quartile** **173**, 210

## M

**major arc** 396
**major segment** 396
map ratio 4
mark-up 83, 84, 85, 90
mean 11, 158–9, 209
median 11, 157, 161, 169–70, 209
members of sets 7
metric units 5, 6, 268
metric/Imperial conversion 5–6
**midclass value** **159**, 209
**minor arc** 396
**minor segment** 396
**mirror line** 10, **289**, 291, 325
mixed numbers 2, 57
modal group 11, 156
mode 11
multiples 1
multiplying 1, 2, 4, 13, 57
**mutually exclusive** events **64**, 65–6, 70, 117–18

## N

**natural number** 191
negative index 5
negative number 4, 14
negative scale factor 285–6, 325
nets, cubes 379, 380
$n$th term of a sequence 191–3, 194, 208
number lines 4, 52, 53–4, 116
number types 1
numerator 1, 57

## O

object 10, 289
obtuse angles 7
octagons 425
one-dimensional (length) 279–80, 282
operations 1
opposite side 342, 344, 362, 432

# Index

outcome 12, 66, 67–8, 118
**outlier 173**

## P

**parabola** 15, **239**, 323
parallel lines 8, 15, 429
parallelogram 6, 9, 418–19, 434
percentage 3, 117
   change 4, 83–4, 85, 92, 117
   compound 96, 117
   to decimals 3
   to fractions 3
   of a quantity 3
**perfect square 225**–6
perpendicular
   bisector 426, 435
   lines 15, 291, 398
   from a point to a line 427, 435
petal design 284, 304
pi ($\pi$) 6
**piece-wise linear graphs** 46, **237**, 239
playing cards 65
point of contact (tangent) 397, 398
polygons 9, 424, 432
polynomial equations 14
**possibility space 67**
powers (indices) 5, 14
prime numbers 1, 401
**principal** (money) **96**–7
**prism** 6, **273**, 275, 380, 381–2, 383
probability 12, 63–6
   adding 65–6, 73, 117–18
   multiplying 67–8, 70–1, 73, 118
**probability trees 70–1**, 73, 118
product of two brackets 128–9, 130, 131
proof 390, 391–3, 399
**proportion, direct** 5, **102**, 106–7, 108, 110, 117
   ratio method 108
   unitary method 107, 108, 109, 110
**proportion, inverse 102**, 111, 117, 251–2, 324
   constant product method 112
   unitary method 112
**pyramid 384**, 386, 432
   surface area 385–6
**Pythagoras' theorem** 10, **357**–8, 382, 385

## Q

$Q_1, Q_2, Q_3$ 173–4, 209, 210
**quadratic equations 220**, 221–2, 226
   solving by factorisation 223–9, 323
   solving graphically 239–41
   solving by **trial and improvement 233**, 234
**quadratic expressions**
   factorising 144–51
   mixed 152

quadrilaterals 8–9, 412
quantity before percentage change 89–90, 91
**quartile 173**, 210

## R

radius 6, 395, 398, 399
range (rounded numbers) 116
range (measure of spread) 11, 155, 157
ratio 4, 103, 104, 105, 108, 117
**raw data 156**
recipe 111
**reciprocal 57**, 58, 111, 116
**reciprocal curve** 246–7, **248**, 251, 324
**reciprocal equation 246**, 324
recognising curves 248
rectangles 6, 9, 421, 433
rectangular numbers 1
**recurring decimals 60**, 116
**reflection** 10, **289**
reflex angles 7
**regular tetrahedron 387**
relative frequency 12
repeated root 225–6
representative fraction 4
rhombus 9, 419, 421, 425, 434
**right circular cone 384**–5
right pyramid 384
right-angled triangles 10, 340, 354, 358, 432
   angles in 354, 358
   **Pythagoras' theorem** 10, **357**–8, 382, 385
   sides of 344, 345, 348–9
**roots 225**
rotation 10, 292, 293
   from enlargement 285, 325
rounding numbers 3, 51–4, 116
running total 161

## S

sale price 88–9
sales tax 91
   *see also* value added tax
**scalar 296**, 324
**scale factor** 11, **285**–6, 305
   to find length 315–16
   negative 285–6, 325
scatter graphs 11, 12
**sector** of a circle **270, 396**
   area of 324, 385
**segment 378, 396**, 435
sequences 192–3, 194, 208
   $n$th term 191–3, 194, 208
sets 7, 76
significant figures 3
**similar figures** 305–**6**, 326
similar triangles 308, 310–14, 317, 326, 342, 404

simplifying 13
**simultaneous equations** 13–14, **198**–9, 200–6, 208–9
**sine of an angle 361**–2, 368, 372, 432
   to find angles 364
   to find sides 362
skeletons 180, 189
**slant height of a cone 385**
SOHCAHTOA 368
solids 379
speed, average 36, 37, 38–9, 40–1, 116
speed/distance/time formula 33, 35, 36, 116
**sphere 385**, 433
spring balance 110
square numbers 1
square roots 4
squares (shapes) 6, 9, 421, 434
**standard form** 3, **61**
standing charge 181
statistics 11–12, 209
**stem-and-leaf diagram 157**, 158
straight line graphs 236–7, 239, 323
   travel graphs 30, 32, 40, 46
**subject of a formula 186**, 208
   changing 185–6, 187, 188, 208
substituting
   one formula into another 190
   numbers into formulas 182, 184
**subtend 378**
subtracting 2, 4, 56
summaries 1–29, 116–27, 207–10, 323–39, 432–48
supplementary angles 7
swimming pool 305
symbols, mathematical 158, 186, 391, 399, 412
symmetry, axis of 10

## T

**tangent of an angle 342**–3, 349, 353, 368, 432
**tangent to a circle 397**, 398–400, 435
tax, income 83, 86, 87
tax, sales 91
tax, value added (VAT) 83, 89
temperature conversion 189
temperature of water 258–9
terms of a sequence 191–2
**tetrahedron** 384, **387**, 432
theory of relativity 62
three-dimensional (volume) 279–80, 282
time/distance/speed formula 33, 35, 36, 116
torus 283
transformations 10–11, 325
   compound 294, 326
   *see also* enlargement, reflection, translation

455

# Index

**translation**   10, **300**–1, 326
transversal   8
**trapezium**   **264**
  area   264–5, 324, 380, 381
  properties   9
travel graphs   30, 32, 40, 46
**tree diagram**   **70–1**, 73, 118
**trial and improvement**   **233**, 234
triangles   6, 8
  congruent   402–10, 412, 415, 416, 433
  constructing   429
  corresponding angles   308
  equilateral   8, 382
  exterior angle of   392
  interior angles of   375, 391, 392
  isosceles   8, 382, 414, 424
  right-angled   10, 340, 344–5, 348–9, 354, 358, 432
  similar   308, 310–14, 317, 326, 342, 404
two-dimensional (area)   279–80, 282

## U

union of sets   7
unitary method, proportion   107, 108, 109, 110, 112
universal sets   7
**upper bound**   **52**, 53–4, 116
**upper quartile**   **173**, 210

## V

value added tax (VAT)   83, 89
**vector**   **296**, 299, 300, 304, 324–5, 326
  representing   297–8
**Venn diagram**   7, **75**–6, 80, 118
vertically opposite angles   7, 314
vertices   310, 379
  of pyramids   384, 432
volume   6–7, 282
  of a cylinder   7, 277
  of a prism   6, 273, 275
  units of   273
**vulgar fraction**   **59**

## W

wage increase   83, 91
water flow   277

## X

$x$-coordinate   14

## Y

$y$-coordinate   14
$y$-intercept   15, 236, 323
$y = mx + c$   15, 323

## Z

zero index   5
zero, multiplying by   221, 323